HYDROGEN STORAGE

Edited by **Lianjun Liu**

Hydrogen Storage
http://dx.doi.org/10.5772/3207
Edited by Lianjun Liu

Contributors
Gabriella Pinto, Simonetta Caira, Marina Cuollo, Sergio Lilla, Lina Chianese, Francesco Addeo, Haiyan Sun, Håvard Jenssen, Teresa Treweek, Marcel Jøhnke, Torben E. Petersen, Tomoko Shimamura, Hiroyuki Ukeda, Bärbel Lieske, Simonetta Caira, Rosa Pizzano, Paolo Polidori, Silvia Vincenzetti, Joanna Barłowska, Anna Wolanciuk, Zygmunt Litwińczuk, Jolanta Król, Gema Lucia Zambrano-Burbano,Yohanna Melissa Eraso-Cabrera, Carlos Eugenio Solarte-Portilla, Carol Yovanna Rosero-Galindo, Massimo Bionaz, Walter Hurley, Juan Loor, Edyta Molik, Genowefa Bonczar, Tomasz Misztal, Aneta Żebrowska and Dorota Zięba

Published by InTech
Janeza Trdine 9, 51000 Rijeka, Croatia

Publishing Process Manager Dragana Manestar
Technical Editor InTech DTP team
Cover InTech Design team

Additional hard copies can be obtained from orders@intechopen.com

Hydrogen Storage, Edited by Lianjun Liu
p. cm.
ISBN 978-953-51-0731-6

For EU product safety concerns:
IN TECH d.o.o., Prolaz Marije Krucifikse Kozulić 3, 51000 Rijeka, Croatia,
info@intechopen.com or visit our website at intechopen.com.

Meet the editor

Dr Jianjun Liu received his Ph. D in Physical Chemistry from the Institute of theoretical chemistry, Jilin University, China, in 2002. He subsequently went on to research activity as a postdoctoral fellow at the Cherry L. Emerson Center for Scientific Computation at the Emory University. Then He became an assistant scientist at the Southern Illinois University, working on computational simulation on complex metal hydrides as hydrogen storage. In 2011, he joined the faculty as an Associate Professor at the State Key Laboratory of High Performance Ceramics and Superfine Microstructure at shanghai Institute of Ceramics, Chinese Academy of Science. He has been active in various organizations and has co-authored over 30 papers and 3 book chapters, as well as has been awarded several national awards and visiting fellows. Dr. Liu's research is focused on computational modeling microscopic structures as well as physical and chemical properties of nanomaterials in hydrogen storage and lithium-ion batteries.

Contents

Preface

Energy storage became a dominant factor in economic development with the widespread introduction of electricity and refined chemical fuels in the later 1800s. Chemical fuels are the dominant form of energy storage, both in electrical generation and energy transportation. Chemical fuels in common use are processed fossil fuels in the form of coal, oil, and natural gas. The world energy consumption has been constantly increasing since the industrial revolution. However, the supply of fossil fuel is limited. Furthermore, the use of fossil fuel puts human health at risk through chemical and particulate pollutants and affects global climate through CO_2 and other greenhouse gas emissions. Hydrogen, as a chemical energy carrier, is widely regarded as a potential cost effective, renewable, and clean energy alternative to petroleum, especially in the transportation sector. Hydrogen storage is becoming a key component of realizing the hydrogen economy for transportation applications.

Emerging nanotechnology has generated a significant influence to many fields. Scientists always think, work, and interact to fabricate and optimize new materials and processes in nanometer scale. In turn, these novel nanomaterials and nanostructures are employed to design real devices, therefore, offering unprecedented efficiency or function compared to conventional bulk phase materials. In the recent years, nanotechnology has extensively applied to develop novel hydrogen storage materials. To reflect the rapid growth of these research in hydrogen storage nanomaterials, this book entitled by "Hydrogen Storage Materials through Nanotechnology" summarize the recent research progress in this field.

In the past years, there are many books already available covering different topics, nanotechnology or hydrogen storage. However, none summarizes the specific topic with emphasis on nanotechnology applied to improve hydrogen storage properties. It is believed that this book can be suitable as a desk reference related to energy storage materials for scientific researchers, government and organizations, as well as industrial R&D labs. This book includes 10 chapters which are divided into two categories, physical and chemical hydrogen storage materials. Each chapter is written by world-leading scientific experts working in nanotechnology and hydrogen storage, the editor made a great effort to ensure the accuracy and appropriateness of the materials collected and any comments and suggestions to improve potential future editions are welcome.

Lianjun Liu
Southern Illinois University
USA

Section 1

Chemical Hydrogen Storage Materials

Chapter 1

Development of Novel Polymer Nanostructures and Nanoscale Complex Hydrides for Reversible Hydrogen Storage

Sesha S. Srinivasan and Prakash C. Sharma

Additional information is available at the end of the chapter

http://dx.doi.org/10.5772/50171

1. Introduction

This book chapter discusses about (i) the characteristics of hydrogen – a clean and renewable fuel, (ii) the grand challenges in hydrogen storage in reversible solid state hydrides, the current technical targets set forth by the US Department of Energy and the FreedomCAR, (iii) current state of hydrogen storage (Broom, 2011), (iv) various types of hydrogen storage methods and modes (Bowman & Stetson, 2010; Sathyapal et. al., 2007). Among the different methods of hydrogen storage, the current book chapter aims to address two important hydrogen storage methods such as physisorbed hydrogen storage via polymer nanostructures and chemisorbed hydrogen storage via complex chemical hydrides. The experimental approaches of synthesizing the solid state hydrides using mechanochemical milling, wet chemical synthesis and electrospinning are discussed. Extensive metrological characterization techniques such X-ray diffraction (XRD), Scanning Electron Microscopy (SEM), Fourier Transform Infrared Spectroscopy (FTIR), Differential Scanning Calorimetry (DSC), Thermogravimetric Analysis (TGA), Thermal Programmed Desorption (TPD), Pressure-Composition-Temperature Isotherms (PCT) are employed to unravel the structural, microstructural, chemical, thermal and volumetric behavior of these materials. The major results based on the structure-property relations are discussed in detail. In summary, a comparative study of various solid state hydrides investigated for reversible hydrogen storage are discussed with potential hydrogen fuel cell applications.

As traditional fossil fuel supplies are dwindling and carbon emissions derived from burning these fuels are being blamed for global weather changes, it is becoming increasingly important to find alternative energy sources. While there are clean and renewable energy production methods, such as wind and solar energy, there is yet to be found a clean and safe means of propelling automobiles. Hydrogen is an ideal candidate since it can easily be refueled in automobiles similar to gasoline. Its onboard storage, however, is a significant barrier for utilizing hydrogen as a fuel. The materials developed as part of this present study provides a significant improvement in the hydrogen storage properties of solid state storage (Jena, 2011).

Two main approaches are investigated as part of this current investigation. The first is to tailor the nanostructure of polyaniline in its emeraldine form, a conductive polymer, so that the surface area of the polymer is increased, by creating nanospheres and nanofibers. The bulk form of polyaniline is also investigated for its hydrogen sorption properties as a comparison. The reason behind using a polymer for hydrogen storage is that polymers contain many hydrogen atoms which allow for the formation of weak secondary bonds between the hydrogen that is part of the polymer and the hydrogen that is meant to be stored. Since polyaniline is easily synthesized and is rather inexpensive, it is an excellent choice as a hydrogen storage material. The alteration of the physical structure of polyaniline into nanospheres and nanofibers allows for an increase in surface area, thereby exposing more material as potential bonding sites for the hydrogen. However, as host materials of hydrogen physical adsorption, it has a weak interaction (quadrupole interaction) with nonpolar hydrogen molecule. Therefore, some positive ions such as Li+ can be added in quinoid and benzenoid to enhance interaction strength. The Lithium doped polymer electrode is recently shown to exhibit higher ionic conductivity (2×10^{-4} S cm^{-1}) at room temperature in addition to its greater charge-discharge cycles at room temperature (Nitanii et. al., 2005; Zhang et. al., 2012). The photoelectron spectroscopic studies and quantum chemical calculations revealed the role of lithium in the polyaniline systems thus enhances the charge transfer; highly localized and energetically favored at N-atomic sites for the bonding of dimers of N-atoms with the Li-atoms (Kuritka et. al., 2006). Moreover, the insertion of Li^+ in the polymer structure affects the electrochemical properties due to their strong interaction with the quinoid and benzenoid rings (Lindino et. al., 2012).

Since polyaniline is composed of quinoid and benzenoid rings and the emeraldine form is terminated with Cl^- ions, additional hydrogen bonding sites, in the form of stronger chemisorption of hydrogen, is made available. The advantage of having both chemisorption and physisorption sites is shown schematically in Figure 1.

Hydrogen, as a molecule, can bond to the material through three main mechanisms. The simplest form is to simply bond weakly to the host material via physisorption as a molecule. This can generally be achieved if the temperature is low enough, and is the main mechanism of storage for physisorption materials, but generally requires temperatures of approximately 77K. If additive materials, such as catalysts, are present, these materials can break up the molecule into ions which then allows for the diffusion through the relatively porous material to then chemically bond with the host material. It is inevitably demonstrated that the vital role of catalyst species such as Ti, Fe, Ni, Pd etc. thus enhances the physico-chemical adsorption of hydrogen atoms on the polyaniline matrices (Skowron´ski & Urbaniak, 2008; Yildirim & Ciraci, 2005). Similarly, hydrogen molecules can also bond the host material. By increasing the surface area as well as the porosity, two main events can occur.

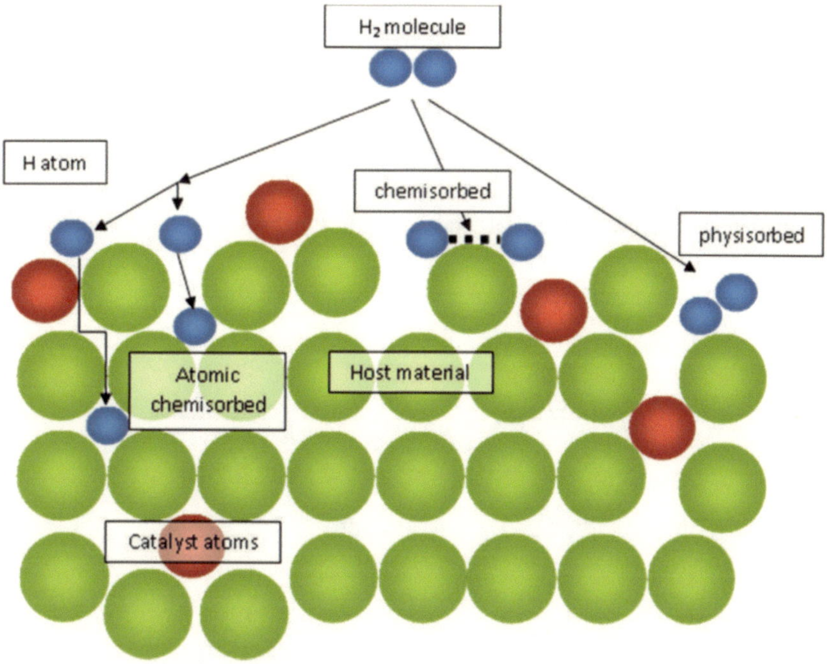

Figure 1. Schematic representation of physisorption and chemisorption with catalyst materials.

The first is that the hydrogen has more bonding sites to the host material. The second is that the diffusion of hydrogen into the material is made possible. Since polyaniline cannot be simply modified to have smaller molecular size, the nanostructure can be altered to increase the surface area and also to increase the hydrogen diffusion pathways. Recently, polyaniline is shown to possess outstanding hydrogen storage properties by carefully modifying the nanostructure of the material. It is found that polyaniline combines both physisorption as well as chemisorption for storing hydrogen, which provides a significant improvement of hydrogen storage properties, since a usable capacity is achieved.

The second mechanism investigated as part of this study is the route of chemisorption, or strong hydrogen atomic bonding, to complex hydride materials. Complex hydrides generally require a high temperature for hydrogen release as the hydrogen bonds are very strong. By reducing the crystallite size of the host material, specifically $LiBH_4$ and $LiNH_4$ and by further destabilizing the structure with MgH_2, the temperature can be reduced to allow for reversible hydrogen storage at a lower temperature (Srinivasan et. al., 2012). The reduction of crystallite size, as well as the exact processing technique is shown to be very important and have a large effect on the storage capacity. Figure 2 shows general hydrogen absorption and desorption of complex hydrides.

The unhydrided material absorbs hydrogen from the outside until it is fully charged with hydrogen. By reducing the pressure on the sample or by increasing the temperature, the hydrogen is then released from the outside first until the material is fully discharged again. When the particles are too large, though, a hydrogen passivation layer can form during the initial hydrogen uptake, thereby reducing any further hydrogenation. Additionally, the kinetics, or rate, of hydrogen sorption is increased with particle size, as this means that the hydrogen has a larger distance to diffuse through.

The effects of particle size on hydrogen storage are shown in Figure 3. By employing mechano chemical milling, the particles are not only reduced in size, but dislocations and vacancies are created, thereby increasing the kinetics and capacity of the material. The interaction of the various particles that make up the hydrogen storage material are also of great importance, as this determines the interaction of the various compounds and either facilitates or hinders successful hydrogen sorption as will be shown.

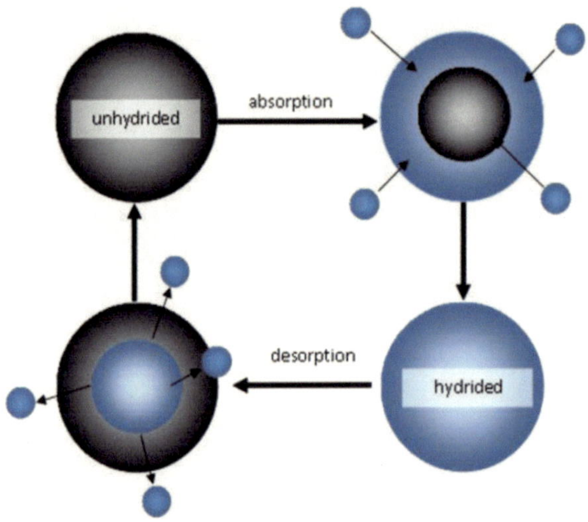

Figure 2. Hydrogen storage mechanism in complex hydrides and the effect of particle size.

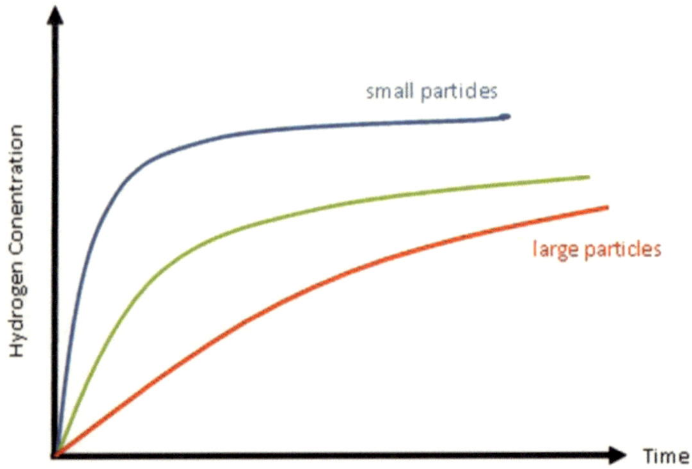

Figure 3. Particle size effects on hydrogen storage.

The complex hydrides that are developed are carefully investigated employing varying processing techniques, a deviation of the traditional means of producing complex hydrides. The nanostructure created with the varying techniques is carefully analyzed and correlated with the material's hydrogen performance. Additionally, the hydrogen storage properties are significantly improved by using optimized quantities of nano sized additives.

2. Hydrogen sorption characteristics of polyaniline nanofibers

The conducting polymer nanostructures combine the advantages of organic conductors and low dimensional systems possessing interesting physicochemical properties (Macdiarmid & Epstein, 1989; Stejskal et. al., 1996; Stejskal & Gilbert, 2002; Trivedi, 1997) and useful applications (Fusalba et. al., 2001; Rossberg et. al., 1998; Virji et. al., 2006). Among the conducting polymers, polyaniline is considered important because of its extraordinary properties of electrical and optical behavior. It was recently reported that polyaniline could store as much as 6 to 8 weight percent of hydrogen (Cho et. al., 2007), which was later refuted. (Panella et al., 2005). Though many controversial results were reported in terms of hydrogen uptake (Cho et. al., 2007; Germin et. al., 2007; Jurczyk et. al., 2007; McKeown et. al., 2007) in polymer nanocomposites, there are still a number of parameters, tailor-made properties, surface morphologies and their correlation with hydrogen sorption behavior to be investigated before these materials can be commercially deployed for on-board hydrogen storage. Similarly, nanotubes (Dillon et. al., 1997; Nikitin et. al., 2008) or nanofibers (Hwang et. al., 2002) have attracted more interest because of their novel properties and wide potential for nanometer-scale engineering applications.

It is known that the nanofibrillar morphology significantly improves the performance of polyaniline in many conventional applications involving polymer interactions with its environment (Wang & Jing, 2008). This leads to faster and more responsive chemical sensors (Sadek et. al., 2005; Virji et. al., 2006), new organic/polyaniline nanocomposites (Athawale & Bhagwat, 2003) and ultra-fast non-volatile memory devices (Yang et. al., 2006). Nanofibers with diameters of tens of nanometers appear to be an intrinsic morphological unit that was found to naturally form in the early stage of chemical oxidative polymerization of aniline. In conventional polymerization, nanofibers are subject to secondary growth of irregularly shaped particles that form the final granular agglomerates. The key to producing pure nanofibers is to suppress secondary growth. Based on this, many methods (interfacial polymerization, rapidly mixed reactions, controlling pH and oxidizing agent) have been developed that can readily produce pure nanofibers with uniform growth, chemical composition and morphology (Ding & Wei, 2007; Rahy et. al., 2008). With this nanofiber morphology, dispensability and processbility of polyaniline are now greatly improved. On the other hand, the template synthesis method is an effective way to grow the nanotubes of various conducting polymers (Huczko, 2000). The preparation conditions and their effect on morphology, size, and electrical properties of nanofibers have been reported elsewhere (Zhang & Wang, 2006). Recently, a novel, simple, and scalable technique to control the formation of the nanofibers of polyaniline and its derivatives via porous membrane controlled polymerization (PMCP) was reported (Chiou et. al., 2008). Through appropriate synthesis conditions, nearly 100% nanofibers are formed with diameters tunable from 20nm to 250nm via the selection of pore diameter, monomer, counter ions, and polymerization conditions. The nanofiber lengths vary from sub-micrometer to several micrometers.

Conducting polyaniline nanofibers were synthesized using chemical templating method followed by electrospun process. These nanofibers have been compared with their standard bulk counterpart and found to be stable up to 150°C. Polyaniline nanofibers prepared by electrospun method reveal high hydrogen uptake of 10wt.% at around 100°C in the first absorption run. However, in the consecutive hydrogenation and dehydrogenation cycles, the hydrogen capacity diminishes. This is most likely due to hydrogen loading into the polymer matrix, chemisorption and saturation effects. A reversible hydrogen storage capacity of ~3-10 wt.% was also found in the new batch of electrospun nanofibers at different temperatures. The surface morphologies before and after hydrogen sorption of these PANI nanofibers encompass significant changes in the microstructure (nanofibrallar swelling effect) which clearly suggest effective hydrogen uptake and release.

Fourier Transform Infrared (FTIR) spectra of PANI-NF-CM and PANI-NF-ES prepared confirmed the formation of PANI in its emeraldine form. The major bonding environment remains unchanged for both structures (see Figure 4). The presence of two bands in the vicinity of 1500cm^{-1} and 1600cm^{-1} are assigned to the non-symmetric C6 ring stretching modes. The higher frequency vibration at 1600cm^{-1} is for the quinoid rings, while the lower frequency mode at 1500cm^{-1} depicts the presence of benzenoid ring units. Furthermore, the peaks at 1250cm^{-1} and at 800cm^{-1} are assigned to vibrations associated with the C-N stretching vibration of aromatic amine out of plane deformation of C-H of 1,4 disubstituted rings. The aromatic C-H bending in the plane (1167cm^{-1}) and out of plane (831cm^{-1}) for a 1,4 disubstituted aromatic ring indicates a linear structure.

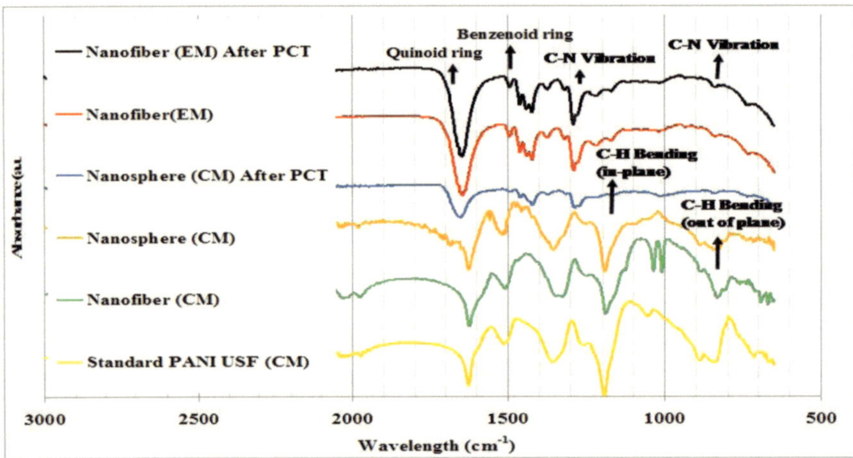

Figure 4. FTIR spectra of polyaniline nanofibers and nanospheres.

Figure 5(a) represents the scanning electron micrograph of PANI-NF-CM and Figure 5(b) shows the structure of the PANI-NF-ES sample. It is clear that the electrospun nanofibers exhibit a smoother surface as opposed to the rough surface of the chemically grown nanofibers. Figure 6 shows the cycle life kinetics of the electrospun PANI nanofibers. No hydrogen uptake was observed until the sample was heated to 100°C. The initial high capacity of 11wt.% is not fully observed when desorbing the hydrogen, as only 8wt.% is released. This occurs in a two-step process characterized by a fast (physisorption) hydrogen release and a slower (chemisorption) hydrogen release step. The full capacity of hydrogen desorbed is then reabsorbed and with each consecutive cycle the amount of hydrogen dwindles until finally no more hydrogen is released of absorbed. This hydrogen behavior is due to hydrogen bonding with both unterminated bonds as well as with the surface of the nanofibers.

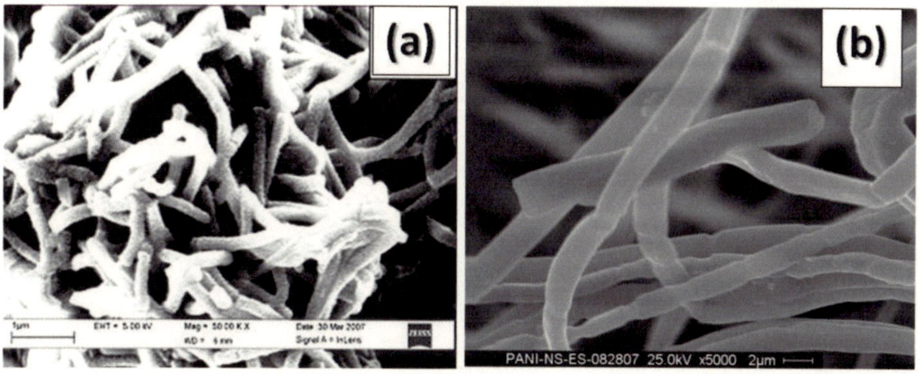

Figure 5. SEM micrographs of polyaniline nanofibers **(a)** chemically grown and **(b)** electrospun.

Pressure-Composition-Temperature (PCT) profiles of the PANI-NF-ES during hydrogen adsorption and desorption were plotted in Figure 7. Hydrogen storage capacity increases with increasing temperature from 50 to 125 °C in various cycles as shown Figure 7(a). At lower temperature of 50 °C, a hydrogen capacity of 3 wt.%, whereas at 100–125 °C, at least two fold increase of capacity (6–8 wt.%) was invariably obtained at various hydrogenation cycles. At the end of each adsorption PCT, desorption PCT experiments were performed by reducing the hydrogen pressure in steps of ΔP = 3 bar, and are depicted in Figure 7(b). A hydrogen storage capacity of 2–8 wt.% was obtained at temperature range of 50–125 °C.

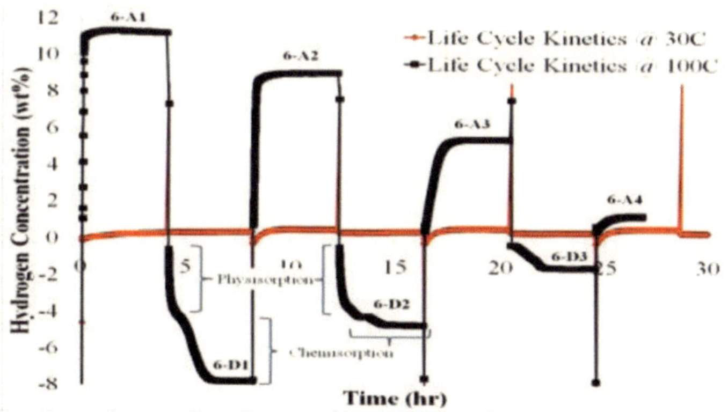

Figure 6. Hydrogen sorption kinetics in PANI-NF-ES.

Figure 8(a) represents the hydrogen absorption kinetic curves of the PANI-NF-ES sample after 55 previous sorption cycles at 30 °C. It can be seen that after approximately 2 h at 80 bar of H_2 pressure close to 5 wt.% of hydrogen is absorbed.

From the same figure, the hydrogen desorption was plotted after 66 cycles at 100 °C and reveals that the kinetics are rather rapid, with most of the hydrogen being released in less than 30 min for a total hydrogen release of close to 6 wt.%. Additionally, the absorption kinetics in the 64th cycle of hydrogen at 80 bar pressure with varying temperature is shown in Figure 8(b). Initially, the hydrogen is absorbed at 125 °C with saturation occurring after approximately 2 h (solid line). After 18 h, the temperature (dotted line) is reduced to 30 °C, whereupon another 3 wt.% of hydrogen is absorbed. This may be due to both chemisorption and physisorption phenomena occur in these materials and the mechanism is yet to be investigated.

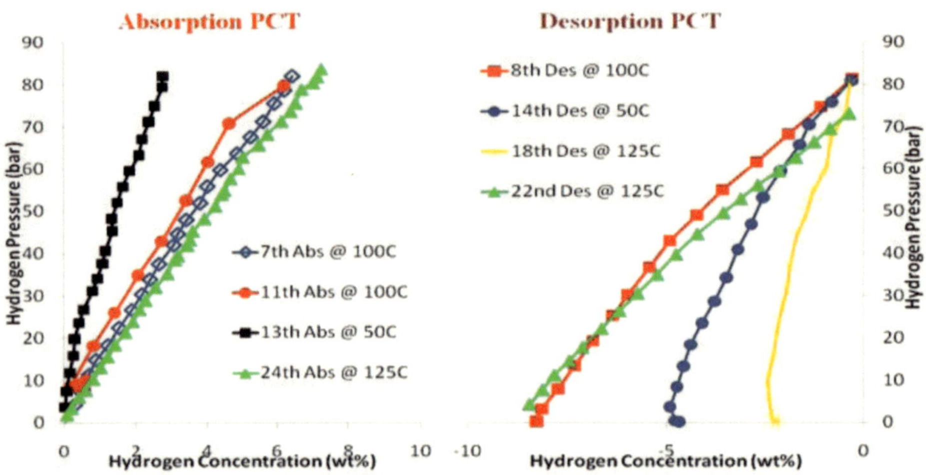

Figure 7. Hydrogen (a) adsorption and (b) desorption PCT curves for the PANI-NF-ES.

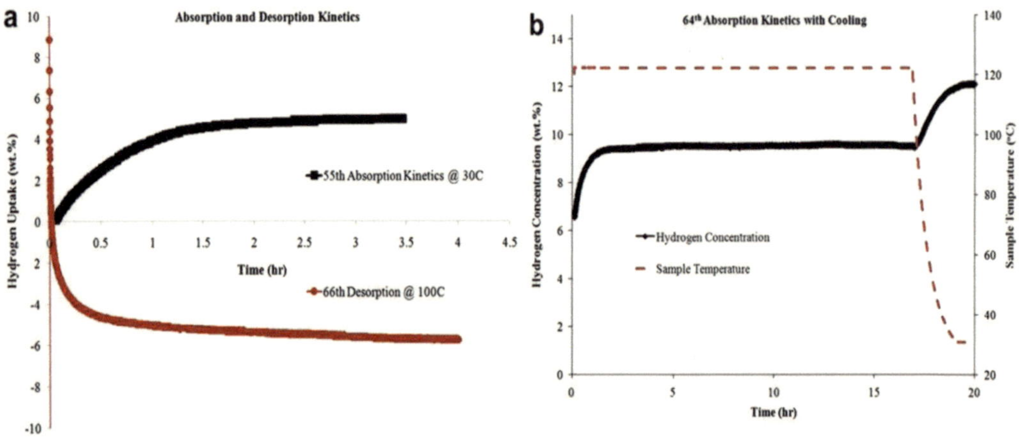

Figure 8. Hydrogen sorption kinetic curves of PANI-NF electrospun samples at different cycles.

The PANI-NF-CM, however, absorbed hydrogen at 30°C and did so reversibly with a capacity of approximately 3wt.%, as seen in Figure 9. Unlike the electrospun nanofibers, the hydrogen was absorbed and released in a one-step process with very rapid kinetics of less than 10min. This fast hydrogen sorption is most likely due to the higher surface area of the chemically grown nanofibers as opposed to the electrospun nanofibers that exhibit a much smoother surface.

When looking at the microstructure of both types of polyaniline nanofibers, it is interesting to note that the nanofibers that were initially present, for the chemically grown samples have disappeared completely, as shown in Figure 10(a). BET surface area measurements, however, have shown that the surface area of the PANI-NF-CM has actually stayed constant, which is explained by the porous nature of the sample of hydrogen cycling. The hydrogen interacted with the PANI-NF-CM and created diffusion pathways for it to pass through the sample reversibly.

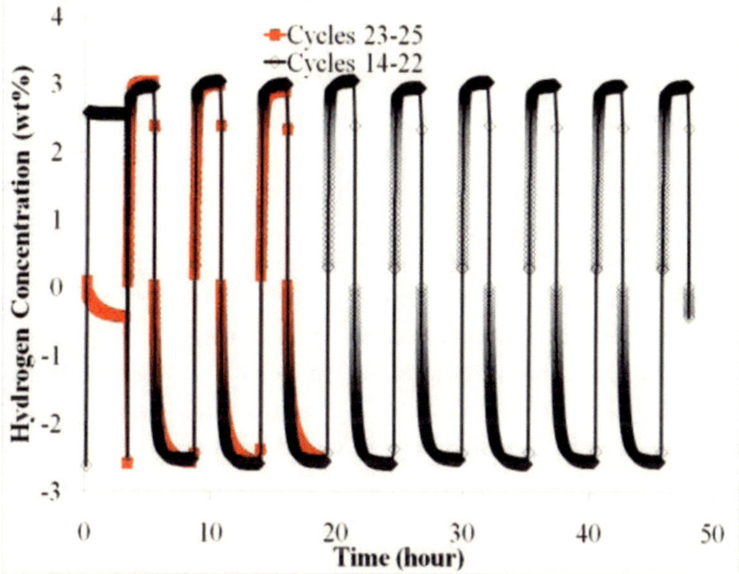

Figure 9. Hydrogen sorption kinetics in PANI-NF-CM.

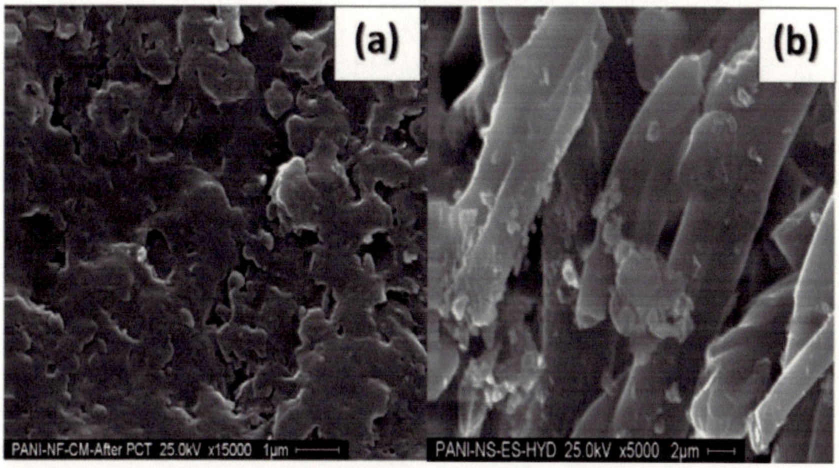

Figure 10. SEM micrographs of polyaniline nanofibers **(a)** chemically grown and **(b)** electrospun after hydrogen cycling measurements.

The microsturcture of the hydrogen cycled PANI-NF-ES, on the other hand, as shown in Figure 10(b), clearly show that the nanofibrallar structure is still intact, although the nanofibers show clear evidence of swelling and also of breaking apart. This further confirms the point that the hydrogen interacted more in a chemisorption manner with the sample, rather than physisorption.

3. Nanoscale complex multinary hydrides for reversible hydrogen storage

Advanced complex hydrides that are light weight, low cost and have high hydrogen density are essential for on-board vehicular storage (Grochala & Edwards, 2004; Schlapbach & Zuttel, 2001; Stefanakos et. al., 2007; US Department of Energy (DOE) Report 2003) Some of the complex hydrides with reversible capacities achieved are Alanates (Bogdanovic & Schwickardi, 1997; Jensen & Zidan, 2002; Srinivasan et. al., 2004), Alanes (Graetz et. al., 2005), Amides (Chen et. al., 2002; Hu & Ruckenstein, 2005) and Borohydrides (Au, 2006; Srinivasan et. al., 2008; Vajo et. al., 2005) magnesium based hydrides (Orimo et. al., 2005; Srinivasan et. al., 2006) and mixed complex hydrides (Jurczyk et. al., 2007) have been recently reported with improved hydrogen storage characteristics. The challenging tasks to design and develop the complex hydrides mandate an optimization and overcoming of kinetic and thermodynamic limitations (Fitchner, 2005; Zuttel, 2004). The enhancement of reaction kinetics at low temperatures and the requirement for high hydrogen storage capacity (> 6.5 wt.%) of complex hydrides could be made possible by either adopting destabilization strategies or catalytic doping. If nanostructured materials with high surface area are used as the catalytic dopants, they may offer several advantages for the physico-chemical reactions, such as surface interactions, adsorption in addition to bulk absorption, rapid kinetics, low temperature sorption, hydrogen atom dissociation and molecular diffusion via the surface catalyst.

The intrinsically large surface areas and unique adsorbing properties of nanophase catalysts can assist the dissociation of gaseous hydrogen and the small volume of individual nanoparticles can produce short diffusion paths to the materials' interiors. The use of nanosized dopants enables a higher dispersion of the catalytically active species (Joo et. al., 2001) and thus facilitates higher mass transfer reactions. Recently, it is claimed that an enhancement of reaction kinetics has been demonstrated with CNT catalyzed $NaAlH_4$ as represented in Figure 14a and 14b (Dehouche et. al., 2005). It was speculated that the repeated interactions of CNT with $NaAlH_4$ during mechanical milling causes an insertion of nanotube bundles (~100-200 nm) in $NaAlH_4$ matrix and accompanying deformation as shown in Figure 14c (Pukazhselvan et, al., 2005).

In addition, the structure of carbon plays an important role on the hydrogenation and dehydrogenation behavior of $NaAlH_4$. Improvement of reaction rate and hydrogen storage capacity has also been reported for the Alanates doped with Ti-nanoparticles (Bogdanovic et. al., 2003). Apparently, the use of nanosized doping agents enables one to achieve a higher dispersion of the catalytically active Ti species than their bulk counterparts. Though the catalytic enhancement of decomposition of sodium aluminum hydride leads to fast sorption kinetics at moderate temperatures, its usable hydrogen storage capacity is rather limited (~5.6 wt%). Lithium borohydride ($LiBH_4$) on the other hand, possesses theoretical hydrogen content of ~18.3 wt%, exhibiting potential promise for on-board applications. However, hydrogen decomposition from $LiBH_4$ starts at an elevated temperature of 380 °C and also shows little or no reversible hydrogenation behavior. It has been reported that catalytically doping SiO_2 lowers the temperature of hydrogen evolution to 300 °C (Zuttel et. al., 2003). In addition to improving the kinetics of $LiBH_4$, tailoring the thermodynamic property is of vital importance for developing an ultimate "holy grail" material for hydrogen storage.

In this direction, the hydrogenation/dehydrogenation enthalpy has been reduced by 25 kJ/mol of H_2 for the ($LiBH4+½ MgH_2$) system (Vajo et. al., 2005). Magnesium hydride destabilizes the $LiBH_4$ structure and forms an intermediate MgB_2 phase (Barkhordarian et. al., 2007; Cho et. al., 2006) during the hydrogen charging and discharging reactions. In a very recent study on the $LiBH_4/MgH_2$ system, it was demonstrated that the hydrogenation and dehydrogenation of $LiBH_4$ occurs via the Li-Mg alloy phase (Yu et. al., 2006) and not through MgB_2 formation. It is generally known that pristine MgH_2 theoretically can store content of ~7.6 wt.% hydrogen (Zaluska et. al., 2001). However, so far, magnesium hydride based materials have limited practical applications because both hydrogenation and dehydrogenation reactions are very slow and hence relatively high temperatures are required (Schlapbach & Zuttel, 2001). Magnesium hydride forms ternary and quaternary hydride structures by reacting with various transition metals (Fe, Co, Ni, etc) and thus improved kinetics (Jeon et. al., 2007). Moreover, the nanoscale version of these transition metal particles offers an additional hydrogen sorption mechanism via its active surface sites (Jeon et. al., 2007; Zaluska et. al., 2001). In a similar way, the synergistic approach of doping nanoparticles of Fe and Ti with a few mol% of CNT on the sorption behavior of MgH_2 has recently been investigated (Yao et. al., 2006). The addition of Carbon Nanotubes (CNT) significantly promotes hydrogen diffusion in the host metal lattice of MgH_2 due to the short pathway length and creation of fast diffusion channels (Wu et. al., 2006).

The dramatic enhancement of kinetics of MgH_2 has also been explored through reaction with small amounts of $LiBH_4$ (Johnson et. al., 2005). Though the MgH_2 ad-mixing increases the equilibrium plateau pressure of $LiNH_2$ (Luo, 2004) or $LiBH_4$ (Rivera et. al., 2006; Vajo et. al., 2005), catalytic doping of these complex hydrides had not been investigated. It is generally believed that the role of the nanocatalyst on either $NaAlH_4$ or MgH_2 is to stabilize the structure and facilitate a reversible hydrogen storage behavior. However, the actual mechanism by which transition metal nanoparticles enhance the dehydrogenation kinetics of complex hydrides is not yet understood very well. Thus, it is desirable to investigate the effects of catalysts, defects, grain boundaries, interface boundaries and impurity atoms on the surface. Based on an extensive literature search, we found no reports on the synergistic effects of transition metal nanoparticles on the complex composite hydrides such as $LiBH_4/MgH_2$, $LiNH_2/LiH/MgH_2$ or $LiBH_4/LiNH_2/MgH_2$ etc.

The present work addresses the grand challenge of hydrogen storage by mechano-chemically milling $LiBH_4$ and $LiNH_2$ with MgH_2 to produce a new complex quaternary Li-Mg-B-N-H structure. This Li-Mg-B-N-H structure possesses storage capacity of more than

10 wt.% at around 150 °C. The parent compounds, $LiBH_4$ and $LiNH_2$, were purchased from Sigma Aldrich with a purity of at least 95%, while MgH_2 was obtained from Alfa Aesar with a purity of 98%. All materials were kept in an inert atmosphere in a glove box and used without further purification. The investigated samples were created in 4g batches with a constant molar ratio of $2LiNH_2$:$LiBH_4$:MgH_2, while taking into account the purity of the parent compounds, by employing high energy ball milling (Fritsch Pulverisette 6) for 5 hours at 300rpm with intermittent hydrogen/argon (5%/95%) purges for 20 minutes before milling and after 2 and 4 hours. This was done to ensure that as little hydrogen as possible was released during the milling process and to reduce the agglomeration of the hydride that occurs when pure hydrogen is used as compared to the hydrogen/argon mixture. The MgH_2 was either added as received or was added as a so-called nano MgH_2. The nano MgH_2 ($nMgH_2$) was created by ball milling the commercial MgH_2 ($cMgH_2$) for 12 hours with intermittent hydrogen/argon purges every 2 hours. This ensured the reduction of particle size as well as the decrease in hydrogen release temperature, as previously reported. The two main processing schemes that were used are shown in Figure 11.

The first processing scheme was to add all parent compounds and mill for 5 or 10 hours using either commercial or nano MgH_2. This is the scheme that is generally used in reported literature and the materials serve as a kind of reference material. The second processing scheme was to first create the quaternary structure LiBNH by milling $LiBH_4$ with $2LiNH_2$ for five hours and then adding either commercial or nano MgH_2, after which the quaternary and the MgH_2 were milled for an additional 5 hours. All milling was carried out in an inert atmosphere and the samples were purged with the hydrogen/argon mixture every 2 hours. In total, five different samples were created. The samples are referred to in this paper according to the naming convention shown in the bold boxes of Figure 11.

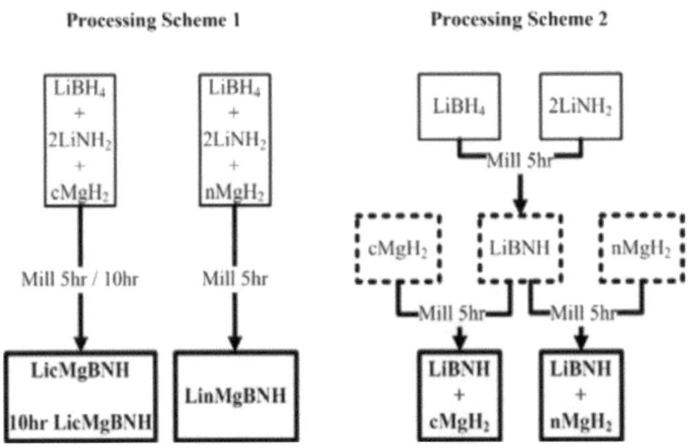

Figure 11. Processing schemes of synthesizing complex hydrides LiBH4/2LiNH2/nanoMgH2 via reactive gas milling.

X-Ray Diffraction. Figure 12 shows the XRD pattern comparing the five differently processed complex hydrides. The parent compounds, $LiBH_4$, $LiNH_2$, as well as both commercial and nano MgH_2 are in the lower half of the figure as a reference. Neither $LiBH_4$ nor $LiNH_2$ peaks are observed in any of the five samples. This confirms that these two materials are fully consumed during the milling process and actually form a new quaternary structure, referred to as LiBNH. The addition of commercial MgH_2 does not cause the formation of a new complex structure, but instead indicates that the quaternary structure is preserved, while the MgH_2 simply intermixes with the material. When the nano MgH_2 is added to $LiBH_4$ and $LiNH_2$ or to the quaternary LiBNH, the MgH_2 peaks are barely picked up by the XRD. This indicates that the small size of the MgH_2 causes the material to intermix and fill voids of the quaternary structure, which results in a nanocrystalline particle distribution, while still preserving the quaternary structure formed by the $LiNH_2$ and $LiBH_4$. All samples are a physical, rather than a chemical, mixture of the quaternary structure LiBNH with MgH_2.

Thermal programmed desorption. Upon producing the complex hydrides, each sample was characterized for its thermal desorption characteristics using TPD with a heating rate of 1, 5, 10, and 15°C/min. As compared to the quaternary structure, the multinary structure containing MgH_2 showed a 3-step hydrogen release mechanism, as is shown in Figure 13. Our TPD analysis of LicMgBNH sample confirms the previously reported data (Yang et al.), but also shows that the processing condition of the material does have an effect on the thermal decomposition characteristics. The first hydrogen release peaks between 157.7°C for 10hr LicMgBNH and 165.2°C for LinMgBNH, which is a relatively small difference in temperature. When investigating the second, or main, peak of the various samples, it is interesting to note that the temperature range for main hydrogen release varies from 287°C for the 10hr LicMgBNH to 306.6°C for LinMgBNH. MgH_2 interacts with the quaternary structure and destabilizes it, thereby releasing hydrogen. In the case of the LinMgBNH and LicMgBNH samples, it takes the entire milling duration (5hr) to form the quaternary phase, thereby giving the MgH_2 little time to allow for the release of hydrogen during milling. In the case of 10hr LicMgBNH, the MgH_2 has 5hr of milling duration (after the formation of the quaternary) to destabilize the material and allow for hydrogen release.

Finally, in the case of LiBNH+cMgH_2, the MgH_2 has to first be reduced in size to allow for the destabilization and the resultant hydrogen release. LiBNH+nMgH_2, however, releases hydrogen during the milling process, as the smaller size of nano MgH_2 allows for immediate destabilization of the already present quaternary. Either prolonged milling duration or reduction in particle size will lead to release of hydrogen in the milling process

Activation energy. The activation energy of each sample was experimentally determined using Kissinger's method, based on TPD data (Figure 13) taken at 1, 5, 10, 15°C/min for the two peaks and correlated to the hydrogen peak release temperature as described by:

$$\frac{d\left(\ln\frac{\beta}{T_m^2}\right)}{d\left(\frac{1}{T_m}\right)} = -\frac{E}{R}$$

where β is the heating rate of the sample in K/min, T_m is the peak temperature in K, E is the activation energy of the hydrogen release, and R is the gas constant. When investigating the first peak, around 160°C, it is interesting to note that the 10hr LicMgBNH sample has the lowest activation energy (109.8 kJ/mole) at 157.7°C as seen in Figure 14. Although all samples exhibit comparable decomposition temperatures, their activation energies vary by ~20 kJ/mole.

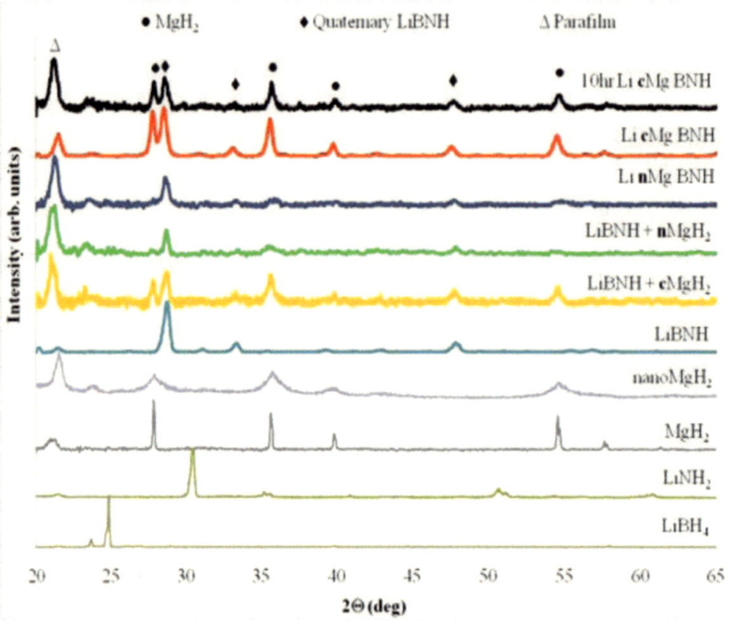

Figure 12. XRD Profile of the five differently processed materials as well as the parent compounds, $LiBH_4$, $LiNH_2$, MgH_2 and nano MgH_2.

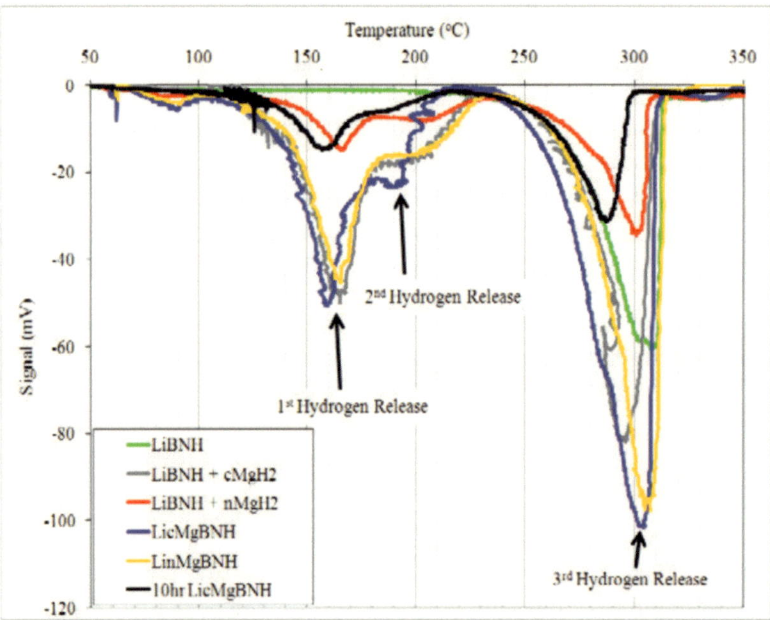

Figure 13. TPD comparison of investigated processing variations showing the two main hydrogen release regions around 160°C and 300°C.

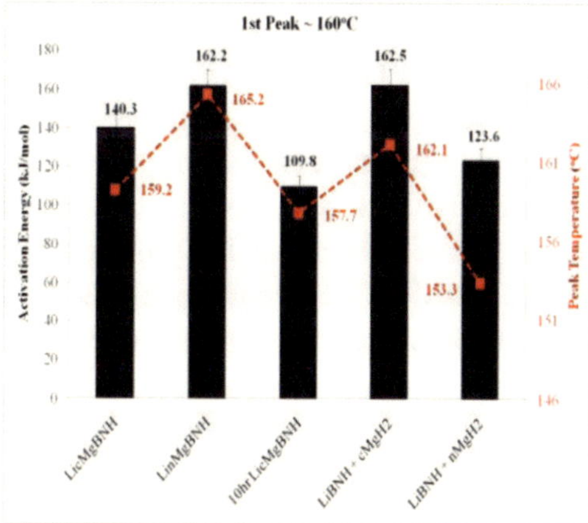

Figure 14.

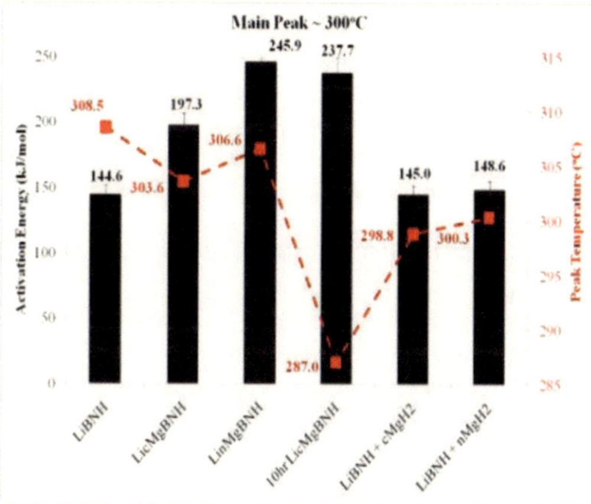

Figure 15. Activation energy, as calculated from the TPD data using Kissinger's method, compared with the first and main peaks of hydrogen release temperature.

A plausible reason is that the reaction pathways of ad-mixing MgH_2 either in the first place (10hr LicMgBNH) or after the quaternary formation (e.g. LiBNH+nMgH_2) proceeds with fine distribution of nanocrystalline MgH_2 in the host matrix of multinary hydrides. It has been recently claimed that the nanocrystallization of MgH_2 has significant impact on lowering the enthalpy of formation and enhancement of the reaction kinetics. The high temperature main hydrogen release peak (300°C) for all the processed materials and the reference LiBNH quaternary hydride are shown in Figure 6. While comparing the activation energies and decomposition temperatures of all the samples, it can be clearly inferred that the quaternary hydrides LiBNH combined with either commercial (LiBNH+cMgH_2) or nanocrystalline MgH_2 (LiBNH+nMgH_2) milled for 5 hours show lower values, e.g. 145-148 kJ/mole at ~300°C. Simply based on this data, it is difficult to justify from both Figures 15 and 6, which sample possesses an optimum hydrogen release characteristic at these two main decompositions. However, it is undoubtedly clear that both the steps occur at two different temperature regimes, namely 160 and 300°C, at which either surface adsorbed or bulk absorbed hydrogen is released.

At the low temperature hydrogen release step, the nano MgH_2 acts as a facilitator in speeding up the reaction; hence both the 10hr LicMgBNH and LiBNH+nMgH_2 materials demonstrate lower activation energies (Figure 14). On the other hand, in the high temperature hydrogen release step (Figure 15), the temperature of 300°C acts as a driving force to release hydrogen from the bulk structures of both LicMgBNH and LinMgBNH milled only for 5 hours. Hence, these materials exhibit lower activation energies which are comparable to pristine LiBNH. Based on the detailed analysis, we draw the conclusion that an additional 5 hours of ball milling, either of the all-in-one hydride (10hr LicMgBNH) or the quaternary/nanocrystalline hydride mixture (LiBNH+nMgH_2), will alter the decomposition characteristics, especially the activation energy which is vital for hydrogen storage.

Pressure-composition-isotherms. Figure 16 represents the PC isotherms of the multinary complex hydrides created with different processing conditions. The PCT studies of the multinary samples are carried out under the following conditions: temperature, T=150-175°C; pressure difference between aliquots, ΔP=3bars (0.3 MPa); absorption pressure limit, P_a=80bars (8 MPa); desorption pressure limit, P_d=0bar (0 MPa); and reservoir volume, V_r=160cm^3. Since all these samples are in hydride phases, the dehydrogenation experiment was followed by the rehydrogenation for at least 10 hours.

The PCT characteristics and their observations are given with respect to the sample processing conditions as follows.

LicMgBNH and LinMgBNH: The multinary complex hydrides processed with either commercial or nanocrystalline MgH_2 and milled all-in-one for 5 hours reveal reproducible hydrogen capacity of 3 to 4wt.%. It is noteworthy to mention that LinMgBNH possesses at least 1wt.% higher capacity and 25°C reduction in temperature as compared to the LicMgBNH counterpart, as also seen from the initial ramping kinetic profiles in Figure 7b. This could be achieved because of the uniform distribution of fine MgH_2 nanoparticles which likely act as hydrogen diffusion enhancement sites, increasing the amount of hydrogen released. Yet another difference between these two processed materials is the tailoring of the plateau pressure (hydrogen/hydride equilibrium region), which is crucial for a hydrogen storage system to be viable for mobile applications. The LinMgBNH material exhibits reduction in the absorption plateau pressure by 20 bars (2 MPa) in contrast to the LicMgBNH due to nanoparticulate formation.

LiBNH+cMgH_2 and LiBNH+nMgH_2:A greater reversible hydrogen storage capacity of 5.3-5.8wt.% was found at temperatures of 150°C to 175°C for the quaternary hydrides LiBNH either milled with commercial or nano MgH_2 for 5 hours. The nano MgH_2 loaded LiBNH outperformed its commercial counterpart with a higher hydrogen capacity of 5.8wt.%. at 150°C as compared to 175°C (refer to Figure 16b). Figure 16a shows that there are plateau pressure regions, which are not as clearly defined as the LicMgBNH and LinMgBNH samples. Moreover, the sorption plateau of these samples resembles greatly the pristine LiBNH (not shown here), which is the precursor material for the multinary hydride formation. Overall, it is unambiguously claimed that LiBNH admixed either with commercial or nano MgH_2 and milled for 5 hours, exhibits a high reversible hydrogen storage capacity of ~6wt.% at temperatures less than 175°C.

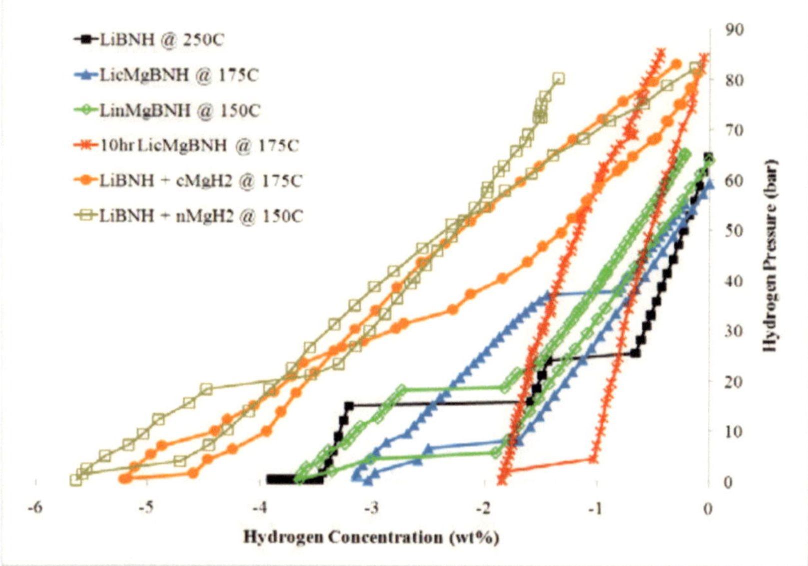

Figure 16. Comparison of the hydrogen sorption characteristics of the various processing conditions at the lowest hydrogen release temperature.

10hr LicMgBNH: The extended milling duration of 10 hours for the three component systems, $2LiNH_2$+$cMgH_2$+$LiBH_4$ show poor hydrogen performance as depicted in Figure16a. A low hydrogen desorption capacity of 2wt.% at a plateau pressure of less than 5bars (0.5 MPa) with little or no reversibility is obtained in this material. The crystallite agglomeration during the prolonged milling is expected to be the limiting factor for the absence of plateau pressure and overall storage capacity. Though this sample exhibits a lower activation energy (~109 kJ/mole) in the first hydrogen release (see Figure 14), the effective hydrogenation needs systematic optimization strategies which are currently under investigation.

Crystallite size effects on hydrogen release characteristics. In order to better understand the hydrogen performance of the differently processed materials, the hydrogen capacity was correlated with the crystallite sizes of the quaternary phase, LiBNH, and the MgH_2. The crystallite sizes were calculated from the XRD data (Figure 12) of each material using Scherrer's method. The initial crystallite sizes of $LiNH_2$, $LiBH_4$, MgH_2, nano MgH_2 and LiBNH were determined to be 138nm, 152nm, 212nm, 27nm, and 60nm, respectively. As seen from Figure 16b, the nano size MgH_2 has a definite effect on the initial hydrogen release temperature. Both samples synthesized with nano MgH_2 release hydrogen at 150°C as compared to 175°C for all the other samples, which were synthesized with its commercial counterpart. The MgH_2 crystallite size for the nano MgH_2 samples are both approximately 10nm, whereas the crystallite size of the commercial MgH_2 samples vary from 35nm to 75nm, as seen in Figure 17. It is important to note that the crystallite size of both MgH_2 and LiBNH are largest for the 10hr LicMgBNH sample, which explains the poor hydrogen performance. This is because of the well known fact that larger particles, and therefore a smaller surface area, can limit hydrogen performance (less than 2wt.% capacity) due to diffusion inhibition and passivation effects.

A milling duration of more than 5 hours is in fact counterproductive and allows for the crystallite size to increase, as both the LiBNH and MgH_2 agglomerate. When looking at the correlation between crystallite size and hydrogen concentration, as shown in Figure 17, it becomes evident that the size of the LiBNH crystallites plays an important role on the amount of hydrogen released.

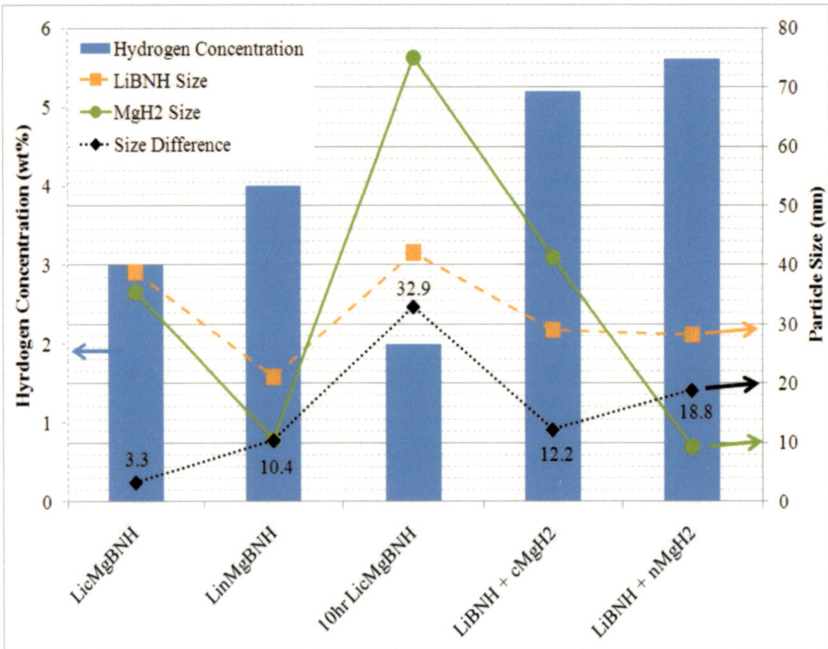

Figure 17. Comparison of hydrogen concentration and crystallite size of the quaternary LiBNH and MgH_2 phases after milling.

If the LiBNH crystallites have a size of approximately 28nm and the MgH_2 crystallites are approximately within 15nm (13nm to 43nm) of this size, the highest possible hydrogen release is achieved (5.5wt.%). When the MgH_2 and LiBNH crystallites are either too similar in size, as in the case for the LicMgBNH sample (3.3nm difference), or if they are too different in size, as in the case for the 10hr LicMgBNH sample (32.9nm difference), the amount of hydrogen released by the sample is reduced. Overall, the new multinary complex hydrides show potential promise on the reversible hydrogen storage characteristics as elucidated in Figure 18. The cyclic kinetics of Li-nMg-B-N-H reveals a reversible hydrogen storage capacity of 6-8wt.% with fast sorption kinetics around 200-250°C which matches or exceeds with the set-forth DOE technical targets. Thus a reversible hydrogen storage capacity of 5-7 wt.% was reproducibly obtained for these materials. With respect to the mass spec analyses, note that the scans did not reveal any evidence for potential parent (or daughter) species (ions); namely, NH_3, BH_3, BH_2NH_2, N_2H_4.

Destabilization of $LiBH_4/LiNH_2/MgH_2$ with Nano Sized Additives. Based on the previous experimental work on the optimization of the processing conditions as well as some previous insight into the role of nano sized additives on the quaternary LiBNH structure, it was decided to systematically investigate the effect of various nano sized additives on the multinary structure LiBNH + $nMgH_2$, as described in the previous section.

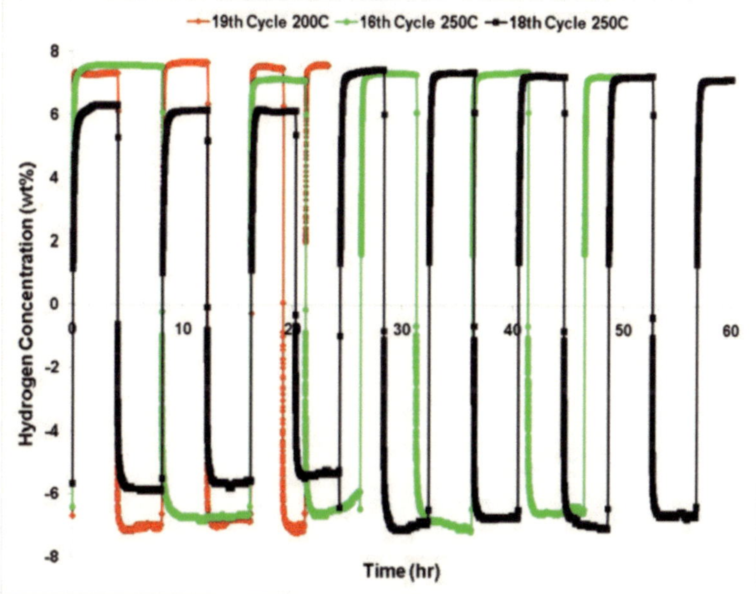

Figure 18. Dehydrogenation and reversible rehydrogenation cyclic kinetics in Li-nMg-B-N-H materials

TPD was used to obtain information about changes in the parent material's hydrogen characteristics. Specifically, the temperature of hydrogen release as well as some general information about the kinetics of hydrogen release can be ascertained from this type of measurement. The peak temperature indicates the optimal hydrogen release temperature, whereas the width of the peak can be used to get insight into the rate at which hydrogen is released, at least qualitatively. A wide peak indicates a low rate of hydrogen release, whereas a narrow and sharp peak indicates rapid hydrogen release. Figure 19 demonstrates the lower temperature release of hydrogen by nanoadditve doping when compared to the pristine complex hydride. As previously described, the parent compound, LiBNH + $nMgH_2$, exhibits a three-step hydrogen release. While the TPD measurements are used for quick-screening the effect of the additives on the hydrogen performance of the material, it can be seen that each additive material either affects the rate of hydrogen release, as depicted by a sharp and narrow peak (especially iron) or significantly lowers the temperature required for hydrogen release.

Since the TPD measurements only give an indication of the hydrogen sorption results, ramping kinetic measurements, where approximately 0.1g of sample are loaded into the PCT and then ramped at a rate of 1°C/min, were performed on all samples. Figure 20 shows the more detailed hydrogen performance of the standard sample, LiBNH+$nMgH_2$ without any additives, as well as with 2mol% of the aforementioned additives.

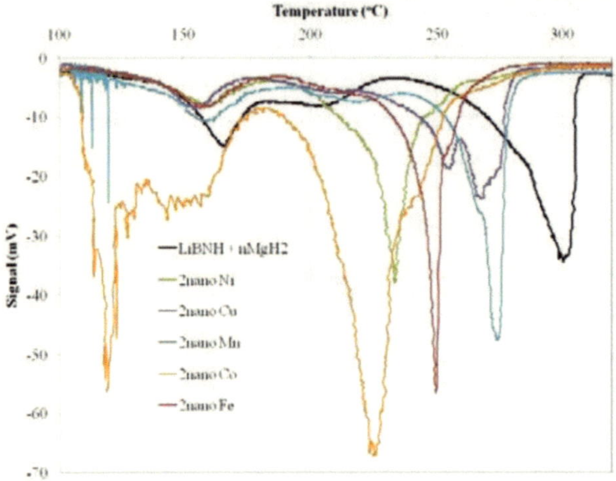

Figure 19. TPD comparison of LiBNH+nMgH₂ without additive and with 2mol% Ni, Cu, Mn, Co and Fe at a constant ramping rate of 1°C/min.

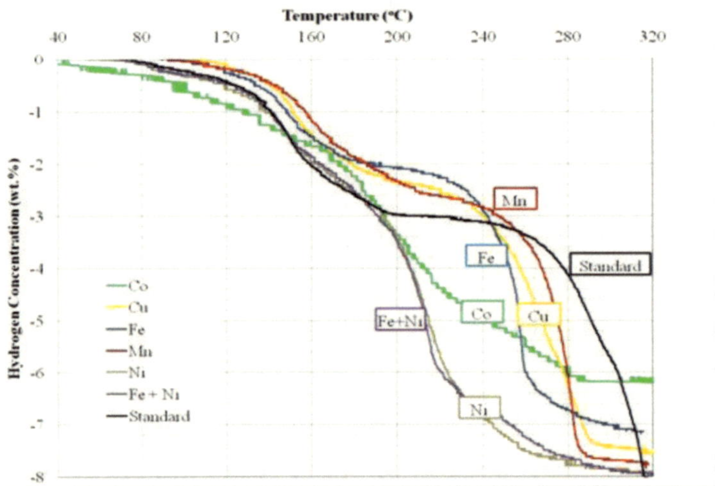

Figure 20. Ramping kinetics measurements of LiBNH+nMgH2 without and with 2mol% nano Mn, Fe, Co, Cu, Ni and Fe+Ni.

It becomes clear that while cobalt seemed promising from the TPD data, it in fact has such slow kinetics, that cobalt is no longer of interest as an additive. The kinetics measurement do confirm the TPD data in that manganese and iron have the fastest kinetics, as indicated by the slope of the desorption curves. Furthermore, the significant reduction in hydrogen release temperature of nickel is confirmed. Figure 21 shows a comparison of the hydrogen release rate and hydrogen release temperature of the standard sample without and with 2mol% of nano sized additives. Since the nano sized nickel showed the lowest hydrogen release temperature of just under 200°C and nano sized iron showed the highest release rate (0.2wt.%/min) at a comparatively low temperature of 245°C, these two additives were chosen to be optimized in terms of their concentration.

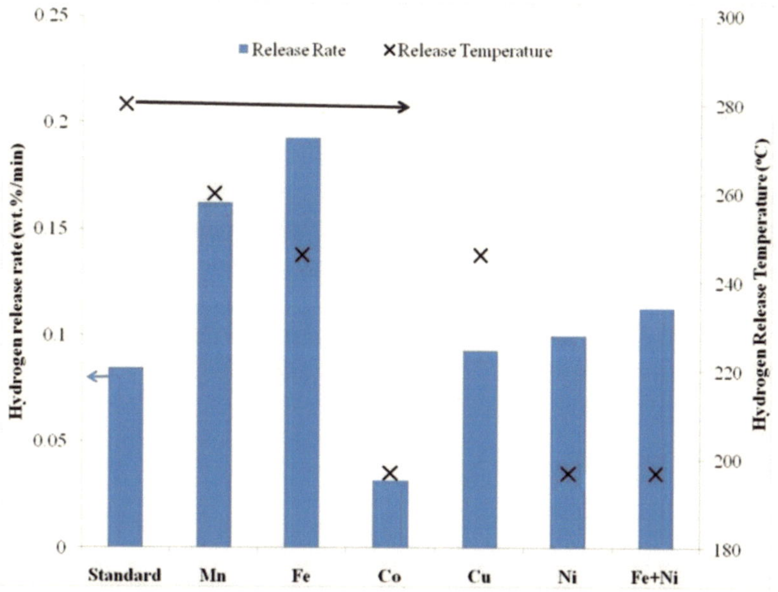

Figure 21. Comparison of hydrogen release temperature and hydrogen release rate of the standard LiBNH+nMgH_2 and LiBNH+MgH_2 with 2mol% of various nano additives.

Material	Capacity	Reversibility	Temperature	Comments
PANI Bulk	0.4 wt%	Small	125°C	
PANI NS-CM	6 wt%	Decreases to 0.5 wt%	30°C	Slow kinetics (hours)
PANI NF-CM	3 wt%	Reversible	30°C	Fast Kinetics (<10min)
PANI NF-ES	10wt%	Reversible decreases with cycle kinetics	100°C (kinetic) 125°C (PCT)	Kinetics combination of physisorption and chemisorption

Table 1. Polyaniline Nanostructures and their hydrogen storage characteristics.

Material	Capacity	Reversibility	Temperature	Activation Energy
LicMgBNH	3.2 wt%	Reversible	175°C	140.3 kJ/mol
LinMgBNH	3.8 wt%	Reversible	150°C	162.2 kJ/mol
LicMgBNH 10hr	1.9 wt%	Reversible	175°C	109.8 kJ/mol
LiBNH + cMgH_2	5.2 wt%	Reversible	175°C	162.5 kJ/mol
LiBNH + nMgH_2	5.6 wt%	Reversible	150°C	123.6 kJ/mol

Table 2. Hydrogen storage capacity of multinary complex hydrides

4. Conclusion

A systematic study of two chemical hydrogen storage systems has been carried out. The first system, polyaniline, was synthesized in its nanostructured forms of nanofibers and nanospheres. This system was investigated for its physisorption of hydrogen through weak secondary atomic bonds. The second main system was investigated for its strong primary atomic bonding of hydrogen and consisted of $LiBH_4$, $LiNH_2$, and MgH_2. Different nano sized additives were added to the complex hydride system, though mainly the complex hydride system was found to destabilize to allow for hydrogen storage at lower temperatures and with faster kinetics. The cumulative results of both the physisorbed and chemisorbed hydrogen storage systems are summarized in Table 1 and Table 2.

Author details

Sesha S. Srinivasan and Prakash C. Sharma
Department of Physics, Tuskegee University, Tuskegee, Alabama, USA

Acknowledgement

Financial support from the Office of Naval Research (ONR-DURIP Grant# N00014-10-1-0721), Florida Hydrogen Initiative (FSEC-DOE Grant# 20126050), and BP Oil Spill grant from Dauphin Island Sea Lab are gratefully acknowledged. Authors wish to thank Dr. Gilbert L. Rochon, President, Dr. Luther S. Williams, Provost and Dr. Shaik Jeelani, VP Research, Tuskegee University, for their encouragement and support in establishing state-of-the-art facilities for hydrogen storage research.

5. References

Athawale, A.A. & Bhagwat, S.V.J.J. (2003). Synthesis and characterization of novel copper/polyaniline nanocomposite and application as a catalyst in the Wacker oxidation reaction. *J. Appl. Polymer Science*, Vol. 89, No. 9, 2412-2417

Au, M. (2006). Destabilized and catalyzed alkali metal borohydrides for hydrogen storage with good reversibility, *U.S. Patent Appl. Publ.*, US 2006/0194695 A1, Aug. 31 (2006)

Barkhordarian, G., Klassen, T., Dornheim M., & Bormann, R. (2007). Unexpected kinetic effect of MgB_2 in reactive hydride composites containing complex borohydrides. *J. Alloys Comp.*, Vol. 440, No. 1-2, pp. L18-L21

Bogdanovic, B., & Schwickardi, M. (1997). Ti-doped alkali metal aluminum hydrides as potential novel reversible hydrogen storage materials. *J. Alloys Comp.*, Vol. 253-254, pp. 1-9

Bogdanovic, B., Felderhoff, M., Kaskel, S., Pommerin, A., Schlichte K., & Schuth, F. (2003). Improved hydrogen storage properties of Ti-doped sodium aluminate using titanium nanoparticles as doping agents. *Adv. Mater.*, Vol. 15, No. 12, pp. 1009-1012

Bowman, R.C., & Stetson, N.T. (2010). DOE Hydrogen and Fuel Cells Program Record, On-Board Hydrogen Storage Systems – Projected Performance and Cost Parameters, Record # 9017. www.hydrogen.energy.gov/pdfs/9017_storage_performance.pdf

Broom, D.P. (2011). *Hydrogen Storage Materials; The Characterization of their Storage Properties*, Springer Energy Technology Series, ISBN 978-0-85729-220-9, London, 1st Edition, XII, p. 258

Chen, P., Xiong, Z., Lou, J., Lin J., & Tan, K.L. (2002). Interaction of hydrogen with metal nitrides and imides. *Nature*, Vol. 420, pp. 302-304

Chiou, N-R., Lee, L.J., & Epstein, A.J. (2008). Porous membrane controlled polymerization of nanofibers of polyaniline and its derivatives. *J. Materials Chem.*, Vol. 18, pp. 2085-2089

Cho, S.J., Song, K.S., Kim, J.W., Kim, T.H., & Choo, K. (2002). Hydrogen sorption in HCl treated polyaniline and polypyrrole, new potential hydrogen storage media. *Proceedings of the American Chemical Society, Division of Fuel Chemistry* Vol.47, No.2, pp. 790-791

Cho, S.J., Choo, K., Kim, D.P., & Kim, J.W. (2007). H_2 sorption in HCl-treated polyaniline and polypyrrole. *Catalysis Today*, Vol. 120, No. 3-4, pp. 336-340

Cho, Y.W., Shim J-H., & Lee, B.-J. (2006). Thermal destabilization of binary and complex metal hydrides by chemical reaction: A thermodynamic analysis. *CALPHAD: Computer Coupling of Phase Diagrams and Thermochemistry*, Vol. 30, pp. 65-69, ISSN: 0364-5916

Dehouche, Z., Lafi, L., Grimard, N., Goyette J., & Chahine, R. (2005). The catalytic effect of single-wall carbon nanotubes on the hydrogen sorption properties of sodium alanates. *Nanotechnology*, Vol. 16, pp. 402-409

Dillon, A.C.; Jones, K.M.; Bekkedahl, T.A.; Kiang, C.H.; Bethune, D.S.; & Heben, M.J. (1997) Storage of hydrogen in single-walled carbon Nanotubes, *Nature (London)*, Vol.386, No.6623, pp. 377-379

Ding, H., Wan M., & Wei, Y. (2007). Controlling the Diameter of Polyaniline Nanofibers by Adjusting the Oxidant Redox Potential. *Advanced Materials*, Vol. 19, pp. 465-469

Fichtner, M. (2005). Nanotechnological aspects in materials for hydrogen storage. *Adv. Eng. Mater.*, Vol. 7, pp. 443-455

Fusalba, F., Gouerec, P., Vellers, D., & Belanger, D. (2001). Electrochemical characterization of polyaniline in nonaqueous electrolyte and its evaluation as electrode material for electrochemical supercapacitors. *J. Electrochem. Soc.*, Vol. 148, No. 1, pp. A1-A6

Germain, J., Frechet, J.M.J., & Svec, F. (2007). Hypercrosslinked polyanilines with nanoporous structure and high surface area: potential adsorbents for hydrogen storage. *J. Mater. Chem.*, Vol. 17, No. 47, pp. 4989-4997

Graetz, J., Lee, Y., Reilly, J.J., Park, S., & Vogt, T. (2005). Structures and thermodynamics of the mixed alkali alanates. *Physical Review B: Condensed Matter and Materials Physics*, Vol. 71, No. 18, pp. 184115-184117

Grochala, W., & Edwards, P.P. (2004). Thermal decomposition of the non-interstitial hydrides for the storage and production of hydrogen. *Chem. Rev.*, Vol. 104, pp. 1283-1315

Hu, Y.H., & Ruckenstein, E. (2005). H_2 Storage in Li_3N. Temperature-Programmed Hydrogenation and Dehydrogenation. *Ind. Eng. Chem. Res.*, Vol. 42, No. 21, pp. 5135-5139

Huczko, A. (2000). Template-based synthesis of nanomaterials. *Appl. Phys. A: Materials Science & Processing*, Vol. 70, No. 4, pp. 365-376

Hwang, J.Y., Hwang, S.H., Lee.; Sim, K.S., & Kim, J.W. (2002). Synthesis and hydrogen storage of carbon nanofibers, *Syn. Metals*, Vol.126, No.1, pp. 81-85

Jena, P. (2011). Materials for Hydrogen Storage: Past, Present and Future. *J. Phys. Chem. Lett.*, Vol. 2, No. 3, pp. 206-211

Jensen, C.M., & Zidan, R.A. (2002). Hydrogen storage materials and. method of making by dry homogenation. *U.S. Patent, 6,471935*

Jeon, K.-J., Theodore, A., Wu, C-Y., & Cai, M. (2007). Hydrogen absorption/desorption kinetics of magnesium nano-nickel composites synthesized by dry particle coating technique. *Int. J. Hydrogen Energy*, Vol. 32, No. 12, pp. 1860–1868

Johnson, S.R., Anderson, P.A., Edwards, P.P., Gameson, I., Prendergast, J.W., Al-Mamouri, M., Book, D., Harris, I.R., Speight J.D., & Walton, A. (2005). Chemical activation of MgH_2; a new route to superior hydrogen storage materials. *Chem. Commun.*, pp. 2823-2825

Joo, S.H., Choi, S.J., & Ryoo, R. (2001). Ordered nanoporous arrays of carbon supporting high dispersions of platinum nanoparticles. *Nature*, Vol. 412, No. 6843, pp. 169-172

Jurczyk, M.U., Kumar, A., Srinivasan, S., & Stefanakos, E. (2007). Polyaniline-based nanocomposite materials for hydrogen Storage. *Int. J. Hydrogen Energy*, Vol. 32, No. 8, pp. 1010-1015

Jurczyk, M.U., Srinivasan, S.S., Kumar, A., Goswami, Y., & Stefanakos, E. (2007). Effects of nano catalysts on a $LiBH_4/LiNH_2$ system for hydrogen storage. *Proceedings of the MS&T Conference*, Detroit, MI, 2007

Kuritka, I., Negri, F., Brancolini, G., Suess, C., Salaneck, W.R., Friedlein, R. (2006). Lithium intercalation of phenyl-capped aniline dimers: a study by photoelectron spectroscopy and quantum chemical calculations. *J Phys Chem B.*, Vol. 110, No. 38, 19023-19030

Lindino, C.A., Casagrande, M., Peiter, A., & Ribeiro C. (2012). Poly(*o*-methoxyaniline) modified electrode for detection of lithium ions, *Quimica Nova*, ISSN:1678-7064, Vol. 35, No. 3, pp. 449-453

Luo, W. (2004). ($LiNH_2$-MgH_2): a viable hydrogen storage system. *J. Alloys Comp.*, Vol. 381, pp. 284-287

MacDiarmid, A.G., & Epstein, A.J. (1989). Polyanilines: A novel class of conducting polymers. *Faraday Discuss. Chem. Soc.*, Vol. 88, pp. 317-332

McKeown, N.B., Budd, P.M., & Book, D. (2007). Microporous polymers as potential hydrogen storage materials. *Macromol. Rapid Commun.*, Vol. 28, No. 9, pp. 995-1002

Niitani, T., Shimad, M., Kawamura, K., Dokko, K., Rho, Y-H., & Kanamura, K. (2005). Synthesis of Li+ Ion Conductive PEO-PSt Block Copolymer, Electrolyte with Microphase Separation Structure. *Electrochemical & Solid-State Letters*, Vol. 8, pp. A385-A388

Nikitin, A., Li, X., Zhang, Z., Ogasawara, H., Dai, H., & Nilsson, A. (2008). Hydrogen storage in carbon nanotubes through the formation of stable C–H bonds. *Nano Lettters*, Vol. 8, No. 1, pp. 162-167

Orimo, S., Nakamori, Y., Kitahara, G., Miwa, K., Ohba, N., Towata, S., & Zuttel, A. (2005). Dehydrideing and rehydriding reactions of $LiBH_4$. *J. Alloys Comp.*, Vol. 404-406, pp. 427-430

Panella, B., Kossykh, L., Dettlaff-Weglikowska, U., Hrischer, M., Zerbi, G., & Roth, S. (2005). Volumetric measurement of hydrogen storage in HCl-treated polyaniline and polypyrrole. *Synthetic Metals*, Vol. 151, No. 3, pp. 208-210

Pukhazselvan, D., Gupta, B.K., Srivastava A., & Srivastava, O.N. (2005). Investigations on hydrogen storage behavior of CNT doped $NaAlH_4$. *J. Alloys Compd.*, Vol. 403, pp. 312-317

Rahy, A., & Yang, D.J., Synthesis of highly conductive polyaniline Nanofibers. *Materials Letters*, Vol. *62*, pp. 4311-4314

Rivera, L., Srinivasan, S., Matthew, S., Wolan, J., & Stefanakos, E. (2006). Destabilized $LiBH_4/MgH_2$ for reversible hydrogen storage. *Proceedings of the AIChE Annual Meeting, Conference Proceedings*, San Francisco, CA, United States, 424d, pp. 6, 2006

Rossberg, K., Paasch, G., & Dunsch, L. (1998). The influence of porosity and the nature of the charge storage capacitance on the impedance behaviour of electropolymerized polyaniline films. *J. Electroanalytical Chem.*, Vol. 443, No. 1, pp. 49-62

Sadek, A.Z., Trinchi, A., Wlodarski, W., Kalantar-zadeh, K., Galatsis, K., Baker, C., & Kaner, R.B. (2005). A room temperature polyaniline nanofiber hydrogen gas sensor. *IEEE Sensors*, Vol. 3, pp. 207-210

Satyapal, S., Petrovic, J., Read, C., Thomas, G., & Ordaz , G. (2007). The U.S. Department of Energy's National Hydrogen Storage Project: Progress towards meeting hydrogen-powered vehicle requirements. *Catalysis Today*, Vol. 120, pp. 246–256

Schlapbach, L., & Zuttel, A. (2001). Hydrogen-storage materials for mobile applications. *Nature*, Vol. 414, No. 15 pp. 353-357

Skowron´ski, J.M.; Urbaniak, J. (2008) Nickel foam/polyaniline-based carbon/palladium composite electrodes for hydrogen storage, Energy Conversion and Management, Vol.49, pp. 2455–2460

Srinivasan, S.S., Niemann, M.U., Goswami, D.Y., Stefanakos, E.K. (2012). Hydrogen-Storage Hydride Complexes, U.S. Patent, 8,153,020, 2012

Srinivasan, S.S., Brinks, H.W., Hauback, B.C., Sun, D., & Jensen, C.M. (2004). Long term cycling behavior of titanium doped $NaAlH_4$ prepared through solvent mediated milling of NaH and Al with titanium dopant precursors. *J. Alloys and Comp.*, Vol. 377, No. 1-2, pp. 283-289

Srinivasan, S., Escobar, D., Jurczyk, M., Goswami, Y., & Stefanankos, E. (2008). Nanocatalyst doping of Zn(BH4)2 for on-board hydrogen storage. *J. Alloys Compd.*, Vol. 462, pp. 294–302

Srinivasan, S., Rivera, L., Stefanakos, E., & Goswami, Y. (2006). Mechano-Chemical Synthesis and Characterization of New Complex Hydrides for Hydrogen Storage, *Mat. Res. Soc. Symp. Proc.*, Vol. 927, pp. EE02

Stefanakos, E.K., Goswami, D.Y., Srinivasan, S.S., & Wolan, J. (2007). Hydrogen Energy, Kutz, Myer (Hrsg.) *Environmentally Conscious Alternative Energy Production, John Wiley & Sons*, 4th volume, Chapter 7, pp. 165

Stejskal, J., Kratochvil, P., & Jenkins A.D. (1996). The formation of polyaniline and the nature of its structures. *Polymer,* Vol. 37, No. 2, pp. 367-369

Stejskal, J., & Gilbert, R.G. (2002). Polyaniline. Preparation of a conducting polymer (IUPAC Technical Report). *Pure and App. Chem.*, Vol. 74, pp. 857-867

Trivedi, D.C. (1997). Polyanilines. *Handbook of Organic Conductive Molecules and Polymers,* H.S. Nalwa (Ed.) Wiley, Chichester, UK 2, pp. 505-572

US Department of Energy. (2003). Report of the Basic Energy Science Workshop on Hydrogen Production, Storage and use prepared by Argonne National Laboratory, May 13-15, (2003)

Vajo, J.J., Skeith, E., & Mertens, F. (2005). Reversible Storage of Hydrogen in Destabilized $LiBH_4$. *J. Phys. Chem. B,* Vol. 109, pp. 3719-3722

Virji, S., Kaner, R.B., & Weiller, B.H. (2006). Hydrogen sensors based on conductivity changes in polyaniline nanofibers. *J. Phys. Chem.* B, Vol. 110, No. 44, pp. 22266-22270

Wang, Y., & Jing, X.J. (2008). Synthesis and hydrogen storage of carbon nanofibers. Synthetic Metals. *J. Phys. Chem. B,* Vol. 112, No. 4, pp. 1157-1162

Wu, C.Z., Wang, P., Yao, X., Liu, C., Chen, C.M., Lu G.Q., & Cheng, H.M. (2006). Hydrogen storage properties of MgH_2/SWNT composite prepared by ball milling. *J. Alloys Comp.,* Vol.420, pp. 278-282

Yang, Y., Ouyang, J., Ma, L., Tseng, R.J., & Chu, C.-W. (2006). Electrical Switching and Bistability in Organic/Polymeric Thin Films and Memory Devices. *Adv. Funct. Mater.*, Vol. 16, No. 8, pp. 1001-1014

Yao, X., Wu, C.Z., Wang, H., Cheng, H.M., Lu, G.Q. (2006). Effects of Carbon Nanotubes and Metal Catalysts on Hydrogen Storage in Magnesium Nanocomposites. *J. Nanoscience and Nanotechnology,* Vol. 6, No. 2, pp. 494-498, ISSN 1533-4880, Online ISSN: 1533-4899

Yildirim, T., & Ciraci, S. (2005). Titanium-Decorated Carbon Nanotubes as a Potential High-Capacity Hydrogen Storage Medium. *Phys. Rev. Lett.,* Vol. 94, pp. 175501-175504

Yu, X.B., Grant D.M., & Walker, G.S. (2006). A new dehydrogenation mechanism for reversible multicomponent borohydride systems—The role of Li–Mg alloys. *Chem. Commun.,* pp. 3906-3908

Yvon, K., & Bertheville, B. (2006). Magnesium based ternary metal hydrides containing alkali and alkaline-earth elements. *J. Alloys Comp.,* Vol. 425, pp. 101-108

Zaluska, A., Zaluski, L ., & Strom-Olsen, J.O. (2001). Structure, catalysis and atomic reactions on the nano-scale: a systematic approach to metal hydrides for hydrogen storage. *Appl. Phys. A,* Vol. 72, pp. 157-165

Zhang, D., & Wang, Y. (2006). Synthesis and applications of one dimensional nano-structured polyaniline: An overview. *Mater. Sci. and Engg. B,* Vol. 134, No. 1, pp. 9-19

Zhang, G., Li, X., Jia, H., Pang, X., Yang, H., Wang, Y., & Ding, K. (2012). Preparation and Characterization of Polyaniline (PANI) doped-$Li_3V_2(PO_4)_3$. *Int. J. Electrochem. Sci.,* Vol. 7, pp. 830-843

Zuttel, A. (2004). Hydrogen Storage Methods. *Die Naturwissenschaften,* Vol. 91, pp. 157-172

Zuttel, A., Wenger, P., Rentsch, S., Sudan, P., Mauron Ph., & Emmeneger, Ch. (2003). $LiBH_4$ a new hydrogen storage material. *J. Power Sources,* Vol. 118, pp. 1-7

Chapter 2

Improvement on Hydrogen Storage Properties of Complex Metal Hydride

Jianjun Liu and Wenqing Zhang

Additional information is available at the end of the chapter

http://dx.doi.org/10.5772/50153

1. Introduction

A large challenging of world economic development is to meet the demand of energy consumption while reducing emissions of greenhouse gases and pollutants [1-5]. Hydrogen, as an energy carrier, is widely regarded as a potential cost effective, renewable, and clean energy alternative to petroleum, especially in the transportation sector [1]. Extensive efforts are being made to develop a sustainable hydrogen economy which is involved by hydrogen production, hydrogen storage, and hydrogen fuel cell in the cyclic system of hydrogen combustion [2, 6]. One key component of realizing the hydrogen economy for transportation applications is developing highly efficient hydrogen storage systems.

Table 1 presents the current available hydrogen storage techniques. Although some basic technical means such as pressurized gas and cryogenically liquefied hydrogen in containers can be used at present, hydrogen capacity is not acceptable in practical applications-driving a car up to 300 miles on a single tank, for example. Therefore, storing hydrogen in advanced solid state materials has definite advantage with regard to a low-cost, high gravimetric and volumetric density, efficiently storing and releasing hydrogen under mild thermodynamic conditions. Over the past decades, many advanced materials such as complex metal hydrides [7, 8], metal hydrides [9], metal-organic framework (MOF) [10-12], and modified carbon nanostructures have been explored to develop efficient hydrogen storage techniques [13-19], but none of them can meet all requirements [20].

			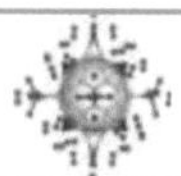	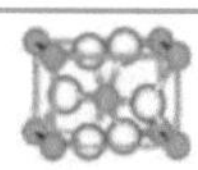	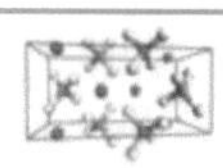
Liquid Hydrogen	Compress Hydrogen	MOF	Nanostru-cture	Metal Hydride	Complex Metal Hydride
-253°C	**25 °C**	**-200 °C**	**25°C**	**330°C**	**>185°C**

Table 1. Available hydrogen storage technologies and corresponding operating temperatures.

Complex metal hydrides (for example, $NaAlH_4$, $LiAlH_4$, $LiBH_4$, $Mg(BH_4)_2$, $LiNH_2$) are currently considered as one of the promising hydrogen storage materials mainly because they have a high hydrogen capacity and are facile to tailor structural and compositional to enhance hydrogen storage performance. The typical structure of complex metal hydrides contains cation alkali metal (M^{n+}) and anion hydrides (AlH_4^-, BH_4^-, NH_2^-) with a closed-shell electronic structure. It should be pointed out that this review focuses on Al- and B-based complex metal hydrides. The bonding characteristics of these complex metal hydrides determine that their dehydriding and hydriding are unfavorable either thermodynamically or kinetically under moderate conditions. As a result, a large obstacle to use complex metal hydrides as on-board hydrogen storage materials is a relatively high hydrogen desorption temperature, a low kinetic rate for hydrogen desorption and adsorption, and a poor reversibility. It is very important to develop the effective chemical and physical methods to improve hydrogen storage properties of these materials.

Herein, take $Mg(BH_4)_2$ as an example. A possible hydrogen desorption process from $Mg(BH_4)_2$ to MgB_2 are depicted by the following equations (1)-(3) [21]:

$$6Mg(BH_4)_2 \leftrightarrow 5MgH_2 + MgB_{12}H_{12} + 13H_2 \tag{1}$$

$$5MgH_2 \leftrightarrow 5Mg + 5H_2 \tag{2}$$

$$5Mg + MgB_{12}H_{12} \leftrightarrow 6MgB_2 + 6H_2 \tag{3}$$

In fact, hydrogen desorption of $Mg(BH_4)_2$ experiences a complicated hydrogen desorption process involving chemical reactions and physical changes such as mass transport and phase separation. Two thermodynamically stable intermediates, $Mg(B_{12}H_{12})$ and MgH_2, are formed in the first step (Equation (1)) with enthalpy and entropy of 39 kJ/mol·H_2 [22]. In 2008, a different value of 57 kJ/mol·H_2 was obtained [23]. The hydrogen desorption reactions of equations (2) and (3) have endothermicity of 75 and 87 kJ/mol·H_2. Therefore, the equations (2) and (3) only occur at a high temperature, 572 K of equation (2) and 643 K of equation (3) [24]. In addition, a stable intermediate usually leads to a thermodynamic pitfall which trap a large amount of hydrogen cannot be cycled. Very recently, Jensen et al. found that at a high condition (~400°C and ~950bar), equations (2)-(3) also can participate hydrogen release/uptake reactions [25]. However, these conditions are unfeasible for practical application.

Promoting the kinetic rates of hydrogen desorption and adsorption of complex metal hydrides play an important role in developing hydrogen storage material. However, because the bonds B–H in BH_4^- and Al–H in AlH_4^- are relatively strong, their dissociations require overcoming a high barrier. Additionally, two processes must be considered to enhance the kinetic rate of hydrogen desorption and adsorption. Firstly, phase transitions coupled with chemical reactions, which sometimes experience a high barrier, slow down the kinetic rate. Secondly, hydrogen diffusion is also important factor to take effect on the kinetic rate of hydrogen desorption and adsorption.

In 1997, Bogdanović et al. demonstrated that a small amount of Ti-compounds doped in $NaAlH_4$ can enhance the kinetic rates of both hydrogen desorption and adsorption of $NaAlH_4$, reduce hydrogen desorption temperature from 210 to 120°C and hydrogen adsorption pressure from 350 to 100 bar, as well as have a good reversibility (80% H) [26]. It stimulated the extensive studies in theory [27-41] and experiment [42-54] to improve the kinetic and thermodynamic properties of hydrogen desorption and adsorption of complex metal hydrides in order to develop the practical hydrogen storage material. More importantly, these studies have extended from doping transition metal to chemical and physical methods such as nanoengineering, and cation substitution. These structural and composition tailor are expected to have strong effects on the thermodynamics of the complex hydrides and the kinetics of hydrogen release and uptake from either the bulk crystalline phase or nanosized particles.

In the past years, there are a few reviews to discuss hydrogen storage materials with different points of view. In this chapter, we focus on improvement on hydrogen storage properties of complex metal hydrides, that is, tailoring thermodynamics and kinetic properties of their hydrogen desorption and adsorption by the various techniques. We do not intend to provide a complete review of the literature about this topic, but rather to emphasize tailoring effect on hydrogen storage properties of complex metal hydrides. The research is mainly categorized into three parts: (i) doping transition metal; (ii) nanoengieering techniques; and (iii) cation Substitution. Finally, we present a conclusive remark for developing complex metal hydrides as hydrogen storage materials by means of altering thermodynamic and kinetic properties.

2. Improving hydrogen storage properties

2.1. Doping transition metal

Catalysts have been widely exploited to hydrogen storage materials to improve the kinetic and thermodynamic properties of hydrogen desorption and adsorption in complex metal hydrides and metal hydrides, following the pioneering work of Bogdanović and Schwickardi [26, 55]. They demonstrated that doping the complex metal hydrides $NaAlH_4$ with a few mol% of Ti lowered the decomposition temperature, improving the kinetics, and, importantly allowed rehydrogenation of the decomposition products. This finding quickly sparks worldwide research activities that aimed at developing catalytically enhanced $NaAlH_4$ and related complex metal hydrides as practical hydrogen storage medium. Then, a great number of experimental and theoretical studies have been devoted to characterize the structures and effect of Ti in $NaAlH_4$. Although many models were proposed to describe (de)hydrogenation of Ti-doped $NaAlH_4$, no clear consensus about structures and catalytic mechanism of Ti in $NaAlH_4$ has been achieved. The only established fact from these studies is a surface-localized species containing a nascent binary phase Ti-Al alloy formed during cyclic dehydriding and rehydriding processes [28, 43, 51, 56-58].

Many experimental studies about the local structure of Ti-doped $NaAlH_4$ showed that highly dispersed Ti in the Al surface plays an important role in hydrogen uptake and release processes. As shown in Figure 1, $TiAl_3$ alloy is the most likely form after dehydriding Ti-doped $NaAlH_4$ [43, 59] . It is consistent with what $TiAl_3$ is thermodynamically the most stable stability in Ti-Al system. The local structure of active species has Ti–Al and Ti–Ti bond distance of 2.79 and 3.88Å, respectively. After mechanical milling, $TiCl_3$ is reduced to zero-state Ti by interaction with $NaAlH_4$. However, $TiAl_3$ doped in $NaAlH_4$ were found to be substantially less effective than $TiCl_3$. Therefore, the catalytic activity of Ti structure may be summarized as "Ti in the Al surface > $TiAl_3$ cluster > crystalline $TiAl_3$" [50].

In fact, determining accurately the local structures in such a complicated system including a dynamic hydriding/dehydriding processes is extremely challenging to many experimental techniques. In this aspect, DFT-based first-principles methods have shown their advantages. Several theoretical studies have been performed with emphasis on substitution of Ti for Al and Na atoms in Ti-doped $NaAlH_4$ bulk and surfaces. Substitution of Ti for Al has been shown theoretically to be the preferred location in bulk $NaAlH_4$. Ĩňiguez *et al.* studied the structure, energetics, and dynamics of pure and Ti-doped $NaAlH_4$, focusing on the possibility of substitutional Ti doping in the bulk. They found that that the doped Ti prefers to substitute for Na and further attract surrounding hydrogen atoms, softening and/or breaking the Al-H bonds. The same group of authors extended their studies to determine the location of Ti. These later results showed that Ti prefers to be on the surface, substituting for Na, and attracting a large number of H atoms to its vicinity. They predicted that a $TiAl_n(n>1)$ structure may be formed on the surface of the sodium alanate [30]. However, Løvvik et al. also suggested that substitution of Ti in bulk $NaAlH_4$ is less favorable than that near surface or defect positions. On the $NaAlH_4$ (001) surface, DFT calculations by Yildirim and Iňiguez showed substitution of Ti for Na is the preferred site [60] whereas Løvvik and Opalka found substitution of Ti for Al is more favorable [40]. The difference has been attributed to the different reference states used in energy calculations.

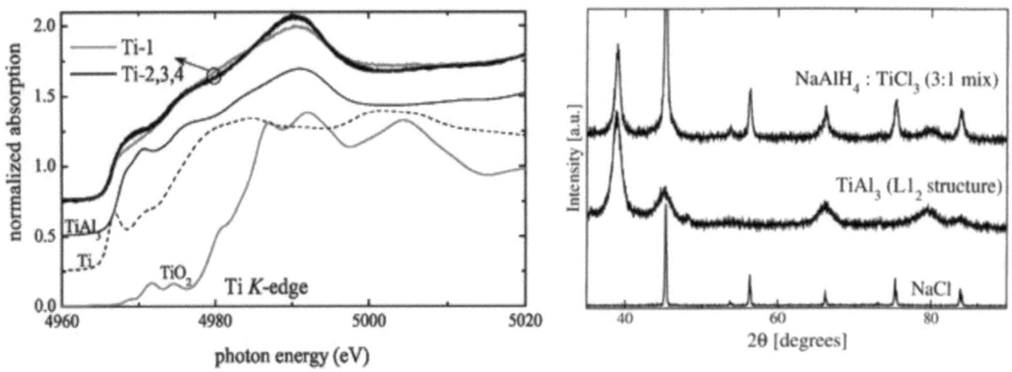

Figure 1. Normaalized XANES spectra for Ti-doped $NaAlH_4$ and the reference compounds (left); X-ray diffraction indicating $TiAl_3$ production in the $NaAlH_4$ system when mechanically milled in a 3:1 ratio with $TiCl_3$. Reproduced from [43] for (left) by permission of The Royal Society of Chemistry, and from [59] for (right) by permission of Elsevier.

However we approached this problem based on a surface model and found a different structure and mechanism. The $TiAl_3H_{12}$ local structure was identified in Ti-doped $NaAlH_4$ (001) and (100) surfaces [41]. Our calculated results show that the hydrogen desorption energies from many positions of $TiAl_3H_x$ are reduced considerably as compared with that from the corresponding clean, undoped $NaAlH_4$ surfaces. Furthermore, we showed that the

$TiAl_3H_{12}$ complex has an extended effect beyond locally reducing the hydrogen desorption energy. It also facilitates hydrogen desorption at a reduced desorption energy by either transferring the hydrogen to $TiAl_3H_x$ or by reducing the hydrogen desorption energy in neighboring AlH_4^- by linking these AlH_4^- units with the complex structure. Our predicted interstitial $TiAl_3H_x$ structure was supported by a recent combined Ti K-edge EXAFS, Ti K-edge XANES, and XRD study of $TiCl_3$-doped $NaAlH_4$ by Baldé et. al [61]. These authors observed that the interstitial structure accounts for more than 70% of all Ti doped in $NaAlH_4$.

Extensive experimental studies have demonstrated that transition metals (TM) can accelerate the kinetic rate of hydrogenation and dehydrogenation reactions in this system. In terms of chemical reactions, TM can weaken Al–H and H–H bonds and thus reduce transition state barriers of hydrogen reactions through electron backdonation interaction from d orbital of TM to σ^* of these bonds [33]. In addition, addition of TM also leads to formation of defect which is also favorable to kinetic improvement of hydrogen diffusion in solid-state materials [31, 62].

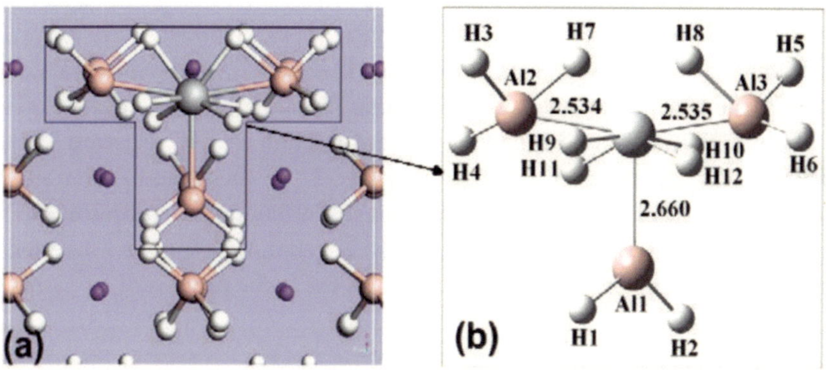

Figure 2. DFT-GGA relaxed structure of Ti-doped $NaAlH_4(001)$ with Ti in the surface interstitial site. (b) Detailed local structure of the $TiAl_3H_{12}$ complex shown in (a). Reprinted from [41].

Few experimental studies on tailoring thermodynamic properties of $NaAlH_4$ by doping TM were performed. Bogdanović and Schüth performed pressure-concentration isotherms for hydrogen desorption of NaAlH4 with different doping levels of Ti [63]. They found Ti doping can significantly alters the thermodynamics of the system, which is demonstrated by the change of the dissociation pressure with doping level. Such a thermodynamic change is mainly attributed to Ti-Al alloy formation.

As mentioned previously, our studies for Ti-doped $NaAlH_4$ found that $TiAl_3H_x$ structure has a significantly effect to reducing hydrogen desorption energy [41, 64]. Such a thermodynamic tuning effect can be explained by the closed-shell 18-electron rule of transition metal structures. In addition, Mainardi et al. performed electronic structure calculations and molecular dynamic simulations for kinetics of hydrogen desorption of $NaAlH_4$ [65]. They found that the rate-determining step for hydrogen desorption was hydrogen evolution from associated AlH_4 species. Ti is predicted to stay on the hydride surface and serves as both the catalytic species in splitting hydrogen from AlH_4^-/AlH_3 groups as well as the initiator Al nucleation sites in Ti-doped $NaAlH_4$ system.

In terms of $NaAlH_4$, an important issue is to select high efficient catalyst for improving thermodynamic and kinetic properties of hydrogenation and dehydrogenation. Anton and Bogdanović studied the hydrogen desorption kinetics of $NaAlH_4$ by different transition metals(TM) and found early TMs have a better catalytic effect for hydrogen desorption kinetics than later TMs [66, 67]. In 2008, we performed DFT calculations for hydrogen desorption mechanism of 3d TM-doped $NaAlH_4$. Similarly, $TMAl_3H_x$ were determined to the most stable structures [33]. In these structures, the electron transfer between hydrogen and Al groups mediated by the d-orbtials of TMs plays an important role in hydrogen release/uptake from analate-based materials.

Only a few publications focus on the theoretical exploration for the mechanism of Ti-catalyzed hydrogenation process [28, 29, 68]. In fact, Ti-catalyzed hydrogenation process includes hydrogen dissociation and the subsequent formation of any hydrogen-containing mobile species from Ti active sites. In 2005, Chaudhuri et. al performed DFT calculations to investigate the position and catalytic mechanism for hydrogenation of Ti in Al (001) surface structure [29]. Two next-nearest-neighbor Ti atoms located on the top of 2×2 Al(001) surface are more favorable to hydrogen dissociation than others positions such as two nearest-neighbor. In this particular local arrangement, the H–H bond can be automatically broken and the dissociated H atoms are connected with Ti and Al. The analysis of electronic structure showed that the bond-breaking process is enhanced by electron backdonation from Ti-3d orbitals to hydrogen σ^* orbitals. However, Ti was believed to promote formation of AlH_3 or NaH vacancies but not included explicitly in the model. Furthermore, NaH was not treated explicitly as the study focused n dehydrogenation. Therefore, a system directly involving NaH is necessary to account for its role in the cyclic process of using $NaAlH_4$ as a hydrogen storage medium.

Recently, we studied hydrogen adsorption process of $TiAl_3H_x$ supported on the NaH(001) surface in order to understand hydrogenation mechanism of Ti-doped NaH/Al [69]. Our results support that $TiAl_3H_x$ gains electronic charge from the NaH hydrides. The hydrided $TiAl_3H_x$ cluster on the NaH surface which dissociates the H_2 molecule at the Ti site in contact with the surface. Furthermore, our DFT-based molecular dynamics simulation (Figure 3) demonstrated that $TiAl_3H_x$ clusters are active for H_2 dissociation after acquiring electrons from the hydride of NaH surface.

Another complex hydride similar to $NaAlH_4$ but having an even higher intrinsic hydrogen capacity is $LiAlH_4$. The decomposition of $LiAlH_4$ is believed to undergo similar steps to $NaAlH_4$. The first decomposition step from tetrahedral $LiAlH_4$ to octahedral Li_3AlH_6 is weakly endothermic [70, 71]. The second decomposition reaction from octahedral Li_3AlH_6 to LiH and Al phase was found to be endothermic with ΔH of 25 kJ/mol·H_2. Its dehydriding was observed to occur at 228-282 °C, likely due to kinetic limiting steps. Apparently, the decomposition temperature is too high for practical purposes. The decomposition of $LiAlH_4$ is very slow without a catalyst [72-75].

Balema et al. found that the mixture of 3 mol% $TiCl_4$ and $LiAlH_4$ under ball milling can cause $LiAlH_4$ to rapidly transform into Li_3AlH_6 [72, 73]. In 2010, Langmi et al. found that

$TiCl_3$ can enhance thermodynamic properties to reduce hydrogen desorption temperature from ~170°C for the first step while melting and 225°C for the second step to 60-75°C below the melting point [76]. These studies indicated that doping $TiCl_3$ can improve thermodynamic and kinetic properties of (de)hydrogenation processes of $LiAlH_4$.

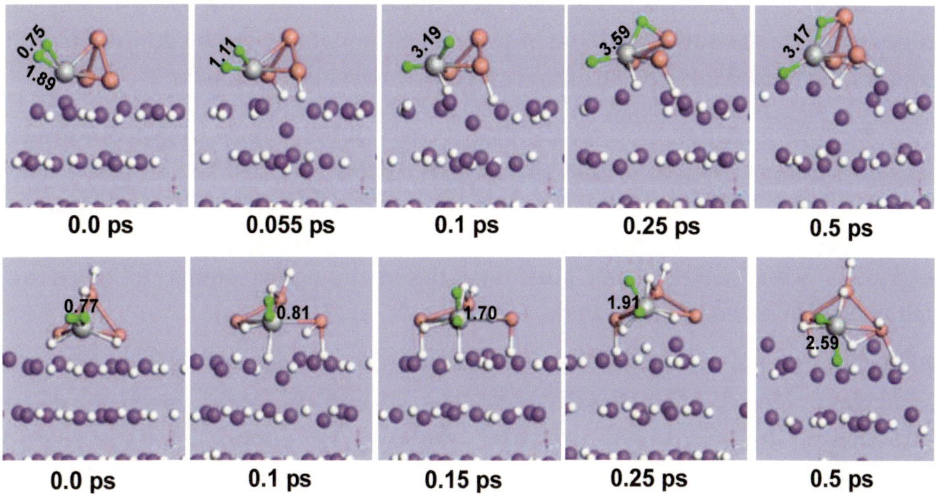

Figure 3. Snapshots from *ab initio* molecular dynamics trajectories for H_2 dissociation on $TiAl_3$ and $TiAl_3H_4$ clusters supported on NaH (001) surface. Purple, white, pink, gray, and green balls represent Na, H, Al, and Ti, and dissociating H_2. Reprinted from [69] by American Chemical Society.

Very recently, Liu et al. directly synthesized $LiAlH_4$ from commercially available LiH and Al powders in the presence of $TiCl_3$ and Me_2O for the first time [77]. However, without $TiCl_3$ or adding metallic Ti, $LiAlH_4$ is not observed in experiment. It suggests that with the presence of $TiCl_3$, $LiAlH_4$ can be cycled, making it a reversible hydrogen storage material. However, the catalytic effect of $TiCl_3$ for enhancing thermodynamic and kinetic properties of $LiH+Al+3/2H_2 \rightarrow LiAlH_4$ still is not studied so far.

Complex metal borohydrides have attracted extensive attention due to due to its intrinsically high gravimetric and volumetric hydrogen capacities (for example, $LiBH_4$, 18.2 wt%, 121 kg/m^3). Unfortunately, the B-H bond in pure $LiBH_4$ material is extremely strong and only liberates 2% hydrogen around the melting point (541-559 K) [1]. Starting from $LiBH_4$, the partial decomposition to $LiH(s)+B(s)+3/2H_2(g)$ has the standard enthalpy of 100.3 kJ/mol·H_2 [78]. The highly endothermic decomposition reaction indicates hydrogen release from $LiBH_4$ must occur at elevated temperatures. The experimental results of Züttel et. al showed that a significant hydrogen desorption peak started at 673 K and reached its maximum value around 773 K [79, 80]. In 2007, Au et al. showed that $LiBH_4$ modified by metal oxides or metal chlorides, such as TiO_2 and $TiCl_3$, could reduce the dehydrogenation temperature and achieve re-hydrogenation under moderate conditions [81, 82]. Modified $LiBH_4$ releases 9 wt% H_2, starting as low as 473 K, which is significantly lower than the hydrogen releasing temperature of 673 K for pure $LiBH_4$. After being dehydrogenated, the modified $LiBH_4$ can absorb 7~9 wt% H_2 at 873 K and 70 bar, a significant improvement from

923 K and 150 bar for pure $LiBH_4$ [80]. Very recently, Fang, et al. reported that a mechanically milled $3LiBH_4/TiF_3$ mixture released 5-6 wt% hydrogen at temperatures of 343~363 K [83]. Similarly, other dopants have been attempted to reduce the hydrogen desorption temperature of MgH_2. Clearly, addition of Ti-compounds (TiO_2, $TiCl_3$, and TiF_3) result in a strong improvement for hydrogen desorption and, to a lesser extent, for re-hydrogenation. On the other hand, the improvement brought by these additives to $LiBH_4$ is not sufficient to make $LiBH_4$ viable as a practical hydrogen storage media

In 2009, we presented our DFT calculations for structures and hydrogen desorption Ti-doped $LiBH_4$ surface [84]. Molecular orbital analysis showed that the structural stability could be attributed to the symmetry-adapted orbital overlap between Ti and "inside" B–H bonds. Several surfaces (001) and (010) can desorb hydrogen in molecular form by high spin state (triplet), while surface (100) must first desorb hydrogen atoms, followed by the formation of a hydrogen molecule in the gas phase.

$Mg(BH_4)_2$ is considered as another promising hydrogen storage materials and it releases approximately 14.9 wt% of hydrogen when heated up to 870K [22, 23, 85-93]. As discussed in Introduction, the dehydrogen process is found to go through multiple steps with formation of some stable intermediates such as $MgB_{12}H_{12}$ and MgH_2 [21]. Therefore, it is very necessary to tune thermodynamic and kinetic properties of hydrogenation and dehydrogenation of $Mg(BH_4)_2$. The addition of $TiCl_3$ into $Mg(BH_4)_2$ was demonstrated to be effective on tuning thermodynamic properties [86]. Hydrogen desorption temperature is reduced to from 870 K to 361 K. However, Ti species gradually convert to Ti_2O_3 and TiB_2 during cycling experiments of hydrogen desorption/adsorption [94], though the catalytic mechanism is still not clear.

2.2. Nanoengineering techniques

Due to size effect and morphology, nanoparticles often display some different physical and chemical properties compared to bulk particles and are applied for instance in catalysis, chemical sensors, or optics [95-98]. A small size of particle can decrease hydrogen diffusion lengths and increase surface interaction with H_2. More importantly, thermodynamics of hydrogen desorption/adsorption of complex metal hydrides usually can be adjusted by controlling particle size [34, 99-103]. Particle size of complex metal hydrides can be usually reduced to ~200 nm by ball milling technique, for $NaAlH_4$ preferably in the presence of TM-based catalysts [57, 104, 105]. Obtaining smaller certain sizes of particles of complex metal hydrides is still challenging. Moreover, with the method of ball milling, the particle size is very difficult to control in an exact value and the size distribution is broad.

In the recent years, a new technique, nanoscaffold, has been extensively used to produce a different size of nanoparticles of complex metal hydrides. However, it should be pointed out that development of controlling nanosize of particle by nanoscaffold technique is really dependent on preparation of porous nanomaterials. Additionally, it is understood that a nanoscaffold technique unavoidably results in a low hydrogen capacity of complex metal hydrides.

By this technique, Baldé et al. synthesized a nanofiber-supported $NaAlH_4$ with discrete particle size ranges of 1-10μm, 19-30nm, and 2-10nm [99]. The experimental measurement on temperature programmed desorption of H_2 for $NaAlH_4$ nanoparticles was presented in Figure 4. The hydrogen desorption temperatures are decreased from 186°C of 1-10μm to 70°C of 2-10nm. More importantly, the activation barriers of hydrogen desorption also change from 116 to 58 kJ/mol correspondingly. It suggests that size reduction of nanoparticle can tailor thermodynamic and kinetic properties of hydrogen desorption/adsorption process of $NaAlH_4$. In addition, they also reported that decreasing particle sizes also lowered the pressures needed for hydrogen uptake. In 2010, Gao et al. confined $NaAlH_4$ into 2-3 nm nanoporous carbon [102]. They observed that H_2 release temperature and rehydrogenation conditions were significantly improved. More importantly, the total reaction is changed to a single step reaction without Na_3AlH_6 formed. The similar studies also exhibited nanosize effect on tuning thermodynamic and kinetic properties for complex metal hydrides.

In 2011, Majzoub et al. presented first-principles calculations for phase diagram of small cluster of Na-Al-H system [34]. They found that decreasing cluster size not only reduces hydrogen desorption temperature but also change reaction path from $NaAlH_4 \rightarrow Na_3AlH_6+Al+H_2 \rightarrow NaH+Al+H_2$ in bulk structure to $NaAlH_4 \rightarrow NaH+Al+H_2$ in a small size of nanoparticles. It should be attributed to the instability of Na_3AlH_6 nanoparticle with a small size. All these studies indicate that controlling nanostructure size provides a practical avenue to tailor thermodynamic and kinetic properties of (de)hydrogenation of complex metal hydrides.

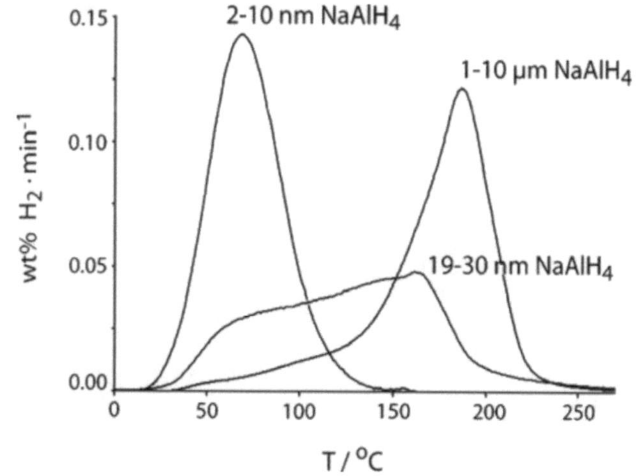

Figure 4. Temperature programmed desorption profile of H_2 for $NaAlH_4$ supported on carbon nanofiber. Reproduced from [99] by permission of American Chemical Society (copyright 2008).

Similarly, decreasing particle size by nanoscaffold technique has also been extended to $LiBH_4$. Vajo and Wang filled $LiBH_4$ into carbon aerogel and AC carbon to form different nanoparitcles [106-108]. They found hydrogen desorption temperature was reduced and kinetic rate was significantly enhanced. Unfortunately, nanosize effect of hydrogen desorption and adsorption of $LiBH_4$ is still not reported so far.

In sum, a small size of nanoparticle of complex metal hydrides can directly result in the change of thermodynamic and kinetic properties for hydrogen adsorption/desorption processes. However, there are two very important questions on nanosize effect of particles of complex metal hydrides. One is to determine the correlation of tuning thermodynamic and kinetic properties with particle size. The other is to establish the hydrogen desorption/adsorption mechanism of complex metal hydrides in a different nanosize.

Except for size effect, nanoengieering also involves the composition of complex metal hydride and nanostructures. Berseth et al. performed joint experimental and theoretical studies for hydrogen uptake and release of $NaAlH_4$ attached on carbon nanostructures such as C_{60}, graphene, and nanotubes [15]. Figure 5 displayed the correlation of hydrogen desorption energies of $NaAlH_4$ with electron affinities of carbon nanostructures. It suggests that that the stability of $NaAlH_4$ originates with the charge transfer from Na to the AlH_4 moiety, resulting in an ionic bond between Na^+ and AlH_4^- and a covalent bond between Al and H. Interaction of $NaAlH_4$ with an electronegative substrate such as carbon fullerene or nanotube affects the ability of Na to donate its charge to AlH_4, consequently weakening the Al–H bond and causing hydrogen to desorb at lower temperatures as well as facilitating the absorption of H_2 to reverse the dehydrogenation reaction.

Similarly, Wellons et al. showed that the addition of carbon nanostructure C_{60} to $LiBH_4$ has a remarkable catalytic effect, enhancing the uptake and release of hydrogen [109]. A fullerene-$LiBH_4$ composite demonstrates catalytic properties with not only lowered hydrogen desorption temperatures but also regenerative rehydrogenation at a relatively low temperature of 350°C. This catalytic effect is probably attributed to C_{60} interfering with the charge transfer from Li to the BH_4 moiety, resulting in a minimized ionic bond between Li^+ and BH_4^-, and a weakened B–H covalent bond. Interaction of $LiBH_4$ with an electronegative substrate such as carbon fullerene affects the ability of Li to donate its charge to BH_4, consequently weakening the B–H bond and causing hydrogen to desorb at lower temperatures as well as facilitating the absorption of H_2.

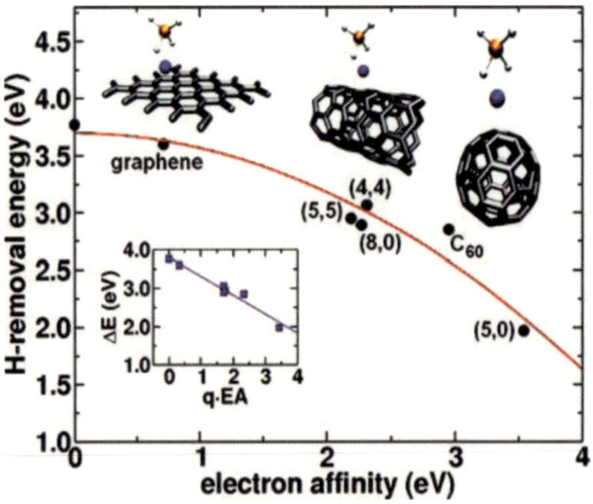

Figure 5. Correlation of the carbon substrate electron affinity and the hydrogen removal energy. Reprinted from [15] by permission of American Chemical Society (copyright 2009).

2.3. Cation substitution

In terms of complex metal hydrides, their stability really depends upon electronic affinity of metal atom. The lower the electronic affinity of metal is, the less stable hydride is. It can further be explained by transferred electron amount from metal atom to hydride. Løvvik, Jensen, Ormio, and Miwa et al. proposed that the metal element with a large electronic affinity can be used to substitute the original metal in order to destabilize reactants, making the enthalpy of the hydrogen release reaction favorable [110-117].

The two cations mixed in one hydride are expected to function synergistically to maintain reasonable stability, and at the same time provide a favorable decomposition enthalpy. Sorby et al. performed an experimental study about dual cation aluminium hydride, $K_2Na(AlH_4)_3$ [118]. Because K has a smaller electron affinity, $K_2Na(AlH_4)_3$ was measured to have a higher hydrogen desorption temperature up to 285°C, which is well consistent with theoretical predict.

Extensive DFT calculations showed that bialkali hexahydrides, such as K_2LiAlH_6, K_2NaAlH_6, KNa_2AlH_6, and $LiNa_2AlH_6$, are stable compared to the pure alanates [110, 111] . In fact, $LiNa_2AlH_6$ has been synthesized experimentally [112, 113]. Mixed aluminohydrides such as $LiMg(AlH_4)_3$ and $LiMgAlH_6$ have also been predicted based on DFT studies and have been synthesized and characterized experimentally [119, 120]. Although their overall hydrogen storage performance was not fully examined, some of these compounds exhibit favorable decomposition temperatures.

Many theoretical and experimental studies on cation modification have been performed to improve thermodynamics and kinetics for borohydrides. Au et al. synthesized a series of bimetallic $M_1M_2(BH_4)_n$ (M_1, M_2=Li, Mg, and Ti) and experimentally measured their hydrogen desorption temperature and hydrogen capacity [121]. They found that dehydrogenation temperature was reduced considerably and the dehydrided bimetallic borohydrides reabsorbed some of hydrogen released, but the full rehydrogenation is still very difficult. In 2010, Fang et al. studied formation of decomposition of dual-cation $LiCa(BH_4)_3$ using X-ray diffraction and thermogravimetry/differential scanning calorimetry/mass spectroscopy techniques [122]. It was found that $LiCa(BH_4)_3$ exhibits improved (de)hydrogenation properties relative to the component phases. In 2011, Jiang et al studied synthesis and hydrogen storage properties of Li-Ca-B-H hydride [123]. They found that the first dehydrogenation temperature is about 70°C, much lower than the pristine $LiBH_4$ and $Ca(BH_4)_2$. All these studies indicate that dual-cation borohydrides have a better thermodynamic property for hydrogen desorption than the single cation borohydride. Therefore, dehydrogenation temperature is significantly improved relative to the single phase.

In addition, some experimental studies on multivalent cation borhydrides such as Al, Sc, and Ti were carried out to reduce hydrogen desorption temperature [115, 117, 124-132]. However, theoretical studies on dehydrogenation mechanism including intermediates and products are desired for further improvement. However, extensive DFT computations have been performed to assess a large number of possible destabilized metal hydrides [133-137]. By assessing the enthalpies of all possible reactions, more than 300 destabilization reactions were predicted to have favorable reaction enthalpies [133]. Wolverton et al. proposed several guidelines to destabilize thermodynamically the metal hydrides in order to design novel hydrogen storage materials [138]. Basically, the enthalpy of the proposed destabilized reaction must be less than the decomposition enthalpies of the individual reactant phases. In addition, if the proposed reaction involves a reactant that can absorb hydrogen, the formation enthalpy of the corresponding hydride cannot be greater in magnitude than the enthalpy of the destabilized reaction.

Vajo et al. examined this strategy by altering the thermodynamics and kinetics of (de)hydrogenation of several metal hydrides [139]. The equilibrium hydrogen pressure and reaction enthalpies can be changed with additives that form new alloys or compound phases upon dehydriding. The formation of new phases lowers the energy of dehydrided state and efficiently destabilizes the component hydrides. A series of experimental explorations have been performed to destabilize the reaction products of $LiBH_4$ and successfully reduce the dehydriding temperatures [139-141].

3. Conclusive remarks

Complex metal hydrides with a high hydrogen capacity have been considered as potential candidates for on-board hydrogen storage materials. However, the high hydrogen desorption temperature and sluggish kinetics prevent them from being applied in practice. It is attributed to unfavorable thermodynamic and kinetic properties of (de)hydrogenation. Over a past decade, a number of efforts have been devoted to improve hydrogen storage properties by altering thermodynamic and kinetic properties of (de)hydrogenation. Doping transition metal in complex metal hydrides can be regarded as a very effective means to tailor thermodynamics and promote kinetics. However, the catalytic mechanism of TM doped in hydrides remains unconfirmed because (de)hydrogenation includes complicated physical and chemical processes. TM may exhibit different structures and catalytic mechanisms in each step. Nanoengineering has extensively been applied to improve thermodynamic and kinetic properties of hydrogen storage materials by means of reducing particle size or mixing with nanostructures. However, some catalytic effect is restricted from some properties such as certain size, as well as chemical and physical properties of nanostructures. Cation substitution to form dual-cation hydride is generally used as a technique to alter thermodynamic property. Although this modification is effective to destabilize reactant, the modified crystal structure does not maintain during reversible processes of (de)hydrogenation.

Author details

Jianjun Liu* and Wenqing Zhang
State Key Laboratory of High Performance Ceramics and Superfine Microstructure,
Shanghai Institute of Ceramics (SIC), Chinese Academy of Sciences (CAS), Shanghai, China

Acknowledgement

Jianjun Liu acknowledge support by the startup funding by Shanghai Institute of Ceramics (SIC), Chinese Academy of Sciences (CAS).

---

* Corresponding Author

4. References

[1] Schlapbach L, Züttel A. Hydrogen-storage materials for mobile applications. Nature. 2001;414:353-8.

[2] Crabtree GW, Dresselhaus MS, Buchanan MV. The hydrogen economy. Physics Today. 2004 Dec;57(12):39-44.

[3] Schultz MG, Diehl T, Brasseur GP, Zittel W. Air pollution and climate-forcing impacts of a global hydrogen economy. Science. 2003 Oct 24;302(5645):624-7.

[4] Schüth F. Challenges in hydrogen storage. European Physical Journal-Special Topics. 2009 Sep;176:155-66.

[5] Dresselhaus MS, Thomas IL. Alternative energy technologies. Nature. 2001;414:332-7.

[6] Ohi J. Hydrogen energy cycle: An overview. J Mater Res 2005 Dec;20(12):3180-7.

[7] Orimo S, Nakamura Y, Eliseo JR, Züttela A, Jensen CM. Complex Hydrides for Hydrogen Storage. Chem Rev. 2007;107(10):4111-32.

[8] Jain IP, Jain P, Jain A. Novel hydrogen storage materials: A review of lightweight complex hydrides. J Alloys Compd. 2010 Aug 6;503(2):303-39.

[9] Graetz J. New approaches to hydrogen storage. Chem Soc Rev. 2009;38:73-82.

[10] Rowsell JLC, Yaghi OM. Strategies for Hydrogen Storage in Metal–Organic Frameworks. Angew Chem Int Ed. 2005;44:4670-9.

[11] El-Kaderi HM, Hunt JR, Mendoza-Cortes JL, Cote AP, Taylor RE, O'Keeffe M, et al. Designed Synthesis of 3D Covalent Organic Frameworks. Science. 2007;316:268-72.

[12] Rosi NL, Eckert J, Eddaoudi M, Vodak DT, Kim J, O'Keeffe M, et al. Hydrogen Storage in Microporous Metal-Organic Frameworks. Science. 2003;300:1127-9.

[13] Niemann MU, Srinivasan SS, Phani AR, Kumar A, Goswami DY, Stefanakos EK. Nanomaterials for Hydrogen Storage Applications: A Review. J Nanomater. 2008.

[14] Zhao Y, Kim Y, Dillon AC, Heben MJ, Zhang SB. Hydrogen Storage in Novel Organometallic Buckyballs. Phys Rev Lett. 2005;94:155504.

[15] Berseth PA, Harter AG, Zidan R, Blomquist A, Araujo CM, Scheicher RH, et al. Carbon Nanomaterials as Catalysts for Hydrogen Uptake and Release in $NaAlH_4$. Nano Lett. 2009;9(4):1501-5.

[16] Li M, Li, Y, Zhou. Z, Shen. P, Chen. Z. Ca-Coated Boron Fullerenes and Nanotubes as Superior Hydrogen Storage Materials. Nano Lett. 2009;9:1944-8.

[17] Yoon M, Yang S, Wang EG, Zhang Z. Charged Fullerenes as High-Capacity Hydrogen Storage Media. Nano Lett. 2007;9:2578-83.

[18] Yildirim T, Ciraci S. Titanium-Decorated Carbon Nanotubes as a potential high-capacity hydrogen storage medium. Phys Rev Lett. 2005;94:175501.

[19] Yoon M, Yang S, Kicke C, Wang EG, Geohegan D, Zhang Z. Calcium as the Superior Coating Metal in Functionalization of Carbon Fullerenes for High-Capacity Hydrogen Storage. Phys Rev Lett. 2008;100:206806.

[20] Felderhoff M, Weidenthaler C, von Helmolt R, Eberle U. Hydrogen storage: the remaining scientific and technological challenges. Phys Chem Chem Phys. 2007 Jun 7;9(21):2643-53.

[21] Li HW, Yan YG, Orimo S, Zuttel A, Jensen CM. Recent Progress in Metal Borohydrides for Hydrogen Storage. Energies. 2011 Jan;4(1):185-214.

[22] Matsunaga T, Buchter F, Mauron P, Bielman A, Nakamori Y, Orimo S, et al. Hydrogen storage properties of $Mg(BH_4)_2$. J Alloys Compd. 2008 Jul 14;459(1-2):583-8.

[23] Li HW, Kikuchi K, Nakamori Y, Ohba N, Miwa K, Towata S, et al. Dehydriding and rehydriding processes of well-crystallized $Mg(BH_4)_2$ accompanying with formation of intermediate compounds. Acta Materialia. 2008 Apr;56(6):1342-7.

[24] Grochala W, Edwards PP. thernal decomposition of non-interstital hydrides for storage and production of hydrogen. Chem Rev. 2004;104:1283.

[25] Severa G, Ronnebro E, Jensen CM. Direct hydrogenation of magnesium boride to magnesium borohydride: demonstration of > 11 weight percent reversible hydrogen storage. Chem Commun. 2010;46(3):421-3.

[26] Bogdanović B, Schwickardi M. Ti-doped alkali metal aluminum hydrides as potential novel reversible hydrogen storage materials. J Alloys Compd. 1997;253-254:1-9.

[27] Aguayo A, Singh DJ. Electronic structure of the complex hydride $NaAlH_4$. Phys Rev B. 2004;69(15):155103.

[28] Chaudhuri S, Graetz J, Ignatov A, Reilly JJ, Muckerman JT. Understanding the Role of Ti in Reversible Hydrogen Storage as Sodium Alanate: A Combined Experimental and Density Functional Theoretical Approach. J Am Chem Soc. 2006;128(35):11404 -15.

[29] Chaudhuri S, Muckerman JT. First-Principles Study of Ti-Catalyzed Hydrogen Chemisorption on an Al Surface: A Critical First Step for Reversible Hydrogen Storage in $NaAlH_4$. J Phys Chem B. 2005;109:6952-7.

[30] Íñiguez J, Yildirim T. First-principles study of Ti-doped sodium alanate surfaces. Appl Phys Lett. 2005;86:103109.

[31] Araújo CM, Li S, Ahuja R, Jena P. Vacancy-mediated hydrogen desorption in $NaAlH_4$. Phys Rev B. 2005;72:165101.

[32] Araújo CM, Ahuja R, Osorio Guillén JM, Jena P. Role of titanium in hydrogen desorption in crystalline sodium alanate. Appl Phys Lett. 2005;86:251913.

[33] Liu J, Han Y, Ge Q. Effect of Doped Transition Metal on Reversible Hydrogen Release/Uptake from $NaAlH_4$. Chem Eur J. 2009;15:1685-95.

[34] Majzoub EH, Zhou F, Ozolins V. First-Principles Calculated Phase Diagram for Nanoclusters in the Na-Al-H System: A Single-Step Decomposition Pathway for $NaAlH_4$. J Phys Chem C. 2011 Feb 17;115(6):2636-43.

[35] Wood BC, Marzari N. Dynamics and thermodynamics of a novel phase of $NaAlH_4$. Phys Rev Lett. 2009;103:185901.

[36] Huang CK, Zhao YJ, Sun T, Guo J, Sun LX, Zhu M. Influence of Transition Metal Additives on the Hydriding/Dehydriding Critical Point of $NaAlH_4$. J Phys Chem C. 2009 Jun 4;113(22):9936-43.

[37] Marashdeh A, Olsen RA, Lovvik OM, Kroes G-J. $NaAlH_4$ Cluster with Two Titanium Atoms Added. J Phys Chem C. 2007;111:8206-13.

[38] Vegge T. Equilobrium Structure and Ti-catalyzed H_2 desorption in $NaAlH_4$ nanoparticles from DFT. Phys Chem Chem Phys. 2006;8:4853-61.

[39] Peles A, Chou MY. Lattice dynamics and thermodynamic properties of $NaAlH_4$: Density-functional calculations using a linear response theory. Phys Rev B. 2006 May;73(18):184302.

[40] Løvvik OM, Opalka SM. Density functional calculations of Ti-enhanced $NaAlH_4$. Phys Rev B. 2005;71:054103.

[41] Liu J, Ge Q. A precursor state for formation of $TiAl_3$ complex in reversible hydrogen desorption/adsorption from Ti-doped $NaAlH_4$. Chemical Communications. 2006(17):1822-4.

[42] Graetz J, Reilly JJ, Johnson J, Ignatov AY, Tyson TA. X-ray absorption study of Ti-activated sodium aluminum hydride. Appl Phys Lett. 2004;85(3):500-2.

[43] Felderhoff M, Klementiev K, Grunert W, Spliethoff B, Tesche B, Bellosta von Colbe JM, et al. Combined TEM-EDX and XAFS studies of Ti-doped sodium alanate. Phys Chem Chem Phys. 2004;6(17):4369-74.

[44] Thomas GJ, Gross KJ, Yang NYC, Jensen C. Microstructural characterization of catalyzed NaAlH4. J Alloys Compd. 2002 Jan 17;330:702-7.

[45] Sandrock G, Gross KJ, Thomas G. Effect of Ti-catalyst content on the reversible hydrogen storage properties of the sodium alanates. J Alloys Compd. 2002;339(1-2):299.

[46] Gross KJ, Thomas GJ, Jensen CM. Catalyzed alanates for hydrogen storage. J Alloys Compd. 2002;330-332:683-90.

[47] Brinks HW, Fossdal A, Fonnelp JE, Hauback BC. Crystal structure and stability of $LiAlD_4$ with TiF_3 additive. J Alloys Compd. 2005;397:291-5.

[48] Balema VP, Balema L. Missing pieces of the puzzle or about some unresolved issues in solid state chemistry of alkali metal aluminohydrides. Phys Chem Chem Phys. 2005;7(6):1310-4.

[49] Bogdanović B, Felderhoff M, Pommerin A, Schuth F, Spielkamp N, Stark A. Cycling properties of Sc- and Ce-doped $NaAlH_4$ hydrogen storage materials prepared by the one-step direct synthesis method. J Alloys Compd. 2009;471:383-6.

[50] Baldé CP, Stil HA, van der Ederden AMJ, de Jong KP, Bitter JH. Active Ti Species in TiCl3-doped NaAlH4. Mechamism for catalyst Deactivation. J Phys Chem B. 2007;111:2797-802.

[51] Brinks HW, Sulic M, Jensen CM, Hauback BC. $TiCl_3$-Enhanced $NaAlH_4$: Impact of Excess Al and Development of the $Al_{1-y}Ti_y$ Phase. J Phys Chem B. 2006;110:2740-5.

[52] Majzoub EH, Herberg JL, Stumpf R, Spangler S, Maxwell RS. XRD and NMR investigation of Ti-compound formation in solution-doping of sodium aluminum hydrides: solubility of Ti in $NaAlH_4$ crystals grown in THF. J Alloys Compd. 2005;394:265-70.

[53] Leon A, Kircher O, Rosner H, Decamps B, Leroy E, Fichtner M, et al. SEM and TEM characterization of sodium alanate doped with $TiCl_3$ or small Ti clusters (Ti_{13}·6THF). J Alloys Compd. 2005.

[54] Herberg JL, Maxwell RS, Majzoub EH. ^{27}Al and ^{1}H MAS NMR and ^{27}Al multiple quantum studies of Ti-doped $NaAlH_4$. J Alloys Compd. 2005;417(1-2):39-44.

[55] Bogdanović B, Brand RA, Marjanovic A, Schwickardi M, Tolle J. Metal-doped sodium aluminium hydrides as potential new hydrogen storage materials. J Alloys Compd. 2000;302:36-58.

[56] Nakamura Y, Fossdal A, Brinks HW, Hauback BC. Characterization of Al–Ti phases in cycled TiF_3-enhanced Na_2LiAlH_6. J Alloys Compd. 2005;416(1-2):274-8.

[57] Brinks HW, Hauback BC, Srinivasan SS, Jensen CM. Synchrotron X-ray Studies of Al1-yTiy Formation and Re-hydriding Inhibition in Ti-Enhanced $NaAlH_4$. J Phys Chem B. 2005;109:15780-5.

[58] Léon A, Yalovega G, Soldatov A, Fichtner M. Investigation of the Nature of a Ti−Al Cluster Formed upon Cycling under Hydrogen in Na Alanate Doped with a Ti-Based Precursor. J Phys Chem C. 2008;112(32):12545–9.

[59] Majzoub EH, Gross KJ. Titanium–halide catalyst-precursors in sodium aluminum hydrides. J Alloys Compd. 2003;356-357(1):363-7.

[60] Iniguez J, Yildirim T. First-principles study of Ti-doped sodium alanate surfaces. Appl Phys Lett. 2005 Mar 7;86(10):103109.

[61] Balde CP, Stil HA, van der Eerden AMJ, de Jong KP, Bitter JH. Active Ti species in $TiCl_3$-doped $NaAlH_4$. Mechanism for catalyst deactivation. J Phys Chem C. 2007 Feb;111(6):2797-802.

[62] Gunaydin H, Houk KN, Ozolins V. Vacancy-mediated dehydrogenation of sodium alanate. Proc Nat Acad Sci USA. 2008 Mar 11;105(10):3673-7.

[63] Streukens G, Bogdanovic B, Felderhoff M, Schuth F. Dependence of dissociation pressure upon doping level of Ti-doped sodium alanate—a possibility for "thermodynamic tailoring" of the system. Phys Chem Chem Phys. 2006;8:2889-92.

[64] Liu J, Ge Q. A First-Principles Analysis of Hydrogen Interaction in Ti-Doped $NaAlH_4$ Surfaces: Structure and Energetics. J Phys Chem B. 2006;110:25863-8.

[65] Dathar GKP, Mainardi DS. Kinetics of Hydrogen Desorption in $NaAlH_4$ and Ti-Containing $NaAlH_4$. J Phys Chem C. 2010;114:8026-31.

[66] Anton DL. Hydrogen desorption kineitcs in transition metal modifed $NaAlH_4$. J Alloys Compd. 2003;356-357(1):400-4.

[67] Bogdanović B, Felderhoff M, Pommerin A, Schuth F, Spielkamp N. Advanced Hydrogen-Storage Materials Based on Sc-, Ce-,and Pr-Doped $NaAlH_4$. Adv Mater. 2006;18:1198-201.

[68] Fang F, Zhang J, Zhu J, Chen GR, Sun DL, He B, et al. Nature and role of Ti species in the hydrogenation of a NaH/Al mixture. J Phys Chem C. 2007 Mar 1;111(8):3476-9.

[69] Liu J, Yu J, Ge Q. Hydride-Assisted Hydrogenation of Ti-Doped NaH/Al: A Density Functional Theory Study. J Phys Chem C. 2011 Feb 10;115(5):2522-8.

[70] Claudy P, Bonnetot B, Lettoffé JM, Turck G. Thermochim Acta. 1978;27:213-21.

[71] Dymova TN, Aleksandrov DP, Konoplev VN, Silina TA, Sizareva AS. J Coord Chem. 1994;20:279-85.

[72] Balema VP, Dennis KW, Pecharsky VP. Rapid solid-state transformation of tetrahedral AlH_4^- into octahedral AlH_6^{3-} in lithium aluminohydride. Chem Commun. 2000:1665-6.

[73] Balema VP, Pecharsky VP, Dennis KW. Solid state phase transformations in $LiAlH_4$ during high-energy ball-milling. J Alloys Compd. 2000;313:69.

[74] Balema VP, Wiench JM, Dennis KW, Pruski M, K. PV. Titanium catalyzed solid-state transformations in LiAlH during high- energy ball-milling. J Alloys Compd. 2001;329:108-14.
[75] Chen J, Kuriyama N, Xu Q, Takeshita HT, Sakai T. Reversible hydrogen storage via titanium-catalyzed LiAlH4 and Li_3AlH_6. J Phys Chem B. 2001 Nov 15;105(45):11214-20.
[76] Langmi HW, McGrady GS, Liu XF, Jensen CM. Modification of the H_2 Desorption Properties of LiAlH(4) through Doping with Ti. J Phys Chem C. 2010 Jun 17;114(23):10666-9.
[77] Liu XF, Langmi HW, Beattie SD, Azenwi FF, McGrady GS, Jensen CM. Ti-Doped LiAlH(4) for Hydrogen Storage: Synthesis, Catalyst Loading and Cycling Performance. J Am Chem Soc. 2011 Oct 5;133(39):15593-7.
[78] Lide DR. CRC Handbook of Chemistry and Physics. Boca Raton: CRC Press; 2004.
[79] Züttel A, Wenger P, Rentsch S, Sudan P, Mauron P, Emmenegger C. $LiBH_4$ a new hydrogen storage material. J Power Source. 2003;118:1-7.
[80] Züttel A, Rentsch S, Fischer P, Wenger P, Sudan P, Mauron P, et al. Hydrogen storage properties of LiBH4. J Alloys Compd. 2003;356-357:515-20.
[81] Au M, Jurgensen A, Zeigler K. Modified Lithium Borohydrides for Reversible Hydrogen Storage (2). J Phys Chem B. 2006;110(51):26482-7.
[82] Au M, Jurgensen A. Modified Lithium Borohydrides for reversible hydrogen storage. J Phys Chem B. 2006;110:7062-7.
[83] Fang ZZ, Ma LP, Kang XD, Wang PJ, Wang P, Cheng HM. In situ formation and rapid decompisition of $Ti(BH_4)_3$ by mechanical milling $LiBH_4$ with TiF_3. Appl Phys Lett. 2009;94:044104.
[84] Liu J, Ge Q. Hydrogen Interaction in Ti-Doped $LiBH_4$ for Hydrogen Storage: A Density Functional Analysis. J Chem Theo Comput. 2009 Nov;5(11):3079-87.
[85] Chlopek K, Frommen C, Leon A, Zabara O, Fichtner M. Synthesis and properties of magnesium tetrahydroborate, $Mg(BH_4)_2$. J Mater Chem. 2007;17(33):3496-503.
[86] Li HW, Kikuchi K, Nakamori Y, Miwa K, Towata S, Orimo S. Effects of ball milling and additives on dehydriding behaviors of well-crystallized $Mg(BH_4)_2$. Scripta Materialia. 2007 Oct;57(8):679-82.
[87] Li HW, Miwa K, Ohba N, Fujita T, Sato T, Yan Y, et al. Formation of an intermediate compound with a $B_{12}H_{12}$ cluster: experimental and theoretical studies on magnesium borohydride $Mg(BH_4)_2$. Nanotechnology. 2009 May 20;20(20):204013-8.
[88] Soloveichik GL, Andrus M, Gao Y, Zhao JC, Kniajanski S. Magnesium borohydride as a hydrogen storage material: Synthesis of unsolvated $Mg(BH_4)_2$. Int J Hydrogen Energy. 2009 Mar;34(5):2144-52.
[89] Yan Y, Li HW, Nakamori Y, Ohba N, Miwa K, Towata S, et al. Differential Scanning Calorimetry Measurements of Magnesium Borohydride $Mg(BH_4)_2$. Mater Trans. 2008 Nov;49(11):2751-2.
[90] Hanada N, Chopek K, Frommen C, Lohstroh W, Fichtner M. Thermal decomposition of $Mg(BH_4)_2$ under He flow and H_2 pressure. J Mater Chem. 2008;18(22):2611-4.

[91] Riktor MD, Sorby MH, Chlopek K, Fichtner M, Buchter F, Zuettel A, et al. In situ synchrotron diffraction studies of phase transitions and thermal decomposition of $Mg(BH_4)_2$ and $Ca(BH_4)_2$. J Mater Chem. 2007;17(47):4939-42.
[92] Matsurtaga T, Buchter F, Miwa K, Towata S, Orimo S, Zuttel A. Magnesium borohydride: A new hydrogen storage material. Renewable Energy. 2008 Feb;33(2):193-6.
[93] Varin RA, Chiu C, Wronski ZS. Mechano-chemical activation synthesis (MCAS) of disordered $Mg(BH_4)_2$ using $NaBH_4$. J Alloys Compd. 2008 Aug 25;462(1-2):201-8.
[94] Choudhury P, Bhethanabotla VR, Stefanakos E. First principles study to identify the reversible reaction step of a multinary hydrogen storage "Li-Mg-B-N-H" system. Int J Hydrogen Energy. 2010 Sep;35(17):9002-11.
[95] Bezemer GL, Bitter JH, Kuipers HPCE, Oosterbeek H, Holewijn JE, Xu XD, et al. Cobalt particle size effects in the Fischer-Tropsch reaction studied with carbon nanofiber supported catalysts. J Am Chem Soc. 2006 Mar 29;128(12):3956-64.
[96] Bell AT. The impact of nanoscience on heterogeneous catalysis. Science. 2003 Mar 14;299(5613):1688-91.
[97] Xi JQ, Kim JK, Schubert EF. Silica nanorod-array films with very low refractive indices. Nano Lett. 2005 Jul;5(7):1385-7.
[98] Tong LM, Lou JY, Gattass RR, He SL, Chen XW, Liu L, et al. Assembly of silica nanowires on silica aerogels for microphotonic devices. Nano Lett. 2005 Feb;5(2):259-62.
[99] Baldé CP, Hereijgers BPC, Bitter JH, de Jong KP. Sodium Alanate Nanoparticles - Linking Size to Hydrogen Storage Properties. J Am Chem Soc. 2008;130:6761–5.
[100] Lohstroh W, Roth A, Hahn H, Fichtner M. Thermodynamics Effects in Nanoscale NaAlH4. ChemPhysChem. 2010;11:789-92.
[101] Fichtner M. Properties of nanoscle metal hydrides. Nanotechnology. 2009;20:204009.
[102] Gao J, Adelhelm P, Verkuijlen MHW, Rongeat C, Herrich M, Bentum PJM, et al. Confinement of $NaAlH_4$ in Nanoporous Carbon: Impact on H_2 Release, Reversibility, and modynamics. J Phys Chem C. 2010;114:4675-83.
[103] Kowalczyk P, Holyst R, Terrones M, Terrones H. Hydrogen storage in nanoporous carbon materials myth and facts. Phys Chem Chem Phys. 2007;9:1786-92.
[104] Bogdanović B, Felderhoff M, Pommerin A, Schüth F, Spielkamp N. Advanced Hydrogen-Storage Materials Based on Sc-, Ce-, and Pr-Doped $NaAlH_4$. Adv Mater. 2006;18:1198-201.
[105] Fichtner M, Fuhr O, Kircher O, Rothe J. Small Ti clusters for catalysis of hydrogen exchange in NaAlH4. NANOTECHNOLOGY. 2003;14:778-85.
[106] Vajo JJ, Olson GL. Hydrogen storage in destabilized chemical systems. Scripta Materialia. 2007 May;56(10):829-34.
[107] Gross AF, Vajo JJ, Van Atta SL, Olson GL. Enhanced hydrogen storage kinetics of $LiBH_4$ in nanoporous carbon scaffolds. J Phys Chem C. 2008 Apr 10;112(14):5651-7.
[108] Zhang Y, Zhang WS, Wang AQ, Sun LX, Fan MQ, Chu HL, et al. $LiBH_4$ nanoparticles supported by disordered mesoporous carbon: Hydrogen storage performances and destabilization mechanisms. Int J Hydrogen Energy. 2007 Nov;32(16):3976-80.
[109] Wellons MS, Berseth PA, Zidan R. Novel catalytic effects of fullerene for $LiBH_4$ hydrogen uptake and release. Nanotechnology. 2009 May 20;20(20).

[110] Lovvik OM, Swang O, Opalka SM. Modeling alkali alanates for hydrogen storage by density-functional band-structure calculations. J Mater Res. 2005 Dec;20(12):3199-213.

[111] Opalka SM, Lovvik OM, Brinks HW, Saxe PW, Hauback BC. Integrated experimental-theoretical investigation of the Na-Li-Al-H system. Inorg Chem. 2007 Feb 19;46(4):1401-9.

[112] Huot J, Boily S, Guther V, Schulz R. Synthesis of Na_3AlH_6 and Na_2LiAlH_6 by mechanical alloying. J Alloys Compd. 1999 Feb 1;283(1-2):304-6.

[113] Brinks HW, Hauback BC, Jensen CM, Zidan R. Synthesis and crystal structure of Na(2)LiAlD(6). J Alloys Compd. 2005 Apr 19;392(1-2):27-30.

[114] Orimo S, Nakamori Y, Kitahara G, Miwa K, Ohba N, Towata S, et al. Dehydriding and rehydriding reactions of $LiBH_4$. J Alloys Compd. 2005;404-406:427-30.

[115] Miwa K, Ohba N, Towata S, Nakamori Y, Orimo S. First-principles study on copper-substituted lithium borohydride, $(Li_{1-x}Cu_x)BH_4$. J Alloys Compd. 2005 Dec 8;404:140-3.

[116] Li HW, Orimo S, Nakamori Y, Miwa K, Ohba N, Towata S, et al. Materials designing of metal borohydrides: Viewpoints from thermodynamical stabilities. J Alloys Compd. 2007 Oct 31;446:315-8.

[117] Hagemann H, Longhini M, Kaminski JW, Wesolowski TA, Cerny R, Penin N, et al. $LiSc(BH_4)_4$: A novel salt of Li^+ and discrete $Sc(BH_4)_4$ complex anions. J Phys Chem A. 2008 Aug 21;112(33):7551-5.

[118] Sorby MH, Brinks HW, Fossdal A, Thorshaug K, Hauback BC. The crystal structure and stability of K_2NaAlH_6. J Alloys Compd. 2005;415(1-2):284-7.

[119] Tang X, Opalka SM, Laube BL, Wu FJ, Strickler JR, Anton DL. Hydrogen storage properties of Na-Li-Mg-Al-H complex hydrides. J Alloys Compd. 2007 Oct 31;446:228-31.

[120] Grove H, Brinks HW, Heyn RH, Wu FJ, Opalka SM, Tang X, et al. The structure of $LiMg(AlD_4)_3$. J Alloys Compd. 2008 May 8;455(1-2):249-54.

[121] Au M, Meziani MJ, Sun YP, Pinkerton FE. Synthesis and Performance Evaluation of Bimetallic Lithium Borohydrides as Hydrogen Storage Media. Journal of Physical Chemistry C. 2011 Oct 27;115(42):20765-73.

[122] Fang ZZ, Kang XD, Luo JH, Wang P, Li HW, Orimo S. Formation and Hydrogen Storage Properties of Dual-Cation (Li, Ca) Borohydride. J Phys Chem C. 2010 Dec 30;114(51):22736-41.

[123] Jiang K, Xiao XZ, Deng SS, Zhang M, Li SQ, Ge HW, et al. A Novel Li-Ca-B-H Complex Borohydride: Its Synthesis and Hydrogen Storage Properties. J Phys Chem C C. 2011 Oct 13;115(40):19986-93.

[124] Blanchard D, Shi Q, Boothroyd CB, Vegge T. Reversibility of Al/Ti Modified $LiBH_4$. J Phys Chem C. 2009 Aug 6;113(31):14059-66.

[125] Cerny R, Ravnsbaek DB, Severa G, Filinchuk Y, D' Anna V, Hagemann H, et al. Structure and Characterization of $KSc(BH_4)_4$. J Phys Chem C. 2010 Nov 18;114(45):19540-9.

[126] Cerny R, Severa G, Ravnsbaek DB, Filinchuk Y, D'Anna V, Hagemann H, et al. NaSc(BH(4))(4): A Novel Scandium-Based Borohydride. J Phys Chem C. 2010 Jan 21;114(2):1357-64.

[127] Lodziana Z. Multivalent metal tetrahydroborides of Al, Sc, Y, Ti, and Zr. Phys Rev B. 2010 Apr 1;81(14).

[128] Yang J, Sudik A, Wolverton C. destabilizing $LiBH_4$ with a metal (M=Mg, Al, Ti, V, Cr, or sc) or metal hydride (MH_2=MgH_2, TiH_2, CaH_2). J Phys Chem C. 2007.

[129] Nickels EA, Jones MO, David WIF, Johnson SR, Lowton RL, Sommariva M, et al. Tuning the decomposition temperature in complex hydrides: Synthesis of a mixed alkali metal borohydride. Angew Chem Int Ed. 2008;47(15):2817-9.

[130] Seballos L, Zhang JZ, Ronnebro E, Herberg JL, Majzoub EH. Metastability and crystal structure of the bialkali complex metal borohydride $NaK(BH_4)_2$. J Alloys Compd. 2009 May 12;476(1-2):446-50.

[131] Kim C, Hwang SJ, Bowman RC, Reiter JW, Zan JA, Kulleck JG, et al. $LiSc(BH_4)_4$ as a Hydrogen Storage Material: Multinuclear High-Resolution Solid-State NMR and First-Principles Density Functional Theory Studies. J Phys Chem C. 2009 Jun 4;113(22):9956-68.

[132] Ravnsbaek D, Filinchuk Y, Cerenius Y, Jakobsen HJ, Besenbacher F, Skibsted J, et al. A Series of Mixed-Metal Borohydrides. Angew Chem Int Ed 2009;48(36):6659-63.

[133] Alapati SV, Johnson JK, Sholl DS. Using first principles calculations to identify new destabilized metal hydride reactions for reversible hydrogen storagew. Phys Chem Chem Phys. 2007;9:1438-52.

[134] Alapati SV, Johnson JK, Sholl DS. Identification of destabilized metal hydrides for hydrogen storage using first principles calculations. J Phys Chem B. 2006 May 4;110(17):8769-76.

[135] Alapati SV, Johnson JK, Sholl DS. Predicting reaction equilibria for destabilized metal hydride decomposition reactions for reversible hydrogen storage. J Phys Chem C. 2007 Feb 1;111(4):1584-91.

[136] Yu XB, Grant DM, Walker GS. A new dehydrogenation mechanism for reversible multicomponent borohydride systems - The role of Li-Mg alloys. Chem Commun. 2006(37):3906-8.

[137] Wolverton C, Siegel DJ, Akbarzadeh AR, Ozolins V. Discovery of novel hydrogen storage materials: an atomic scale computational approach. J Phys Condense Matt. 2008 Feb 13;20(6).

[138] Siegel DJ, Wolverton C, Ozolins V. Thermodynamic guidelines for the prediction of hydrogen storage reactions and their application to destabilized hydride mixtures. Phys Rev B. 2007 Oct;76(13).

[139] Vajo JJ, Skeith SL, Mertens F. Reversible storage of hydrogen in destabilized $LiBH_4$. J Phys Chem B. 2005 Mar 10;109(9):3719-22.

[140] Piinkerton FE, Meyer MS, Meisner GP, Balogh MP, Vajo JJ. Phase boundaries and reversibility of $LiBH_4/MgH_2$ hydrogen storage material. J Phys Chem C. 2007 Sep 6;111(35):12881-5.

[141] Barkhordarian G, Klassen T, Dornheim M, Bormann R. Unexpected kinetic effect of MgB_2 in reactive hydride composites containing complex borohydrides. J Alloys Compd. 2007 Aug 16;440(1-2):L18-L21.

Chapter 3

Enclosure of Sodium Tetrahydroborate ($NaBH_4$) in Solidified Aluminosilicate Gels and Microporous Crystalline Solids for Fuel Processing

Josef-Christian Buhl, Lars Schomborg and Claus Henning Rüscher

Additional information is available at the end of the chapter

http://dx.doi.org/10.5772/50186

1. Introduction

The development of new materials for production and high storage capacities is most essential for an efficient use of the future energy source "hydrogen". Besides cryogenic and high pressure storages several chemical alloys like metal hydrides, carbon nanotubes or clathrates have been discussed [1-5]. New metal-organic framework compounds (MOFs) have been developed and proved to be outstanding hydrogen storage materials [6-9]. Besides these new materials also the well known hydride salt sodium tetrahydroborate ($NaBH_4$) has recently been moved into new centre of interest as a possible hydrogen source according to its large hydrogen capacity, 5.3 wt% H_2, which could be used to gain 2.4 l H_2/g $NaBH_4$ in the reaction with water [10-13].

Our recent studies succeeded in an easy and safe way of handling $NaBH_4$ salt in strong alkaline aluminate and silicate solutions. Brought together gelation occurs immediately which could be further solidified by drying [14]. There remains a heterogeneous solid containing $NaBH_4$ crystals and sodalite-type nanocrystals which are "glued" together in a matrix formed by short range ordered Si-O-Al (sialate) bonds. This new compound exhibits a high capacity of up to 72 wt% of $NaBH_4$ which could easily be handled in moisture atmosphere without any segregation or loss of $NaBH_4$ for weeks [15]. The complete amount of $NaBH_4$ inserted during synthesis could be used for the hydrogen production controlled by pH-value with the addition of weak acid solution. Details about synthesis of the gel, alteration during solidification and quantification of hydrogen storage capabilities will be described here in section 2.

Section 3 follows another idea of the enclathration of $NaBH_4$ into a zeolite framework structure which prevents the BH_4-anion from hydrolysis and offers a safe and specific way of hydrogen release in a rather controlled way. In [16] Barrer suggested an impregnation of pre-formed zeolites like X and Y with boronhydride salts like $Al(BH_4)_3$ or $NaBH_4$. Some experiments in this direction will also be demonstrated here, showing, that the BH_4-anion cannot be stabilized in such types of matrixes (section 3.1). Contrary to this the incorporation of BH_4-anions into the small sodalite cages during the formation of the sodalite crystals succeeded in a direct way of hydrothermal synthesis [17-21]. By this method $NaBH_4$-sodalites with aluminosilicate, gallosilicate and aluminogermanate framework compositions could be prepared (section 3.2) and a detailed understanding of their structure could be worked out (3.3). Further investigations succeeded in variations of crystal sizes between typically obtained microcrystals and nanocrystalline material also showing details of their hydrogen release reactions [22, 23]. The state of the art of a control of hydrogen release reactions of the BH_4-anions in the sodalite crystals in consecutive reaction steps with water will be outlined for the microcrystalline aluminosilicate sodalite (section 4.1). Indications of back reaction of the pre-reacted BH_4-anion in the sodalite cage were also reported [24, 25]. The realization of direct reinsertion of hydrogen in reacted $NaBH_4$ is ruled out so far, which makes its global use as energy storage so problematically. However, this problem could be overcome for the BH_4-anion in the sodalite cages. Thus we report in section 4.2 our results realizing first steps of hydrogen reinsertion into pre-reacted BH_4-sodalite.

2. Synthesis and characterisations of $NaBH_4$ crystals grown in an aluminosilicate gel

2.1. Solidified $NaBH_4$ aluminosilicate gel and its partial crystalline secondary products

The new material is a gel, formed from sodium-tetrahydroborate ($NaBH_4$), sodium-silicate, sodium-aluminate and water with the chemical composition 3 Na_2O : 2 SiO_2 : Al_2O_3 : 9-18 $NaBH_4 \cdot x\ H_2O$; $0 < x < 112$. The material can be obtained by a two-step reaction process. Firstly an aluminosilicate mixture under addition of high portions of $NaBH_4$ has to be prepared at room temperature. Secondly this mixture has to be dried by heating between 80 and 110°C between 0.5 and 4 h. During this procedure the gel undergoes a partial stepwise alteration resulting in solidification of the final product. As an example a typical synthesis batch can easily be prepared by dissolving $NaAlO_2$ (Riedel-de Haen 13404) in 1.5 ml H_2O before high $NaBH_4$ amounts (between 100 - 850 mg) have to be added to this solution and dissolved under stirring until a clear solution arises (solution I). A second solution is prepared from 310 mg Na_2SiO_3 (Fluka 2299129) and 1.5 ml water. After total dissolution of the silicate the same amount of $NaBH_4$ as used for preparation of solution I is added and the mixture has to be stirred too, until a clear solution has formed (solution II). Afterwards gel precipitation starts by the dropwise addition of solution II to solution I. A pasty liquid results from this alkaline gel-boranate mixture. After this the gel is exposed to drying procedure in an oven as described above.

Directly after precipitation of the $NaBH_4$ gel the product has the state of an amorphous sodium aluminosilicate, containing the whole amount of BH_4-anions from the inserted sodium-tetrahydroborate. As a result of separation of NaOH during gel precipitation and alteration process, the alkalinity remains very high, thus preventing the tetrahydroborate from decomposition by hydrolysis. The subsequent process of drying at 80°C up to 110°C between 0.5 h and 4.0 h causes rapid gel hardening. According to this solidified aluminosilicate gels are converted into secondary products with a high content of $NaBH_4$. Gel precipitation and alteration during drying at 110°C up to two hours was followed by X-ray powder diffraction as shown in Fig. 1. The powder diagram of the pure salt $NaBH_4$ is inserted in Fig. 1 for comparison.

It can be seen that without further drying at enhanced temperature only short range order could be present revealing a very broad peak around 30° 2 Theta (d = 2.97 Å) and a shoulder around 48° 2 Theta. The broad peak could be related to short ranged ordered

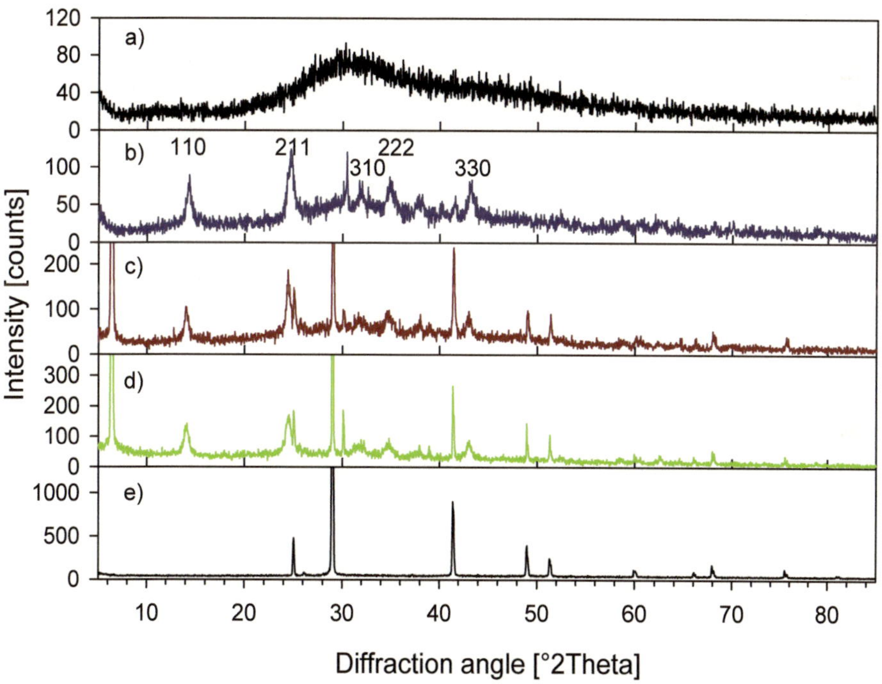

Figure 1. X-ray powder patterns of $NaBH_4$ - gel in dependance oft the drying period during drying at 110°C: directly after gel precipitation (a) and after drying for 0.5 h (b), 1.0 h (c) and 2.0 h (d). The pattern of the pure $NaBH_4$ salt dried at 110°C, is inserted for comparison (e).

Si-O-Al (sialate) type bonds which are typically also observed in geopolymers, i.e. when a mixture of water glass and metakaolin is aged [26]. A significant crystallization can be seen after 30 minutes of drying, reducing significantly the amount of short ranged ordered Si-O-

Al bonds. The four most intensive peaks can uniquely be indexed as (110), (211), (310), (222) and (330) within a sodalite framework. Further aging at 110°C does not alter the sodalite type peaks, which remain rather broad (Fig. 1 c, d). However, the intensity related to short range ordered sialate bonds becomes strongly reduced simultaneously with the appearance of a rather sharp diffraction peak at 6.5° 2 Theta (d = 13.66 Å). The initial crystallisate in the aluminosilicate could be identified as sodalite type nanocrystals. The strong and sharp peak at 6.5° 2 Theta (13.66 Å) in the XRD pattern (Fig. 1) could be related to a "disordered" sequence obtaining a 1.5· a superstructure, a = typical lattice parameter of the sodalite. The structure may not be seen as an intermediate between sodalite and cancrinite [27]. Parallel to this growth of a sodalite type aluminosilicate the $NaBH_4$ phase re-crystallizes. This process of $NaBH_4$-recrystallization under the strong alkaline conditions within the solidified aluminosilicate gel can be seen according to the evaluation of the $NaBH_4$ peaks in agreement with data of PDF-9-386. Thus the complete material can be regarded as a composite material containing $NaBH_4$ and sodalite-type phase embedded in or glued together by sialate bonds

A SEM photograph of the $NaBH_4$ gel, exposed to open conditions for 4 weeks is given in Fig. 2. Some more general features may be described for the handling of the $NaBH_4$-gel under open conditions. XRD pattern taken in a series up to 4 weeks are shown in Fig. 3. Compared with the powder pattern taken directly after precipitation and drying for two hours, after 10 days held under open conditions no remarkable decomposition occurs. According to the alkalinity of the sample some sodium carbonate was formed by uptake of carbon dioxide from the air. This can be seen by weak additional lines in the powder pattern, consistent with Na_2CO_3 PDF-18-1208. After 4 weeks under open conditions the intensity of all diffraction peaks becomes reduced and the samples gain a paste-like character.

Figure 2. Scanning electron microscopy of $NaBH_4$ gel held for 4 weeks under open conditions and EDX area analysis.

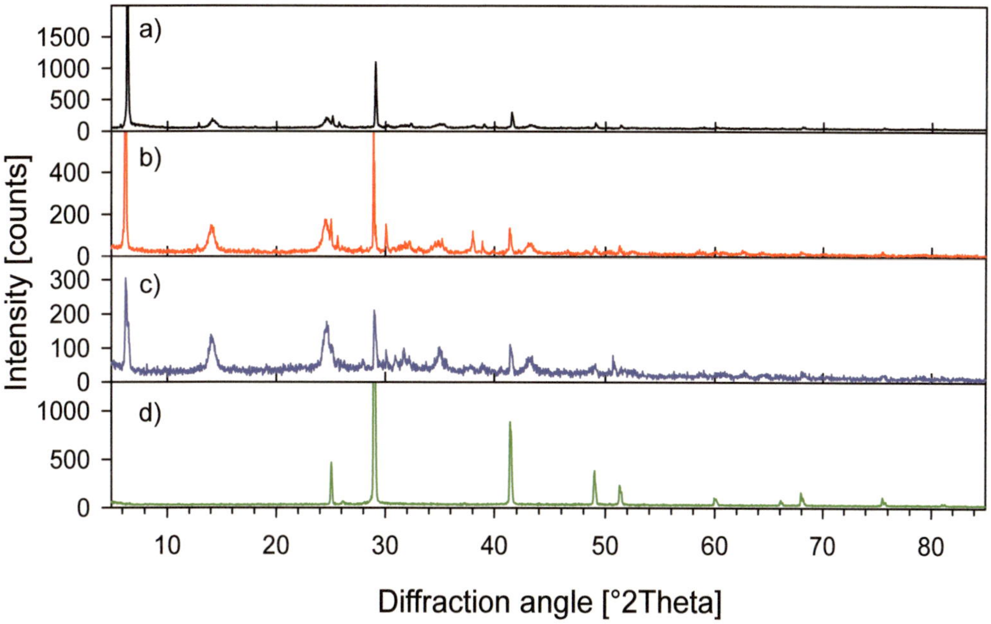

Figure 3. XRD investigation of the stability of $NaBH_4$-gel under open conditions: a) directly after precipitation and drying for two hours, b) after 10 days, held under open conditions and c) after 4 weeks under open conditions. The pattern of pure $NaBH_4$-salt (dried at 110°C) is included for comparison (d).

The samples were further analysed by infrared (IR) absorption spectroscopy (KBr method). The tetrahedral BH_4-anion groups of the $NaBH_4$ crystals could be identified by strong vibration modes at 1143 (ν_4), 2286 (2 ν_4), 2241 (ν_3) and 2390 ($\nu_{2}+\nu_4$) ([28-31], and more recently [32]) as shown in Fig. 4. By comparison with the spectrum of $NaBH_4$ the peaks related to the $NaBH_4$ in the aluminosilicate gel can be identified at the same positions in spectra a-d.

A significant uptake of water molecules to the gel could be seen as indicated by the increasing intensity of the peak at 1630 cm^{-1} (H_2O bending) and the main peak around 3500 cm^{-1} together with the broad shoulder towards lower wavenumbers. Despite this uptake of water no sign of hydrolysis reaction accompanied by formation of other borate species [33-35] at the expense of $NaBH_4$ can be derived from the spectra. This is a further indication for the stability of the $NaBH_4$ in the composite material. The peak at around 1450 cm^{-1} and the smaller one at 880 cm^{-1} indicate the presence of CO_3^{2-}-anions [36, 37]. As already mentioned, formation of sodium carbonate is the result of reaction with CO_2 from the air under alkaline conditions on the gel surface. Indications of the presence of sodalite framework are given by the small peaks at 436 and 469 cm^{-1}, the triplicate peak at 668, 705 and 733 cm^{-1} and the contribution at 990 cm^{-1} as could be realized by comparison with a typical sodalite spectrum Fig. 4. The sodalite spectrum appears to be superimposed to what has been called geopolymer type matrix, i.e. the alumosilicate gel. Similar spectra were observed during in situ investigations of sodalite crystallization from appropriate alumino-silicate gel in a KBr matrix [38].

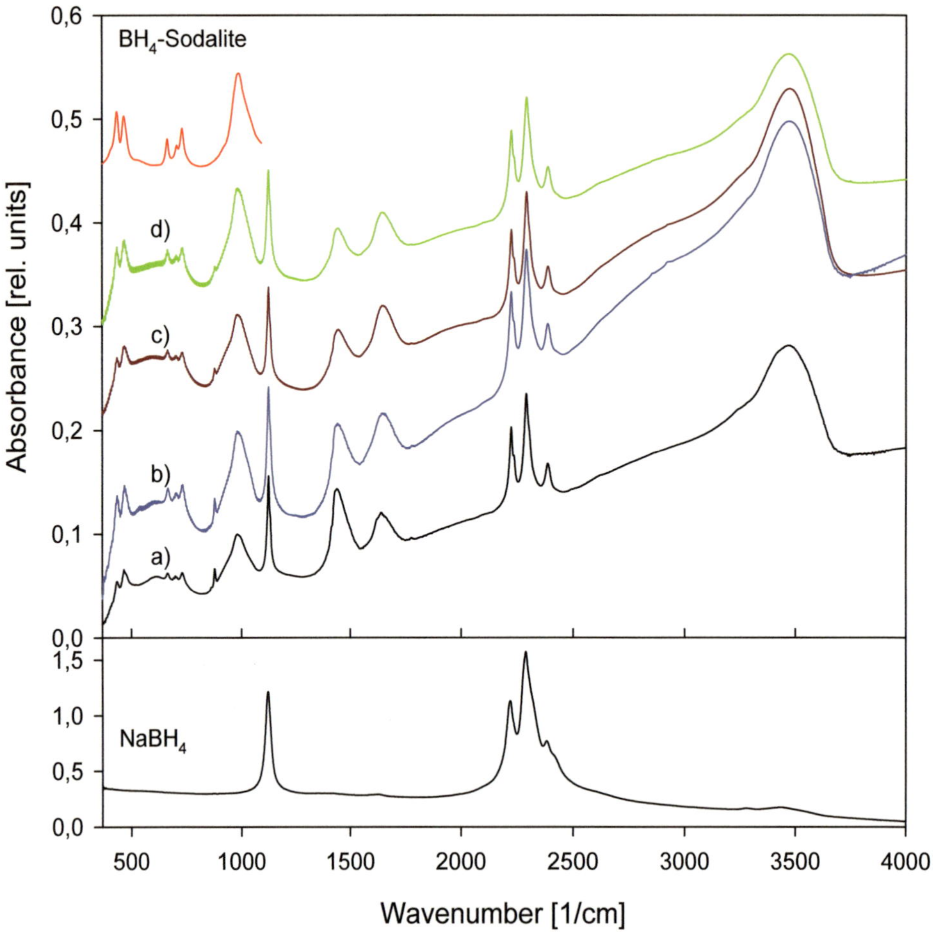

Figure 4. IR-absorption spectra of a series of $NaBH_4$-gel: directly after precipitation and drying for 2 h at 110°C (a); sample after 1 week (b), 3 weeks (c) and 4 weeks (d). Spectra b-d were normalized to spectrum a related to the BH_4-absorption intensity for better comparison. An example of typical as prepared BH_4-Sodalite is also shown.

Further tests of thermal stability of the new composite material including $NaBH_4$ crystals were carried out for heating the sample in a muffle furnace under open conditions in air. Spectra obtained before the treatment and treated for 2 h at 100°C steps up to 500°C are given in Fig. 5 (a-e). It can be seen that the $NaBH_4$ in the gel largely remains stable up to 300°C and then starts to decompose into borate species at temperatures between 300 and 400°C in air. A simple test of hydrogen release could be given by burning the sample initiated with a pocket lighter. The sample is burning and a glassy-like mixture of aluminosilicate and sodiumborate remains as could be identified in the IR absorption spectra, also shown in Fig. 5 (f). A more precise way for hydrogen release is the reaction in acid solutions [39]. The investigation of hydrogen release from solidified aluminosilicate gels by wetting with diluted acid is described in the following.

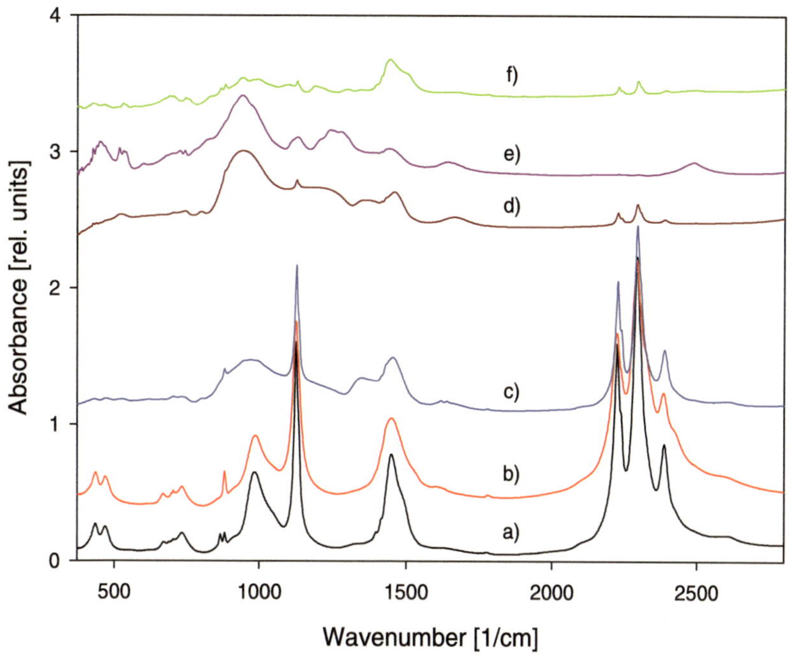

Figure 5. IR-absorption spectra of $NaBH_4$-gel heated in a muffle furnace for 2 h at 100°C (a), 200°C (b), 300°C (c), 400°C (d), 500°C (e) and of a sample after burning (f).

2.2. Hydrogen release experiments and further optimizations of storage capability

Hydrogen release experiments and optimizations of the starting compositions and solidification conditions for the hydrogen storage capability were carried out using an apparatus made from glassware as shown in Fig. 6. It consists of a 100 ml bulb, a gas syringe which measures the volume expansion related to the gas release, and the possibility of injection of liquid reactants. The bulb could also be heated and the temperature inside the bulb could be measured during the experiment. A stability check of the $NaBH_4$ enclosed in the aluminosilicate gel may be given in comparison to the raw $NaBH_4$-salt. Below 40°C $NaBH_4$ creates a stable hydrated form $NaBH_4 \cdot 2H_2O$ in contact with water. This species dehydrates at about 40°C to water and $NaBH_4$ which leads to the uncontrollable reaction with water [32]. For a further check of the reactivity of $NaBH_4$-salt with water an amount of 26.2 mg of $NaBH_4$ was wetted with 10 ml water in the bulb and heated to 70°C. This reveals an increased gas volume to about 58.7 ml in the reaction with water. For 10 ml water in the repeated experiment the gas volume increased by about 21 ml as a reference value. Since the amount of 26.2 mg $NaBH_4$ could release 58.66 ml of hydrogen, as also verified in further experiments below Fig. 7, it could be estimated that about 58% of the $NaBH_4$ reacted at 70°C under such conditions. Repeating the same experiment for an example of aluminosilicate gel (the $NaBH_4$-gel_0.47, Fig. 7, Tab. 1) reveals that only 11% of the $NaBH_4$ enclosed in the gel reacted. This shows a significant increase in protection.

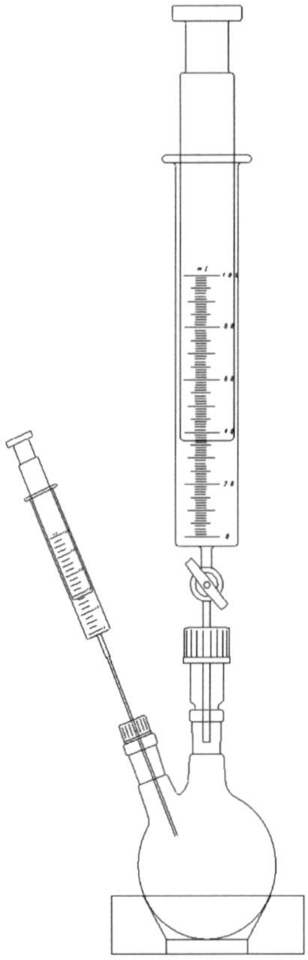

Figure 6. Apparatus as used for measuring gas release

A controlled hydrogen release from the $NaBH_4$ enclosed in the aluminosilicate gel could be achieved by lowering the pH-value by adding diluted acid. Using an injection needle as sketched in Fig. 6 diluted acid was added through a pierceable rubberplug, so the apparatus remained gas tight. The released gas was identified as hydrogen using the hydrogen-oxygen-reaction. Due to the high alkaline character of the $NaBH_4$-gel CO_2 from the surrounding air is absorbed in small amounts as CO_3^{2-} (compare IR-Spectra Fig. 4). By adding the diluted acid the CO_2 is also released. A gas check showed, however, that the concentration of CO_2 was below the detection limit of 1000 ppm and therefore negligible. As diluted acid a 1 % solution of hydrochlorid acid was chosen. It showed the best compromise between needed volume for entire hydrogen release and reaction velocity. If higher concentrations of acid were used the gas is released too fast. To get reliable results for every gel composition at least 5 different sample masses were investigated. The volumes of added acid were subtracted from the shown volume at the gas syringe to get the pure released gas volumes. The volumes obtained were plotted for the different samples against the used sample mass. With a linear regression the volume of released hydrogen per 100 mg sample could be calculated and compared to the pure $NaBH_4$ salt Fig. 7, Tab. 1.

$NaBH_4$/solid ratios between 0.26 and 0.75 were investigated. Tab. 1 depicts the synthesis masses of the investigated samples. The experiments show a linear trend between used sample mass and released hydrogen volume (Fig. 7). The more $NaBH_4$ is enclosed in the aluminosilicate gel the more hydrogen can be stored. Using a $NaBH_4$/solid ratio of 0.75 at the synthesis the released hydrogen volume is equivalent to 72 % of the pure $NaBH_4$-salt. More added $NaBH_4$ during the synthesis lowers the protection ability of the aluminosilicate gel after solidification. The hydrogen content approaches a saturation with further added $NaBH_4$ mass during synthesis. Some higher amount could still be enclosed by decreasing the solidify temperature. This was investigated using $NaBH_4$/gel ratios of 0.47 and 0.59. At 80°C solidified samples the 0.59 $NaBH_4$-gel releases about 15 % more hydrogen compared to the identical synthesis, solidified at 110°C. This higher amount of hydrogen storage capacity is reached, however, at the expense of longer solidify time up to 48 hours. Drying temperatures below 40°C are not able to solidify the $NaBH_4$-gel even after 96 hours which makes these temperatures inefficient. Below 40°C the hydrogen release per sample mass is very low because main fractions of the sample mass consist of water.

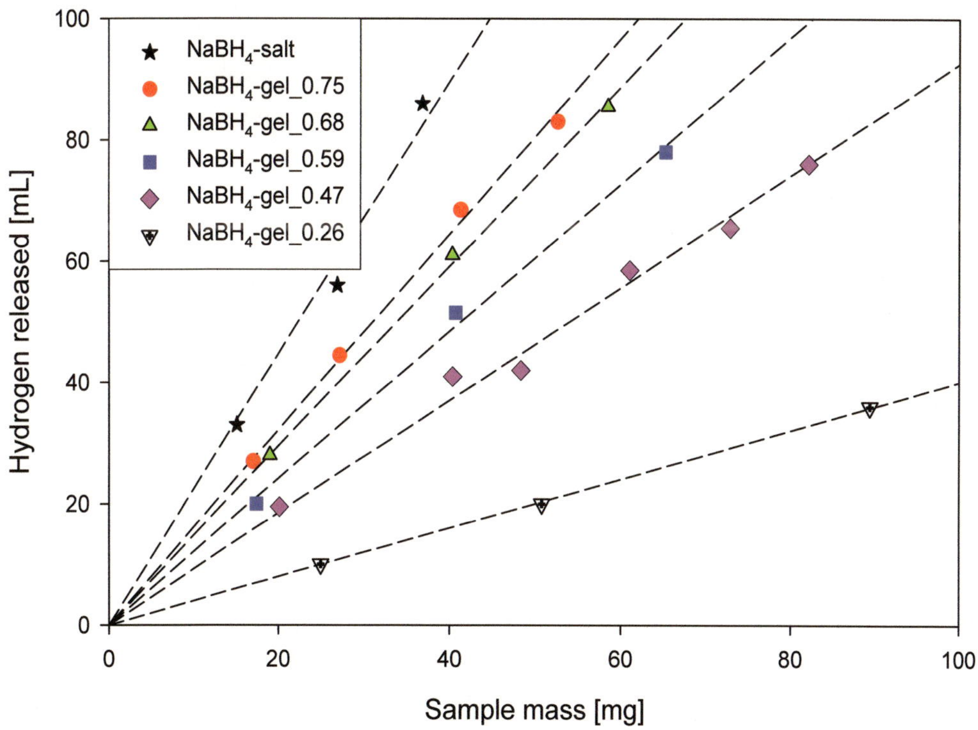

Figure 7. Released gas volume related on sample mass by the reaction of $NaBH_4$-gel of various ratios $NaBH_4$/gel as denoted with diluted hydrochloric acid. The results using pure $NaBH_4$-salt are also shown. Dashed lines result from linear regressions to the data.

Sample name	$NaBH_4$ [mg]	Na_2SiO_3 [mg]	$NaAlO_2$ [mg]	R	ST [°C]	H_2/100mg [ml]	H_2/H_2 from $NaBH_4$-salt [%]
$NaBH_4$-salt	-	0	0	1	-	224(exp.) 240 (lit.)	100
gel_0.26	200	310	250	0,26	110	40	17,86
gel_0.47	500	310	250	0,47	110	92	41,07
gel_0.47_80	500	310	250	0,47	80	106	47,32
gel_0.47_40	500	310	250	0,47	40	94	41,96
gel_0.47_18	500	310	250	0,47	18	55	24,55
gel_0.59	800	310	250	0,59	110	121	54,02
gel_0.59_80	800	310	250	0,59	80	143	63,84
gel_0.68	1200	310	250	0,68	110	148	66,07
gel_0.75	1700	310	250	0,75	110	161	71,88

Table 1. Used reactants and results of hydrogen release. Column 1-3: the used amounts of reactants (solids) for gel preparation (in mg); column 4: the ratio R =$NaBH_4$/solid; column 5 solidify temperature ST (°C) ; column 6: H_2/100 mg = hydrogen released per 100 mg sample (in ml) from linear regression to data Fig. 7); column 7: H_2/H_2 from $NaBH_4$-salt.

It can be concluded that the hydrogen content of the solidified aluminosilicate gels can be varied with the amount of added $NaBH_4$ during the synthesis. Till now the highest ratio of $NaBH_4$ per solid reactants in the synthesis is about 0.75 of gels solidified at 110°C. Some higher amount could still be enclosed decreasing the solidify temperature, however, on extension of the solidification time.

3. The BH_4-anion enclosed in cages of the sodalite

In addition to the materials like carbon nanotubes or MOFs as well as the aluminosilicate gel, zeolites could be suitable matrices for inclusion of hydrogen because of their open framework structures. Hydrogen loading into zeolite cavities under high pressure has been discussed. It could be shown that the sodalite structure could exhibit a high storage capacity but requires a loading temperature of 300°C and a pressure of 10 MPa [40]. A completely different way has been discovered more recently by direct enclosure of the BH_4-anion in the sodalite cage during soft chemical synthesis under hydrothermal conditions [17, 18]. The enclathration of one BH_4-anion into each of the sodalite cages prevents the anion from hydrolysis and offers a safe and specific way of hydrogen release as will be described in detail below. Barrer proposed in his outstanding work [16] an impregnation of pre-formed zeolites like A, X and Y with boronhydride salts like $Al(BH_4)_3$ or $NaBH_4$. However, the incorporation of hydride-anions into the small sodalite cage type units in a post-synthesis step seems impossible due to the diameter restrictions of the six ring window. Some primary attempts of direct synthesis of zeolites like LTA with $NaBH_4$-filled toc-subunits may be shown to be unsuccessful and LTA can only be obtained in mixtures with BH_4-sodalite.

3.1. Primary steps

The three-dimensional structure net of the sodalites is known for more than 80 years [41]. It was found as basic structure type of many zeolite related compounds up to date [42]. The general sodalite composition is $Na_8[T^1T^2O_4]_6X_2$ where T^1 is a trivalent cation (usually Al^{3+}) and T^2 a tetravalent cation (usually Si^{4+}) but others like Ga^{3+} and Ge^{4+} can be built in during synthesis [18, 43-45]. The sodalite framework is built up by a space filling package of truncated octahedral cages ("toc-units") formed by tetrahedral TO_4 units. Each cage is filled by a $[Na_4X]^{3+}$ -complex with X representing a monovalent anion or anion group as for example the BH_4-anion. Those guests are enclathrated during synthesis according to the chlathralite like properties of sodalites [46]. As known for other salt-filled sodalites the thermal reactivity of the enclathrated guests can differ from the behaviour of the pure salt according to special interactions of the guest-complex with the sodalite host-framework [47-50]. The toc-unit of the sodalite structure is a common building unit in other zeolites like LTA, LSX, X and Y. Fig. 8 gives a schematic view on the sodalite framework and the framework of zeolite LTA.

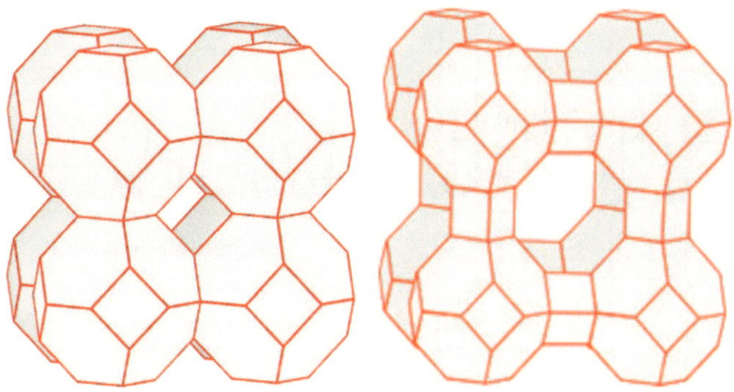

Figure 8. Structure scheme of frameworks of sodalite (left) and zeolite LTA (right).

With the aim to clarify major questions connected with wet chemistry of $NaBH_4$ and zeolites with wider open frameworks than sodalites, experiments on impregnation possibility by interaction of zeolite A (LTA) and $NaBH_4$ and on direct synthesis of $NaBH_4$-LTA using the gel method under addition of $NaBH_4$ were carried out. The impregnation experiments were performed using commercial zeolite Na-LTA (Fluka-69836) and $NaBH_4$. The zeolite was stirred in 2 M NaOH-$NaBH_4$ solution at room temperature with a solid:liquid ratio of 1:20. After a treatment period of 60 minutes the solution was filtered. NaOH residues were carefully washed out of the solid followed by drying at 80°C over night. However the IR absorption spectrum does not show the presence of any BH_4-anions in the otherwise typical LTA zeolite signatures. Therefore, it can be concluded that the $NaBH_4$ salt or BH_4-anions do not enter or cannot be stabilized neither in the supercages (grc) nor in the toc units of LTA.

Because of this failure of BH_4-incorporation into the cavities of pre-formed zeolite LTA another experimental series was performed to test possibilities of direct formation of $NaBH_4$-LTA. The common gel method was used under addition of $NaBH_4$ salt to the gel during its precipitation. Sodium metasilicate and sodium aluminate were used for gel formation and $NaBH_4$ salt was added to both of the starting solutions before they were mixed to form the gel. Whereas the sodalite crystallization occurs under high alkalinity which prevents rapid hydrolysis of the inserted BH_4-anion, LTA formation needs lower alkaline solutions. Thus the experiments were performed under the Na_2O:H_2O- ratio of 1:20 M. Under those conditions of low alkalinity a partial decomposition of $NaBH_4$ cannot be excluded. Therefore the crystallization time was shortened neglecting an otherwise necessary further gel aging step in order to prevent $NaBH_4$ decomposition choosing a temperature of 100°C for 2-4 h. This higher temperature for LTA formation was selected to also accelerate the crystallization process within the short crystallization time interval.

The X-ray powder pattern of a typical synthesis product is shown in Fig. 9. The powder patterns of common zeolite LTA (Fluka) and $NaBH_4$-sodalite are inserted in this figure for comparison. Beside sharp peaks, consistent with the powder pattern of zeolite LTA, broad lines of sodalite can be distinguished from the powder pattern of the synthesis product.

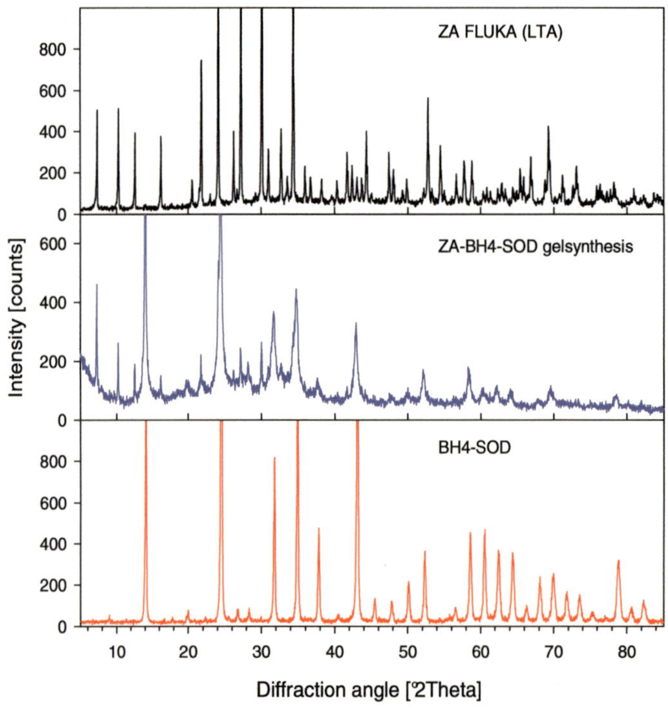

Figure 9. X-ray powder patterns of: product of direct synthesis in the LTA-$NaBH_4$ system at 100°C and 4 h synthesis time (pattern in the middle) and of common zeolite LTA-Fluka (on top) and $NaBH_4$-sodalite (bottom).

According to this, the product can be regarded as a mixture of a lower amount of zeolite LTA and mainly nano-sized sodalite beside some short range ordered aluminosilicate units, indicated by the broad peak in the range 20° to 40° 2 Theta. The formation of the two phases zeolite LTA and sodalite can also be observed in the IR absorption spectrum in comparison to spectra of the $NaBH_4$-sodalite and zeolite LTA (Fluka) Fig. 10. In particular the sodalite can be identified by the triplicate sodalite "fingerprint".

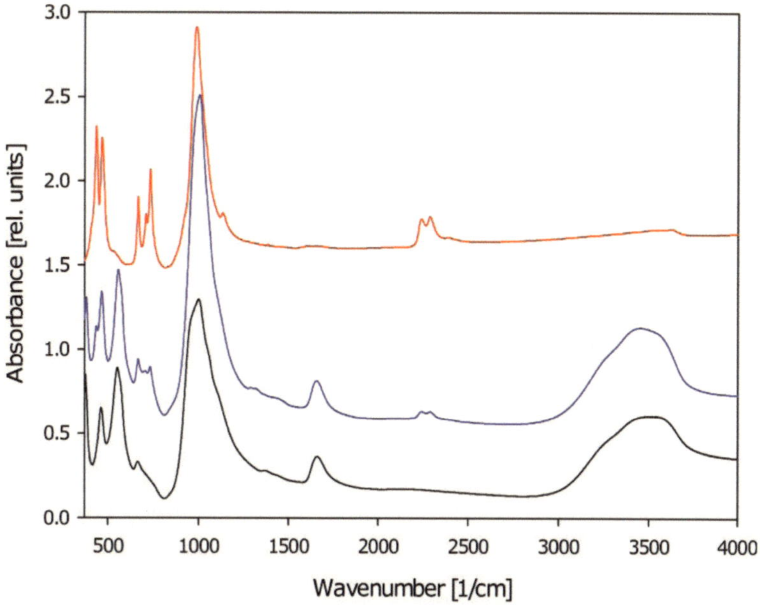

Figure 10. FTIR-spectra of BH_4-SOD, two phase product of the direct synthesis experiment and the spectrum of zeolite LTA-FLUKA; (spectra from top to bottom).

Further experiments under variation of the alkalinity, the solid/liquid ratio as well as the time of syntheses all failed to obtain $NaBH_4$ zeolite LTA. In each case a two phase product was observed, consisting of $NaBH_4$ sodalite and $NaBH_4$-free zeolite LTA. The reaction parameters mentioned here showed only a small influence on the mass ratio of both of these phases. The results are a hint that even a direct crystallization of $NaBH_4$-zeolite LTA seems to be impossible under the conditions of gel-crystallization usually used in zeolite chemistry.

3.2. Enclosure of the BH_4-anion in micro- and nano-crystalline sodalites

The route of synthesis follows certain rules in order to include the BH_4-anion in the sodalite cage and not to obtain just co-crystallization of $NaBH_4$ and the sodalite within the aluminosilicate matrix as described in section 2. Synthesis of $NaBH_4$-aluminosilicate sodalite in microcrystalline form was performed under mild hydrothermal and strong alkaline conditions (NaOH) using kaolinite as Si-Al-source. An excess of $NaBH_4$ salt has to be added to this solution. From wet chemical reaction behaviour of pure sodium tetrahydroborate in water it is known that the kinetics of decomposition are highly influenced by the alkalinity [51, 52, 10]. Further parameters of synthesis like the solid to liquid ratio, temperature and reaction time had to be optimized for sodalite synthesis with hydrolysis sensitive BH_4-anions. 50 ml Teflon coated steel autoclaves were used for synthesis.

After screening experiments the amounts of 1 g of kaolinite, 2 g of sodium tetrahydroborate salt and 10 ml of 16 M sodium hydroxide solution were selected for preparation of the favored reactant mixture. Crystallization was performed at a temperature of 110°C for 24 hours reaction time. The final products were washed with water and dried at 80°C for 24 hours [17]. Tetrahydroborate-sodalite nanoparticles were successfully synthesized even at lower temperature hydrothermal conditions (60°C) from high alkaline aluminosilicate gels and $NaBH_4$ salt [22, 23]. Preparation of basic hydrosodalite by this very simple method was first described by [53, 54] during experiments on zeolite A crystallization at very low temperatures under superalkaline conditions. Fine tuning of this gel method by [55] also yielded basic-hydrosodalite nanoparticles. Gel conditions are suitable for precipitation of salt-filled sodalites and cancrinites, too, as recently demonstrated by [56, 57] for the nitrate sodalite-cancrinite system. The use of similar gels at low temperatures under superalkaline conditions was shown to be a suitable method for $NaBH_4$-sodalite nanoparticle formation [22, 23] as the hydrolysis reaction of the highly moisture sensitive $NaBH_4$ salt is retarded under low temperature strong alkaline conditions [10, 51, 52, 58].

Details for obtaining BH_4-sodalite nanoncrystalline samples are batch compositions 13 Na_2O : 2 SiO_2 : 1,5 Al_2O_3 : 5 $NaBH_4$: 220 H_2O prepared from analytical grade chemicals sodium-metasilicate, sodium aluminate, $NaBH_4$ and sodium hydroxide solution. Syntheses were performed in a Teflon coated steel autoclave at 60°C. The final products were washed with 500 ml water and dried at 110°C for 48 h [22, 23]. A heating period of 12 h was proved to be an optimal reaction time for nanoparticle formation of suitable size and sufficient crystallinity.

The gallosilicate tetrahydroborate enclathrated sodalite was prepared by alkaline hydrothermal treatment of a solid mixture of gallium oxide, sodium silicate, $NaBH_4$ in 10 ml of a 6 M NaOH at 110°C for 24 h, using teflon lined autoclaves. The final product was washed with water and dried at 80°C for 24 hours [18]. Synthesis of the aluminogermanate phase was performed from a beryllonite analogous $NaAlGeO_4$ following [59]. This starting material was obtained from GeO_2, γ-Al_2O_3 and Na_2CO_3 heated for 12 h at 1200°C before quenched to room temperature and crystallized at 800°C for 48 hours. The sodalite was subsequently synthesized by treatment of the $NaAlGeO_4$ in 10 ml 4 M NaOH at 110°C for 24 hours, again using a Teflon lined autoclave and same washing procedure as for the gallosilicate sodalite [18].

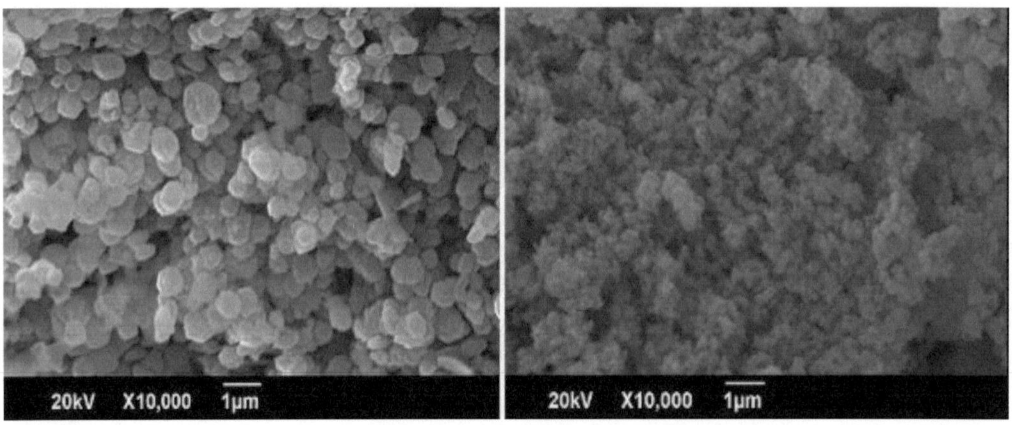

Figure 11. SEM-image of the microcrstalline sample (left) and the nanocrystalline product of the 12h experiment (right).

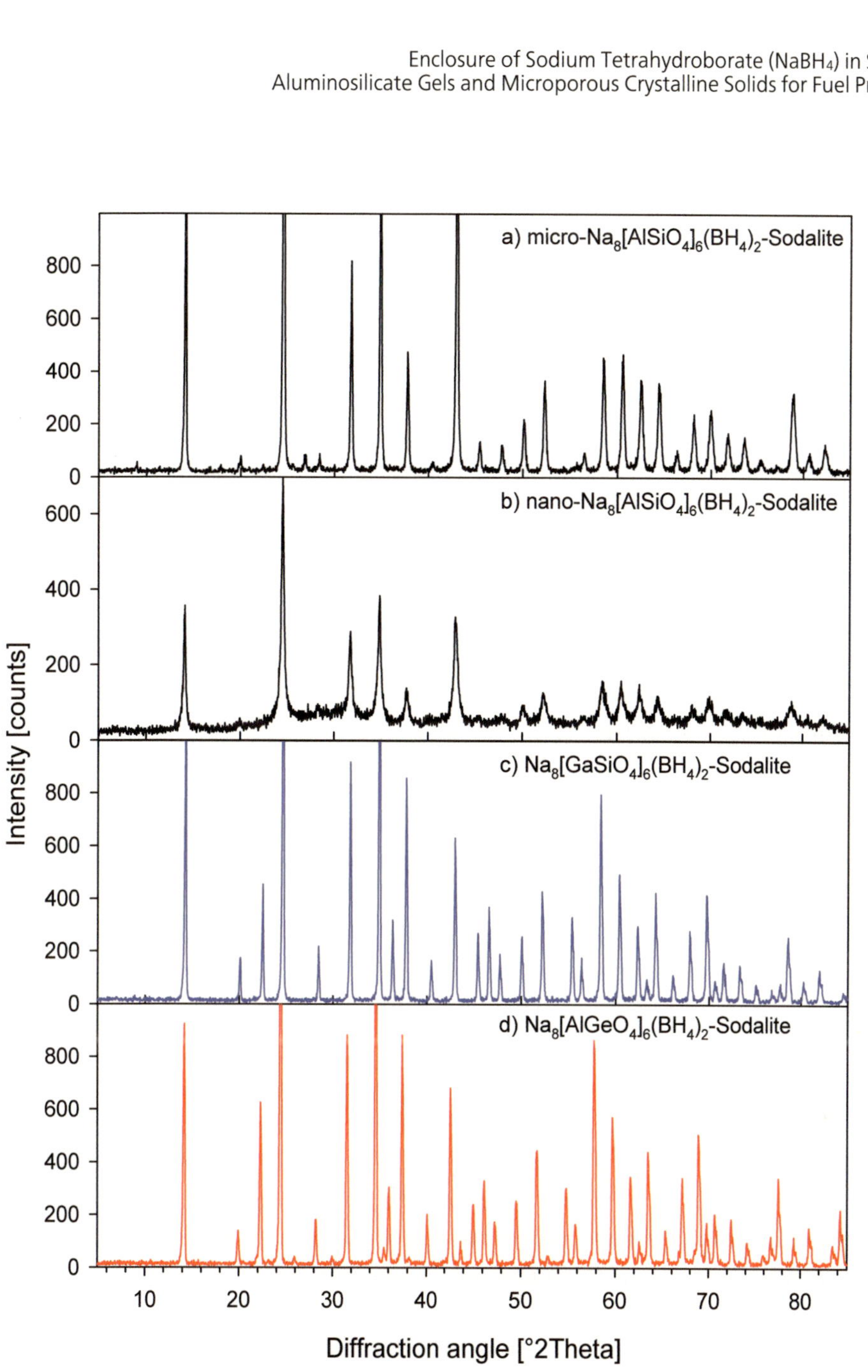

Figure 12. XRD pattern of the $NaBH_4$-sodalites: micro- (a) and nano- (b) crystalline aluminosilicate sodalite, gallosilicate sodalite (c), aluminogermanate sodalite (d).

Formation of crystals of good quality with an average size > 0.5 μm could be stated fom SEM investigations of microcrystalline tetrahydroborate sodalite $Na_8[AlSiO_4]_6(BH_4)_2$ (Fig. 11, left side). For the nanocrystalline sample, obtained after 12 h reaction time (Fig. 11, right side) nanocrystals are "glued together" by amorphous material to larger spherical agglomerates of about 100 nm size [22, 23].

The X-ray powder patterns of the micro- and nano-crystalline aluminosilicate-sodalite, gallosilicate sodalite and aluminogermanate sodalites are shown in Fig. 12. All diffraction peaks of the X-ray powder pattern could uniquely be indexed to pure phase sodalite within P-43n.

Rietveld refinement of the XRD-data for the micro crystalline aluminosilicate sodalite sample showed an amount up to 10 % related to a broad "background contribution" which could be peaked in the range around 30° 2 Theta. This contribution is of importance for hydrogen release from the sodalite as it contains and transports the amount of available water to the sodalite cages as shown below (section 4.1). This contribution which is related to short range ordered sialate type matrix to the sodalite crystals increases in the nanocrystalline sample. This can be seen in Fig. 12 (b) for the nanocrystalline sodalite compared to the XRD pattern of the microcrystalline sample. For this sample, obtained after 12 h reaction time, refinement of lattice constant reveals a = 8.9351(8) Å, being slightly enlarged compared with the microcrystalline phase, a = 8.9161(2) Å. An average crystal size of 25 nm was calculated for this nanocrystalline sodalite and the amount of short range ordered aluminosilicate material could be estimated to about 50 % using the "TOPAS" software [22].

The enclathration of BH_4-anions inside the cages of the micro- and nanocrystalline aluminosilicate sodalite can be seen in the IR-absorption spectra of the samples, Fig. 13. For the microcrystalline sample the aluminosilicate sodalite framework related vibrations, i.e. the six typical peaks can be seen: at about 436 and 469 cm^{-1}; "triplicate peaks" or "sodalite fingerprint" at 668, 705 and 733 cm^{-1} and asymmetric Si-O vibration at 987 cm^{-1} [60]. The BH_4-related vibrations can be seen at 1134 cm^{-1} and the characteristic triplicate peaks at 2240, 2288 and 2389 cm^{-1} by direct inspection in comparison to the spectrum of $NaBH_4$. These peaks are very slightly shifted compared to the peak positions in $NaBH_4$ which could be seen only in an enlarged scale. There are also indications for H_2O contributions around 1630 cm^{-1} (bending of H_2O) and in the range between 3000 and 3600 cm^{-1} (H_2O stretching) related to the water content in the short ranged ordered aluminosilicate. This contribution is significantly enlarged in the nanocrystalline sample. According to this the intensities of the sodalite framework and BH_4^- contribution appear smaller. The IR spectra of gallosilicate and aluminogermanate $NaBH_4$ sodalite are also shown in Fig. 13, too. In both spectra the enclathrated BH_4-anions can be seen by intense absorption bands of the BH_4^- tetrahedral group as compared with the spectrum of the pure salt. The spectrum of the gallosilicate-sodalite framework shows two clear resolved maxima at 922 cm^{-1} and 945 cm^{-1} for the asymmetric T-O-T vibrations. For the ν_s modes two very close adjacent signals with vibrations at 642 cm^{-1} and a shoulder at 624 cm^{-1} as well as a peak at 556 cm^{-1}can be seen from Fig. 13 (c) and finally the framework bending mode was found as one sharp signal at 457 cm^{-1}. The spectrum of the aluminogermanate sodalite Fig. 13 d shows one asymmetric T-O-T mode at 858 cm^{-1}, two symmetric T-O-T vibrations (609 cm^{-1} and 636 cm^{-1}) and a T-O deformation mode at 387 cm^{-1} [18].

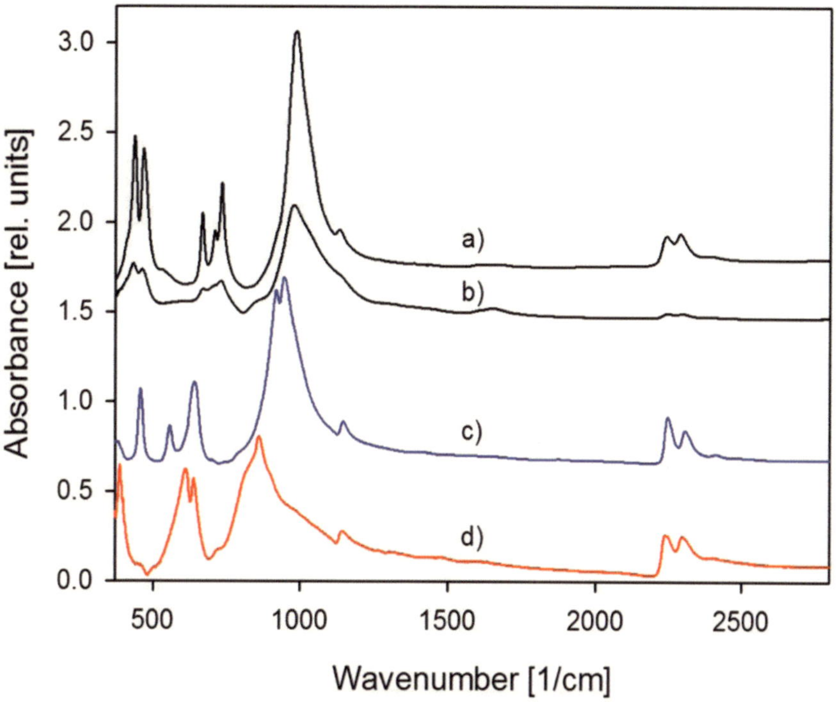

Figure 13. FTIR spectra of $NaBH_4$ aluminosilicate sodalites as synthesized: a) microcrystalline phase, b) nanocrystalline phase and the spectrum of $NaBH_4$ salt (after [22]). Also shown are the spectra of gallosilicate NaBH4-sodalite (c) and aluminogermanate $NaBH_4$-sodalite (d) ([18]).

3.3. Crystal structure of $NaBH_4$ sodalites: X-ray diffraction and MAS NMR study

3.3.1. XRD

The X-ray powder patterns of the microcrystalline aluminosilicate, gallosilicate as well as aluminogermanate $NaBH_4$ sodalites were analyzed by Rietveld method [17]. The atomic parameters of NaCl sodalite were taken as starting values [61] with the BH_4^- group with boron in the centre of the sodalite cage instead of the Cl-anion. Hydrogen was refined on x,x,x positions, restrained to distances of 116.8 pm as found in $NaBD_4$ [62].

The refined positional-, displacement- and occupancy- parameters together with R-values [63], cell constant and cell volume, are collected in Tab. 2 (standard deviation for occupation factors of B and Na: 3%). In the aluminosilicate sodalite the Si-O and Al-O distances are 163(3) pm and 174(3) pm, respectively $l_1 \cdot a$, $l_2 \cdot a$, a = lattice parameter, Fig. 14 a. For example with the formulas as given by [50] some sodalite structure specific parameters as shown in Fig. 3.7 (a) are the tetragonal tetrahedral distortions α'_{Si} = 112.6°, α''_{Si} = 107,9°, α'_{Al} = 110.8°, α''_{Al} = 108.8°, the tilt angles ϕ_{Si} = 23.7° and ϕ_{Al} = 22.3°, and the Al-O-Si angle γ = 139°. The "cage filling" configuration could be worked out as illustrated in Fig. 14 b. The Na atoms have three oxygen atoms at 234(2) pm as well as three hydrogen atoms at 267(5) pm in an octahedral arrangement as nearest neighbours. Positional disorder of the hydrogen atoms according to dynamic averaging of orientational disorder is suggested here. Recent ab-initio calculations could rather closely fit to the experimentally determined structural parameters [64].

Atom	P-4 3 n	occup.	x	y	z	B /10²nm³
$Na_8[AlSiO_4]_6(BH_4)_2$: (SG P-43n: a = 8.9161(2) Å, V = 708.79(2) Å³, R_{WP} = 0.042, R_P = 0.032, R_I = 0.023, R_e = 0.031, d = 0.951)						
Na	8e	1.02(3)[1]	0.1834(15)	x	x	2.3(7)
Al	6d	1.0	¼	0	½	1.5(5)[2]
Si	6c	1.0	¼	½	0	1.5[2]
O	24i	1.0	0.1391(28)	0.1487(29)	0.4390(21)	1.1(7)
B	2a	1.1[1]	0	0	0	2.2(34)[3]
H	8e	1.1[1]	0.424(16)	x	x	2.2[3]
$Na_8[GaSiO_4]_6(BH_4)_2$: (SG P-43n: a = 8.9590(1) Å, V = 719.09(3) Å³, R_{WP} = 0.097, R_P = 0.074, R_B = 0.023, R_e = 0.043)						
Na	8e	1.02(2)[1]	0.1731(3)	x	x	2.7(2)
Ga	6d	1.0	¼	0	½	1.1(2)[2]
Si	6c	1.0	¼	½	0	1.1(2)[2]
O	24i	1.0	0.1336(3)	0.1513(3)	0.4310(4)	1.9(2)
B	2a	1.1[1]	0	0	0	3.2(8)[3]
H	8e	1.1[1]	0.4371(38)	x	x	3.2(8)[3]
$Na_8[AlGeO_4]_6(BH_4)_2$: (SG P-43n: a = 9.0589(2) Å, V = 743.40(6) Å³, R_{WP} = 0.112, R_P = 0.082, R_B = 0.026, R_e = 0.043)						
Na	8e	1.02(3)[4]	0.1740(4)	x	x	2.5(2)
Al	6d	1.0	¼	0	½	1.4(2)[5]
Ge	6c	1.0	¼	½	0	1.4(2)[5]
O	24i	1.0	0.1425(4)	0.1441(4)	0.4296(5)	1.6(2)
B	2a	1.1[4]	0	0	0	3.2(8)[6]
H	8e	1.1[4]	0.4401(38)	x	x	3.2(8)[6]

[1-6]Parameters with the same number were constrained to each other

Table 2. Atomic parameters of sodalites $Na_8[AlSiO_4]_6(BH_4)_2$ [17] and $Na_8[GaSiO_4]_6(BH_4)_2$, $Na_8[AlGeO_4]_6(BH_4)_2$ [18]

In the case of gallosilicate sodalite the Ga-O and Si-O distances are 161.1(3) pm and 181.9(3) pm, respectively. The boron atom is located at the centre of the sodalite cage. In a static statistical model of positional disorder the sodium atoms have three oxygen atoms at 234(4) pm as well as three hydrogen atoms at 253(1) pm as nearest neighbours i.e. in principal the same arrangement, as in aluminosilicate sodalite shown in Fig. 14 b. Positional disorder of the hydrogen atoms according to dynamic averaging of orientational disorder is suggested here, too [18]. For the aluminogermanate sodalite the Ge-O and Al-O distances read 173.0(3) pm and 174.9(3) pm. The sodium atoms in this phase have three oxygen atoms at 234(4) pm and three hydrogen atoms at 257(5) pm as nearest neighbours.

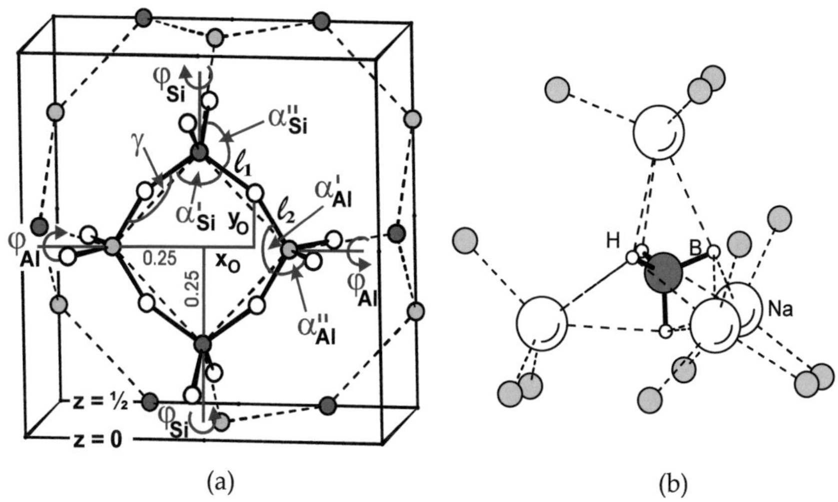

Figure 14. a, left) Perspective view of one half of the sodalite cage (after [50]) depicting sixring and fourring windows (dashed lines) and denoting bond lengths (l_1, l_2) and tetragonal tetrahedral distortions (α', α''), tilt angles (ϕ) and oxygen coordinates of the framework (x_0, y_0, z_0). Note $z_0 > 0$ determines $\phi > 0$ and z coordinate shifted by ½ compared to values given in Tab. 2. **b, right)** Coordination of the non-framework atoms in the microcrystalline $NaBH_4$ sodalite together with three oxygen atoms of the framework around each sodium atom [17].

3.3.2. MAS NMR

MAS NMR measurements of nuclei the ^{1}H, ^{11}B and ^{23}Na were performed for the microcrystalline $NaBH_4$ sodalites with aluminosilicate-, gallosilicate- and aluminogermanate framework on a Bruker ASX 400 spectrometer. Further informations on orientational disorder, dynamics and host-guest interaction between framework atoms and BH_4-anions inside the sodalite cages can be derived from MAS NMR spectroscopy [17, 65]. The ^{11}B MAS NMR results of the $NaBH_4$ sodalites are shown in Fig. 15 and in Tab. 2. All spectra were recorded at 128.38 MHz (pulse duration: 0.6 μs; 100 ms pulse delay, 10000 scans were accumulated at a spinning rate of 12 kHz). Chemical shifts were determined using $NaBH_4$ ($\delta = -42.0$ ppm from $BF_3{\cdot}Et_2O$) as an external reference [65]. As a result of boron nitrite of the probehead a broad line is found in all spectra, beside a sharp narrow signal. Thus spectra after subtraction of the broad background are included in Fig. 15. The sharp signals with isotropic chemical shifts (δ_{iso} (^{11}B) around –49.08 ppm (see Tab. 3) are typical for boron tetrahedrally coordinated by four hydrogen atoms [66]. The full-width at half-maximum (FWHM) of the sharp lines are included in Tab. 3.

^{11}B MAS NMR				
Sodalite	δ_{iso} [ppm]	FWHM [ppm]	C_Q [kHz]	η_Q
AlSi	–49.08	1.70	–	–
GaSi	–49.91	1.28	–	–
AlGe	–49.01	1.11	–	–
^{23}Na MAS NMR				
AlSi	–6.61	4.43	8.82	0
GaSi	–1.31	2.32	6.41	0
AlGe	–1.60	1.88	6.75	0.22

Table 3. Isotropic chemical shift (δ_{iso}), quadrupolar coupling constant (C_Q), asymmetry parameter (η_Q), Lorentzian/Gaussian broadening (FWHM) of the ^{11}B MAS NMR spectra of $NaBH_4$ aluminosilicate (AlSi), gallosilicate (GaSi), aluminogermanate (AlGe) sodalite [65].

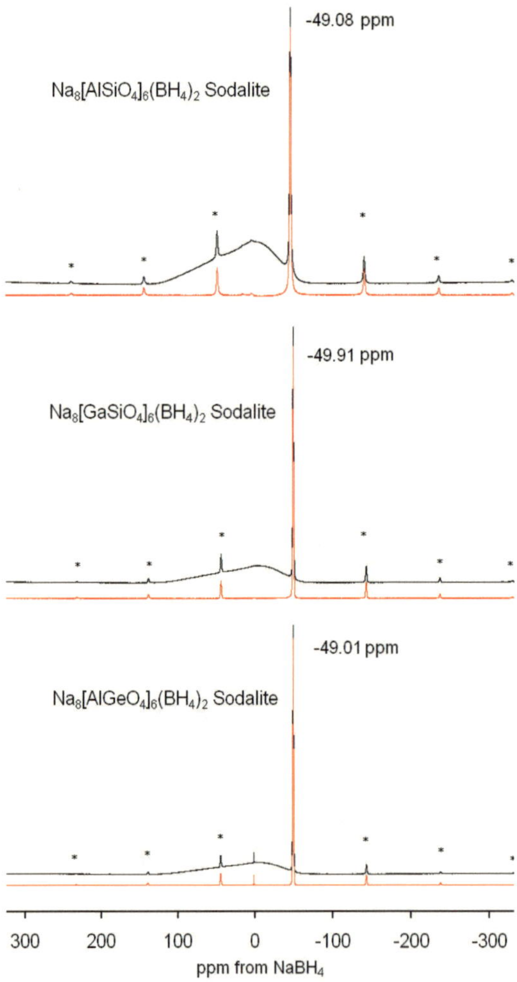

Figure 15. ^{11}B MAS NMR spectra of the microcrystalline $NaBH_4$ sodalites; asterisks mark the spinning side bands, lower spectra are given after the subtraction of the probe background (after [65]).

Compared with pure $NaBH_4$ ($\delta^{11}B$ = -42.0 ppm) a slight downfield shift of the signals can be found for all three sodalites as a result of the matrix effect of the surrounding framework atoms. The nearly equal values of the chemical shifts for the three sodalites indicate no significant influence of framework composition on the arrangement or the shape of the BH_4-anions inside the sodalite cages [65]. Threefold-coordinated non-framework boron or four-coordinated framework boron can also be excluded from the distinct chemical shift values of the sharp signal in all three cases, because ^{11}B of BO_3-units resonates in the range between 12 and 25 ppm, and BO_4-groups exhibit these shifts in the approximate range between –4 ppm and –6 ppm [66, 67].

Following [65] further informations can be derived from the signals. The sharp and narrow lineshape of the signals show almost no quadrupolar interactions due to the discreet BH_4-unit possessing a highly symmetrical environment for boron in the sodalite-cage as well as possibly fast dynamic site exchange of hydrogen. Because of a relatively low C_Q (~ 0.501 MHz) value expected for BO_4-groups, and a larger value (C_Q ~ 2.5 MHz) expected for BO_3-units [66-68] the present quadrupole coupling parameters rule out any contribution of BO_3- or BO_4-units. A small contribution at +2 ppm can be seen in the spectrum of aluminosilicate sodalite after subtracting of the probehead signal. This signal could be attributed to a small amount of $B(OH)_4$-anions in the sodalite cages. A similar signal at about 1.7 ppm is known in the ^{11}B MAS NMR spectrum of $NaB(OH)_4$-aluminosilicate sodalite [69].

The ^{23}Na MAS NMR signals were recorded at 105.84 MHz (pulse duration: 0.6 μs, pulse delay: 100 ms; accumulation of 5000 scans; spinning rate: 12 kHz; external reference: solid NaCl salt). The "dmfit2003" software was used for peak fitting [70]. The ^{23}Na MAS NMR spectra are given in Fig. 16 (left). Further informations on chemical shifts, FWHM and quadropolar parameters C_Q (quadrupole coupling constant) and η_Q (asymmetry parameter) are summarized in Tab. 3. The signals exhibit a narrow line shape typical for a single type of sodium coordination indicating nearly no quadrupole interactions according to the very small quadrupole parameters (C_Q and η_Q in Tab. 3). This implies well defined position probably caused by highly symmetric and cubic orientation of the BH_4-anions on the one side and the framework oxygen on the other side but also enables dynamic fast motion of the guest atoms within each sodalite cage leading to motional narrowing.

The ^1H MAS NMR spectra were obtained at 400.13 MHz with a single pulse sequence duration of 1.5 μs (90 degree pulse length of 6.5 μs) and a recycle delay of 5-30 s. 100 scans were accumulated with a spinning rate of 14 kHz, and tetramethylsilane (TMS) was used as an external reference. The high spinning speed helps to remove any residual dipolar broadenings. The ^1H MAS NMR spectra of the sodalites, given in Fig. 16 (right) exhibit a single sharp intense line at a chemical shift of about –0.6 ppm for all three sodalites assigned to the hydrogen atoms of the BH_4-groups. This indicates that proton chemical shift is not influenced by the chemical composition of the sodalite framework. A shift in high field direction, compared with $NaBH_4$ salt ($\delta(^1H) \approx 1.0$ ppm) results from the enclathration of single BH_4-groups in contrast to their incorporation in the NaCl-type structure of the salt and its strong heterovalent bonds [65]. A further but weak signal at ~ 5.0 ppm in each spectrum could be related to water molecules enclathrated in distorted sodalite cages [71] or following the interpretation due to water incorporation in short range ordered sialate bonds. The nature of the shoulder of the main signal at about 0.6 ppm could be related to enclathrated water molecules or to $B(OH)_4$-impurities as observed in ^{11}B NMR signals, too. A third less intense signal at 1.2 ppm appears only in the spectrum of the aluminogermanate sodalite. It could indicate small amounts of ($H_3O_2^-$)-impurities in a few sodalite cages [71, 65].

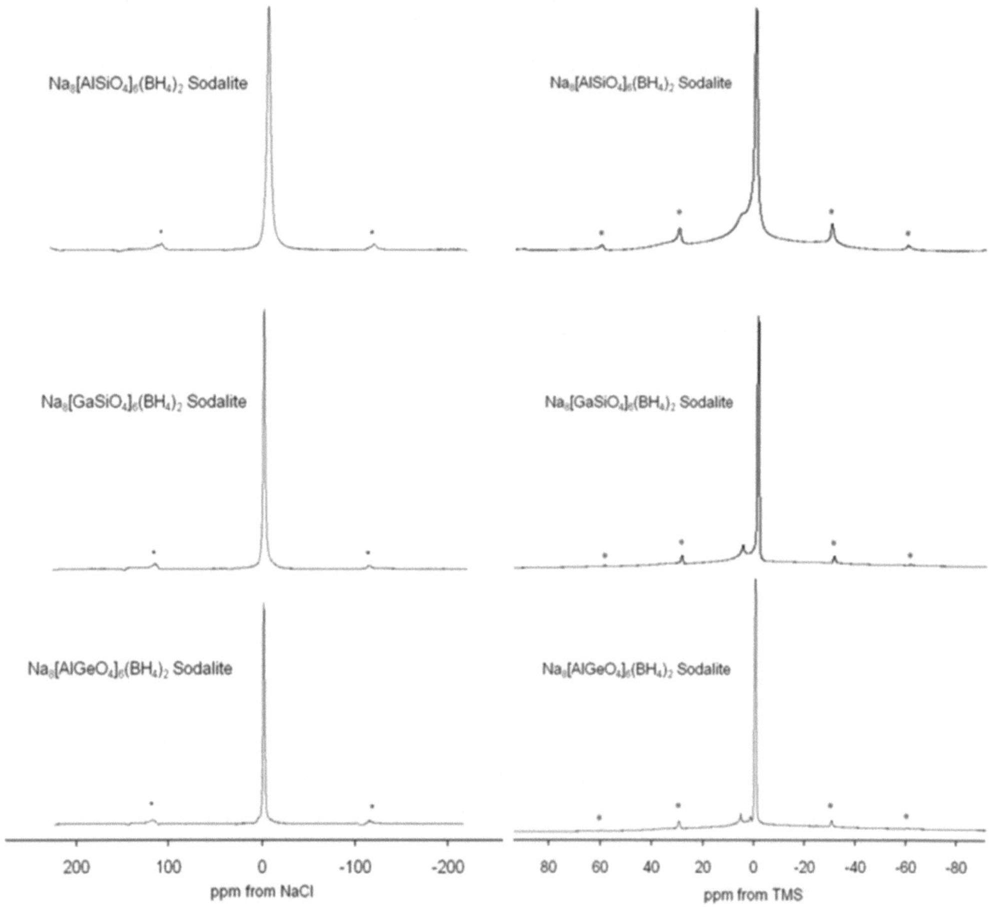

Figure 16. (left) ^{23}Na MAS NMR spectra of the microcrystalline $NaBH_4$ sodalites; asterisks mark the spinning side bands (after [65]). **Fig. 16 (right)** ^{1}H MAS NMR spectra of the microcrystalline $NaBH_4$ sodalites; asterisks mark the spinning side bands (after [65]).

4. Hydrogen release and reinsertion experiments with the BH_4-sodalite samples

This section shows that the BH_4-anion enclosed in the sodalite cage enables a controlled release of hydrogen with water in consecutive steps when heated (4.1). Experiments of regeneration of the pre-reacted BH_4-anion in the sodalite cage are presented (4.2). A basic understanding could be achieved using temperature dependent infrared (TIR) absorption experiments [38]. Since this method is used intensively in this chapter it may be described here briefly. In TIR experiments typically around 1 mg of sample are diluted in 200 mg NaCl or KBr powders and pressed into pellets. The pellets are fixed in an Ag tube (Fig. 17) which can be heated in a furnace to the desired temperature. In this arrangement transmitted IR light (I) can be monitored in situ as absorbance = -lg(I/I_0), where I_0 is the transmitted intensity through a reference pellet. The measurements were conducted under vacuum in an appropriate IR sample chamber (FTIR Bruker IFS 66v) to avoid disturbances by variations in the outer atmosphere during the measurements. With increasing temperature dehydration effects of the matrix (KBr, NaCl) could occur if not dried sufficiently before its use. As shown below, additives to the matrix as for example the addition of KNO_3, could be used in order to tracer the effect of hydrogen release by reduction. Thermogravimetric (TG) and differential thermal analytical (DTA) measurements were carried out using a commercial instrument (Setaram Setsys evolution 1650). Flowing atmospheric conditions of various gases were used with the sample carried in corundum crucibles. Further hydrogen reloading experiments were performed with controlled hydrogen pressures up to 200 bars and temperatures up to 400°C using in house built autoclaves with equipment as shown in Fig. 18. The pressure was checked permanently during heating by the manometer. For the experiments the samples were filled into Au capsules which were only closed by slightly pressing the ends together for realizing a throughput for gas. Temperature calibrations of the autoclave were carried out between 100 and 400°C obtaining a precision within 5°C for the absolute value. The heating rate was 4°C/min. After the chosen holding time at a fixed temperature, the autoclave was cooled down rapidly, e.g in 10 minutes from 300°C.

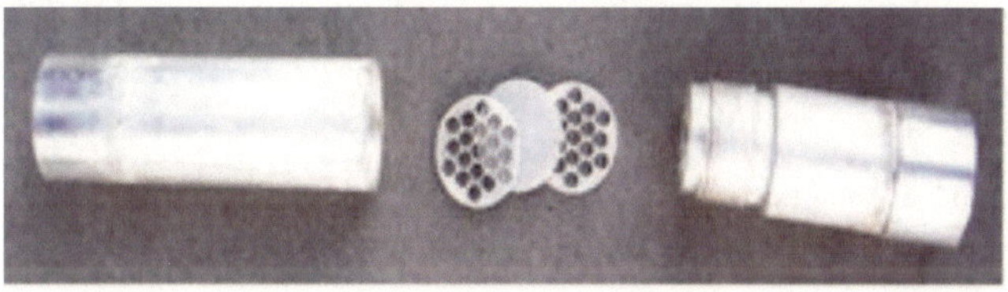

Figure 17. Ag tube, a pressed pellet (diameter 13 mm), and two silver nets for supporting the pellet in the tube and improve the thermal contact during the temperature dependent infrared (TIR) measurements.

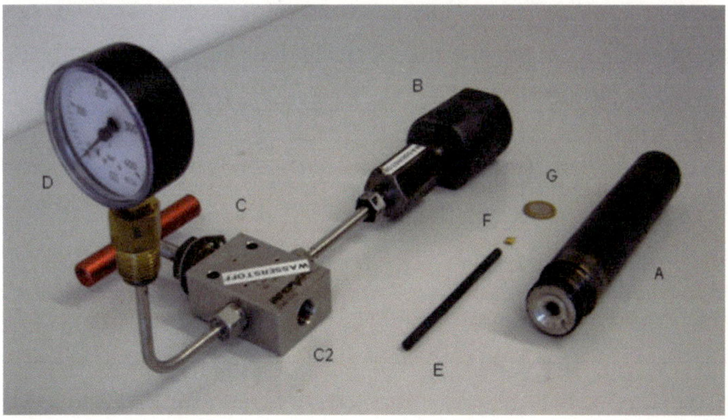

Figure 18. Equipment for hydrogen pressure experiments: Autoclave (A), connection piece (B) between autoclave and closing device with three-way switch (C) with opening to the outer hydrogen bottle (C2) and to the manometer (D). Filling piece (E) and Au capsule (F).

4.1. Variation of cage anions of $Na_8(AlSiO_4)_6(BH_4)_2$ when heated

The effect of various heating rates and heating temperatures in ex situ TG experiments as well as in situ TIR measurements in dry and wet NaCl environments has been investigated for the microcrystalline $NaBH_4$-sodalite. Heating with a constant rate of 2°C/min to 300, 400 and 500°C (He-flowing conditions, 20 ml/min) reveals weight losses of about 0.2 %, 0.6 % and 0.9 %. This weight loss is related to dehydration and to hydrogen release. The dehydration corresponds obviously to water contained in some short range ordered sialate bonds as explained in section 3.2 (compare also below peaks denote HOH and D in Fig. 19). In experiments with heating rates of 2°C/min, 4°C/min and 6°C/min to 300°C the weight loss reads 0.2 %, 0.4 % and 0.7 %, respectively. Thereby the IR absorption intensity of BH_4 related peaks decreases about 11 % at heating rate 2°C and 7 % at heating rate 6°C/min. This shows that the faster the heating rate the lesser the loss of BH_4-anion in the sodalite cage. Therefore, the faster the heating rate the more water leaves the sample without use for hydrogen release and, accordingly, the higher the weight loss. The spectra of the samples taken after cooled down to room temperature (rate 2°C/min) from 300, 400 and 500°C, exposed to atmospheric conditions and pressed into KBr pellets are shown in Fig. 19. Compared to the unheated sample the BH_4 related intensity has decreased by about 15.5 %, 13 % and 11 % when heated to 500, 400 and 300°C, respectively. Related with the decrease in BH_4-anion concentration new peaks appear denoted in the spectra by A, B, B′ and C. This notation follows that given earlier [17, 22] relating peak A, B, B′ to anions H_3BOH^-, $H_2B(OH)_2^-$ and $HB(OH)_3^-$, respectively. According to this the following reactions could be seen:

$$H_4B^- + H_2O = H_3B(OH)^- + H_2;\ \text{peaks A (A1, A2)} \tag{1}$$

$$H_3B(OH)^- + H_2O = H_2B(OH)_2^- + H_2\ ;\ \text{peak B} \tag{2}$$

$$H_2B(OH)_2^- + H_2O = HB(OH)_3^- + H_2\ ;\ \text{peak B}' \tag{3}$$

A further reaction of hydrogen release followed by two steps of dehydration which finally leads to peaks C could be suggested as:

$$HB(OH)_3^- + H_2O = B(OH)_4^- + H_2 \tag{4}$$

$$B(OH)_4^- = BO(OH)_2^- + H_2O \tag{5}$$

$$BO(OH)_2^- = BO_2^- + H_2O\ ;\text{peaks C} \tag{6}$$

The step eq. 4 which reveals the anion species $B(OH)_4^-$ is not seen directly due to fast dehydration above about 400°C as known for the $B(OH)_4$-sodalite [72, 38]. For better comparison appropriate spectra of microcrystalline $Na_8(AlSiO_4)_6(B(OH)_4)_2$ and those obtained by dehydration at 200°C for $Na_8(AlSiO_4)_6(BO(OH)_2)_2$ and at 400°C for $Na_8(AlSiO_4)_6(BO_2)_2$ are shown in Fig. 20. The BO_2-anion in the sodalite cages can markedly be observed by means of the peaks at about 1958 and 2029 cm^{-1} denoted as peaks C here. The anion species $BO(OH)_2^-$ shows characteristic peaks in the range 1490-1520 cm^{-1} (denoted C′ further below) and around 1150 cm^{-1}, beside the OH-stretching at about 3610-3620 cm^{-1} compared to the OH-stretching of the $B(OH)_4$-anion at 3620-3640 cm^{-1} [38]. According to this BO_2-anion species in Fig. 19 can readily be identified indicating that reactions eq. 4, 5 and 6 occurred.

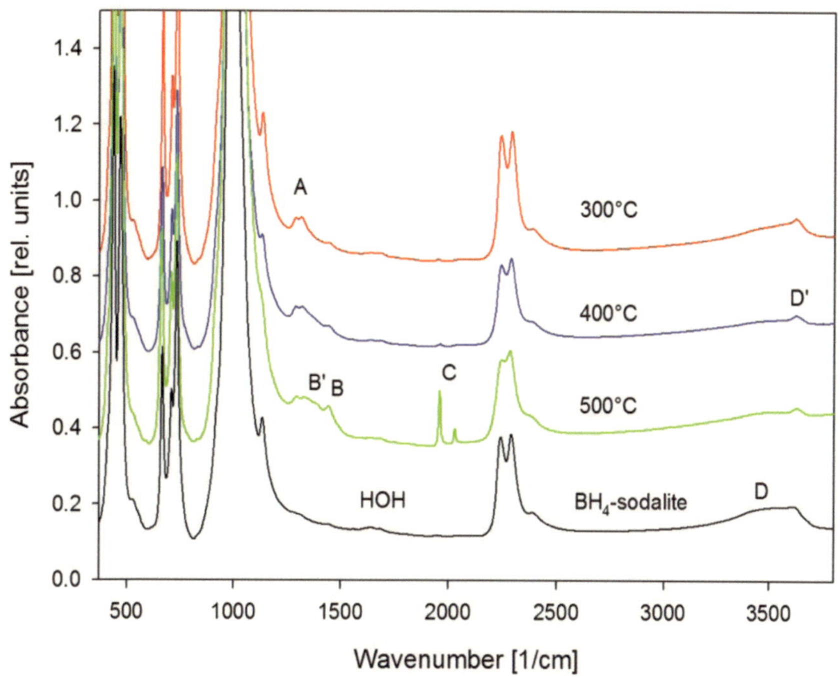

Figure 19. IR-absorption of BH_4-sodalite before and after heated to 300, 400 and 500°C. For peaks denoted A, B, B′, C, D, D′ and HOH see text.

The microcrystalline sodalite shows a sharper peak which remains present during heating (see below) with maximum at about 3620 cm^{-1} denoted as D′. As discussed by [17, 22] D′ could be related to an OH-anion in the sodalite, indicating basic or hydro-hydroxo sodalite. D′ could also be related to the formation of B-OH forms or to superimposition effects of various species. A clear distinction is hard to obtain. Another important contribution is the presence of water molecules indicated by H-O-H bending vibrations at about 1640 cm^{-1} and OH stretching at 3000-3600 cm^{-1} called D in Fig. 19. It could be observed that these contribution becomes partly dehydrated during heating and are in some extend related to the hydrogen release reactions. If the sample is dehydrated the hydrogen release reaction invariably stops. It could be shown that all the sodalite with all formed species remains stable below 600°C also at invariably long time. Heating above 630°C immediately leads to framework destruction. These thermal instabilities were investigated in detail by [19, 20].

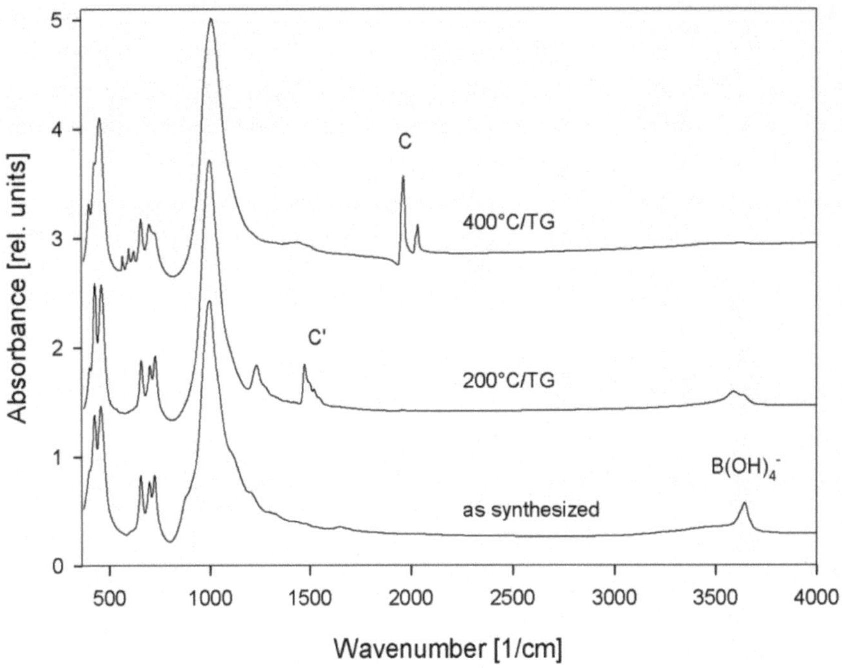

Figure 20. IR-absorption of $B(OH)_4$- sodalite (as synthesized), after dehydration at 200°C to $BO(OH)_2$-sodalite and at 400°C to BO_2- sodalite.

The effect of rehydration could be used to further proceed the hydrogen release reactions as shown in Fig. 21. The sample preheated at 500°C was taken out of the thermobalance, exposed to atmospheric conditions and run again up to 500°C under He conditions in 7 further cycles. Spectra obtained after each run show a gradual decrease of BH_4-absorption intensity and increase in BO_2^- content (Fig. 21). Similar to the C peaks, the B peak also becomes a bit more pronounced, whereas the peaks B′ and A become broader and more unspecific. The framework vibrations remain largely unchanged and may depict only the changes due to the changed borate species, i.e no significant destruction of the framework is observed. Thus these experiments show the high stability of the sodalite during several heating cooling cycles. A certain amount of water content related to the matrix consisting of short range ordered sialate bonds of the sample is reloadable and can be used to continue the hydrogen release reactions. An exponential decay of the ratio BO_2^-/BH_4^- could be observed (inset in Fig. 21). There is a significant increase in intensity in the range of peak B above about 400°C, too. There is further work to do to find out whether here intracage reactions occur or if borate species outside the sodalite cages are formed which could be related with a destruction of the framework. Explanation could, however, also be given with an other anion in the cage, e.g. H_2BO^- (compare below, eq. 7, 8).

The reaction sequence of hydrogen release may be followed in more detail for the microcrystalline $NaBH_4$-sodalite in TIR experiments as shown in Fig. 22. In the lower part spectra are shown when cooled down to room temperature after heating to 200, 300, 400 and 500°C in NaCl pressed pellet. In the upper part the spectra are taken at denoted temperatures. It can be seen that peaks A grow in intensity followed by B and B′ with increasing temperature. The formation of $B(OH)_4$-anions may not be detected as they become dehydrated to $BO(OH)_2$-anions as indicated by peak C′ at temperatures between 250 and 450°C. Above this temperature the formation of BO_2-anions is indicated by the peaks C (C1, C2). In TIR experiments of the 500°C pre-reacted sample it was observed that the C peak intensitity (BO_2^-) decreases from about 200°C and increases again in intensity above about 370°C. Complementary to this the intensity related to ($BO(OH)_2^-$) (called C′, Fig. 22, right) increases and decreases. Additionally the BH_4^- intensity decreases above about 370°C. This indicates that in a first step the reacting water must rehydrate the BO_2^- species and finally effectively reaches the BH_4-anions in the centre of the sodalite crystals. This implies that the reactions are controlled by the diffusion through the sodalite cages. The diffusing species might mainly be OH^- and H^- via defect formation in the cages. Thus water disproportion and H_2 formation may occur at the crystal surface.

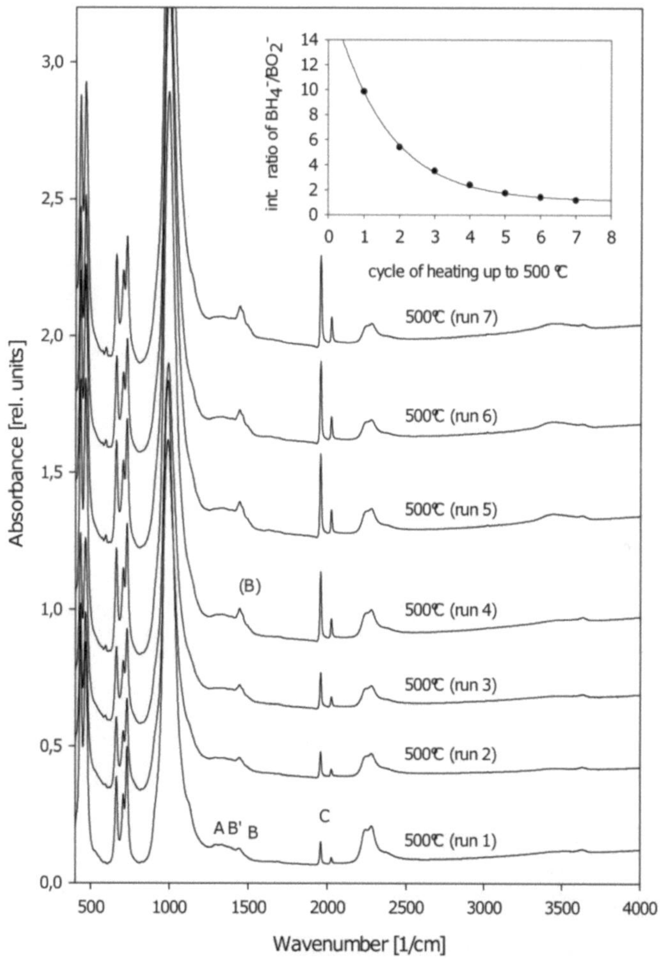

Figure 21. IR-absorption spectra of BH_4-sodalite sample of nominal composition $Na_8(AlSiO_4)_6(BH_4)_2$ (as synthesized) after thermal treatment at 500°C (heating rate 2°C/min, 0.5 h holding time, flowing He 20 ml/min). Spectra were taken after each cycle and exposure the sample to atmospheric conditions for 24 h by KBr method for a part of the sample. The remaining sample was given to the next thermal treatment.

A closer view of the characteristic sodalite fingerprint peaks ν_1, ν_2, ν_3 is shown in Fig. 22 (left). A significant shift of ν_1, ν_2, ν_3 towards lower wavenumbers occurs with increasing temperature which is almost completely reversible with decreasing temperature (Fig. 23). The small deviations can be related to the changes in the cage filling species. TIR experiments were also carried out using KBr pressed pellets to demonstrate the effect of an exchange of Na-cations from the sodalite with K-cations from the matrix as indicated by the strong deviation in temperature dependance of ν_1, ν_2, ν_3 from the effect observed using NaCl pellets (Fig. 23). A significant exchange occurs above about 250°C which could be similarly demonstrated for other sodalites, too [38]. This indicates that the Na-cations become highly mobile as could also be observed in temperature dependent investigations using MAS NMR [73] and XRD structure refinements [50] on related sodalite compositions. It can be concluded that above 250°C the jump rate of Na^+ in the cage and through the sixring windows becomes very significant which implies that the out-in jump rates of other ions could also increase.

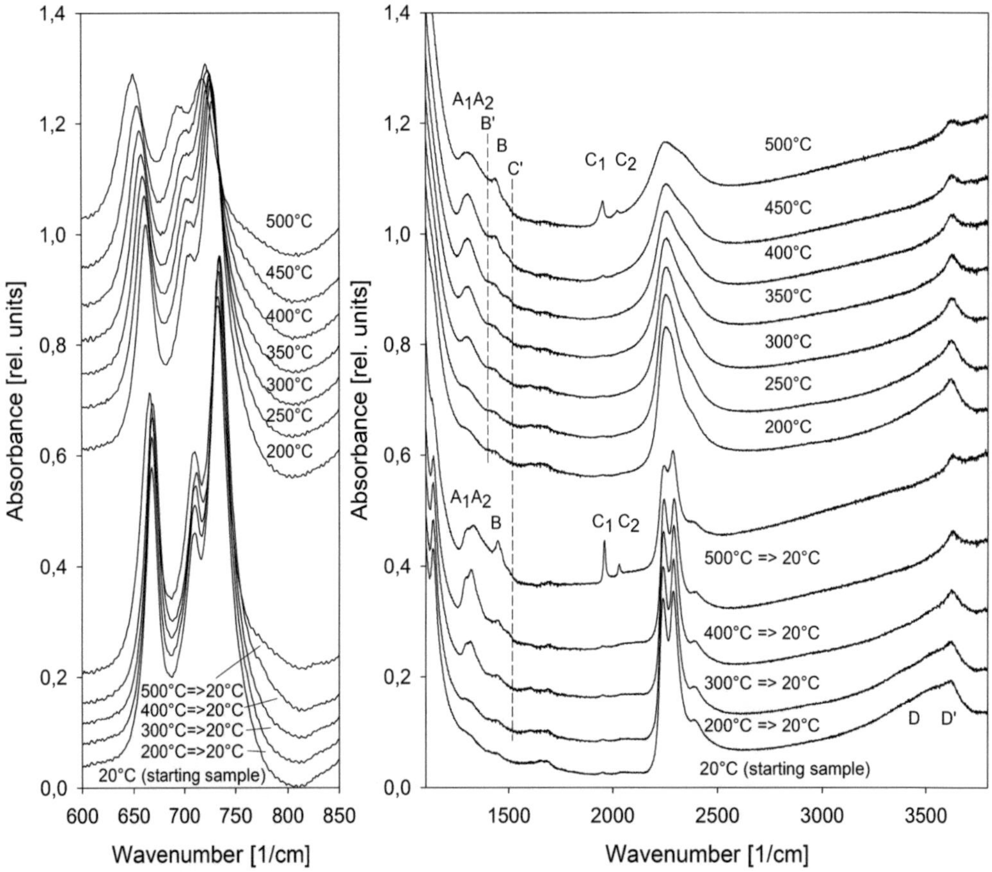

Figure 22. TIR absorption spectra of BH_4-sodalite sample of nominal composition $Na_8(AlSiO_4)_6(BH_4)_2$ (as synthesized) of the starting sample at 20°C and taken in the heating up run at temperatures as denoted (upper part) and at 20°C when cooled down from temperatures as denoted (lower part) for the range of the characteristic sodalite fingerprint of framework vibrations ν_1, ν_2, ν_3. (left) and the cage filling species (right).

The change in peak intensities obtained in the NaCl related TIR experiment (Fig. 22) for peaks A, B, C′ and C could be evaluated in more detail as shown in Fig. 24. The intensity of A peaks increases significantly above about 220°C, crossing over in an effective temperature independent behaviour above about 320°C. About at that temperature peak B and also C′ start to increase in intensity. The later effect shows that dehydration occurs as described by eq. 5. This implies that reaction eq. 4 occurred rather fast which could not be resolved. Species B′, i.e. eq. 3, could not be considered separately because of superposition in the spectra with peaks A and peak B. A decrease in intensity of peaks A, C′ and also B′ is observed above about 450°C where peaks C start to increase strongly in intensity. This shows that above 450°C strong dehydration occurs leading finally to the BO_2-anion in the sodalite cage. It can be suggested that part of this water is used for a further effective hydrogen release with reactions as given by eqs. 1-4.

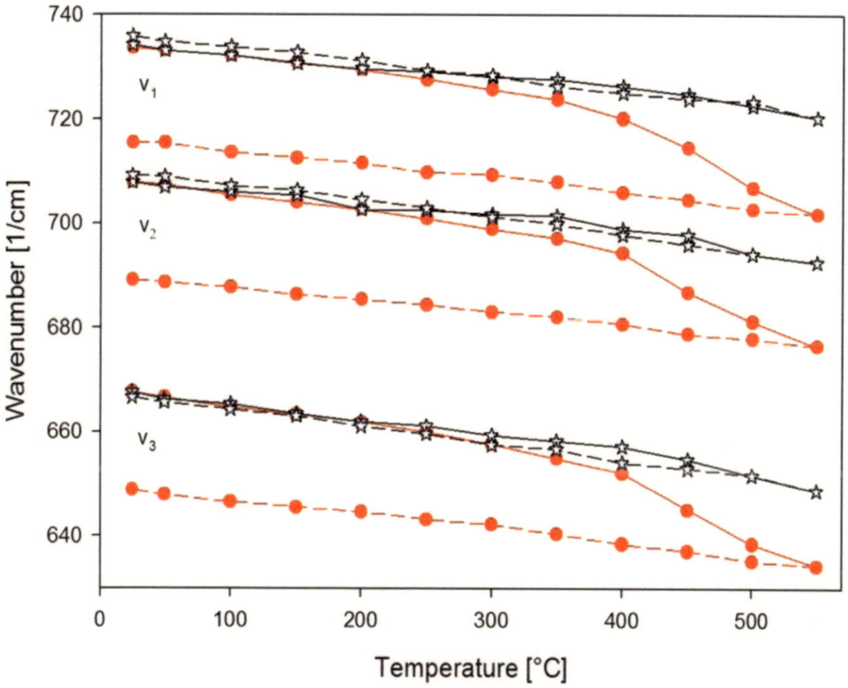

Figure 23. Temperature dependence of framework peaks ν_1, ν_2, ν_3 when heated in NaCl pellet (open stars) and in KBr pellet (closed circles) taken in the heating up run (connected by thin solid line) and when cooled down (connected by dashed line).

One interesting point is that the peak intensity of peak B also re-increases above about 450°C. If this effect is related to further hydrogen release reaction or to a dehydration reaction or a dry hydrogen release of the type eq. 7 or 8 [17, 22] requires further investigations.

$$HB(OH)_3^- = H_2BO^- + H_2 \tag{7}$$

$$H_2B(OH)_2^- = H_2BO^- + H_2O \tag{8}$$

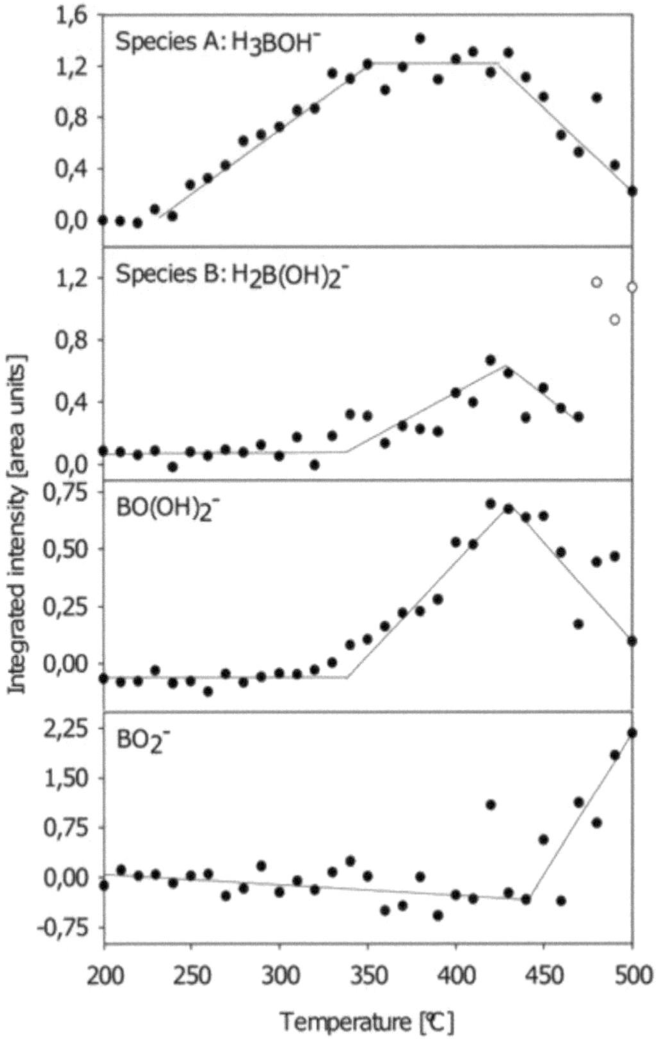

Figure 24. Temperature dependence of intensities of characteristic peaks of intra cage anion species as denoted. Thin solid lines are guide to the eyes. Open circles in the species B panel mark the re-increase of intensity when heated above 475 °C (compare text).

A proof that hydrogen is released is given using a nitrate tracer reaction in TIR experiments [22, 24, 74] as shown in Fig. 25. Here some amount of KNO_3 has been added to the sample pellet which could be identified as an additional peak in the spectra (see inset in Fig. 25). With increasing temperature this peak strongly decreases in intensity above about 250°C. Repeating the TIR experiment with only KNO_3, it shows a slight decrease in intensity of the KNO_3 related peak above about 400°C due to a gradual thermal decomposition. Therefore, it can be concluded that hydrogen is released from the BH_4-sodalite sample leading to a redox reaction with KNO_3. Since this reaction cannot be achieved by hydrogen molecules (compare below, Fig. 27) it can also be concluded that hydrogen is set free in an activated form.

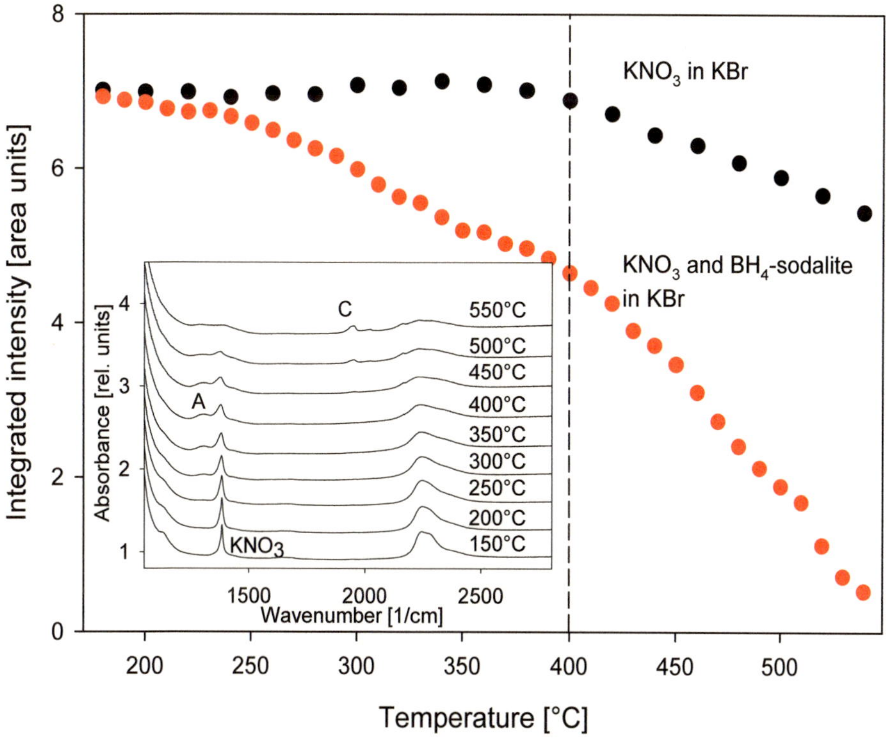

Figure 25. TIR experiment with the addition of KNO_3. Main figure shows the temperature dependence of the integrated intensity of the KNO_3 related peak in the pellet with the BH_4-sodalite (compare spectra given in the inset) compared to the behavior observed without. Dashed vertical line marks temperature of decomposition known for KNO_3.

Another interesting point is the effect of using "wet NaCl" for the pressed pellet in the TIR experiment as this increases the amount of H_2O available for the hydrogen release reaction from the embedded BH_4-sodalite. This even leads to a complete decrease of the BH_4-content with increasing temperature. The loss in BH_4-intensity in "dry NaCl" and "wet NaCl" are compared in Fig. 26 demonstrating again the effect of the different amounts of available H_2O content for the hydrogen release reaction. Included is the observed effect for nanocrystalline BH_4-sodalite as well, also using dry NaCl and wet NaCl. Here the difference is marginal distinct since the nanocrystalline sample contains a much higher amount of water related to the higher contribution of short range ordered sialate bonds. Moreover the reaction path for hydrogen release is found at significantly reduced temperature due to the much smaller crystal size.

4.2. Experiments on regeneration of pre-reacted BH_4-sodalite

Based on the observation that the hydrogen release reactions of the BH_4-anion in the sodalite cage with H_2O could be carried out in consecutive steps it seems likely that all reactions according to eqs. 1-4 or at least some are reversible. It can be concluded that hydrogen release from BH_4-sodalite crystals is a diffusion controlled process were OH-anions

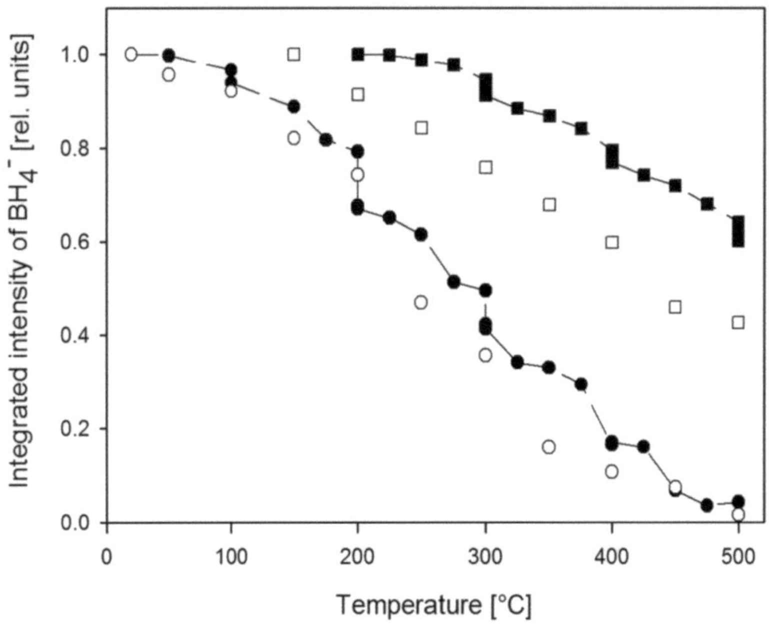

Figure 26. Integrated intensity of the BH_4-anion absorption. Closed and open squares for micro BH_4-sodalite in dry and wet NaCl, respectively. Closed and open diamons for nano BH_4-sodalite in dry and wet NaCl.

diffuse inward and H-anions outward. Following this, the reinsertion reaction seems only be possible if appropriate H^-/OH^- gradients could be realized which should govern the efficiency of "reactor regeneration". This assumption could be supported by the "nitrate reduction reaction" observed in Fig. 25. Following reference [75] such type of low temperature nitrate reduction can not be achieved by hydrogen molecules. It has been shown that hydrogen molecules need to be activated by the presence of special catalysts, e.g. Pt or Pd. This requirement could be demonstrated in TG/DTA experiments with heating/cooling runs with as received commercial KNO_3 (Merck) with and without the addition of Pt powder (Fig. 27). Using pure KNO_3 the effect of the structural phase transition can be seen at 140°C and 130°C in the heating up (20°C/min) and cooling down (20°C/min) run as endothermic and exothermic peak, respectively. There is a negligible weight loss of less than 0.5 %, which could be related to dehydration effect. The experiments were carried out using forming gas (10%H_2/90%N_2). In a second run using the same conditions but about 10 wt% of Pt powder was added to the KNO_3 sample. A weight loss of about 13 % for the KNO_3 content is observed. There is a strong exothermic peak centered around 150°C due to the reaction related to the weight loss. The endothermic peak due to the structural phase transition appears only as a very small minima in the heating up run and the reversal to the low temperature phase is absence indicating the complete chemical reaction of KNO_3. The exothermic peaks at 220°C and 180°C in the cooling run can be related to the chemical products. The results are in agreement with observations reported by [75]. Two important conclusions can be drawn: 1. The hydrogen leaving the sodalite crystals are in an activated form which is able to reduce the nitrate even in some distance to the crystals in the pressed pellet (KBr/NaCl) as used in the TIR experiments. 2.

Hydrogen can be brought into an activated form at temperature already as low as 150°C in the presence of Pt. Therefore this effect could be used to increase the appropriate partial pressure of activated hydrogen to stop or even reverse the reactions eq. 1-4. Conclusively series of experiments were carried out using Pt powder in addition to the sample either brought in simply as additive to the powdered sample or even as additive in the synthesis route. Some main results may be outlined in the following.

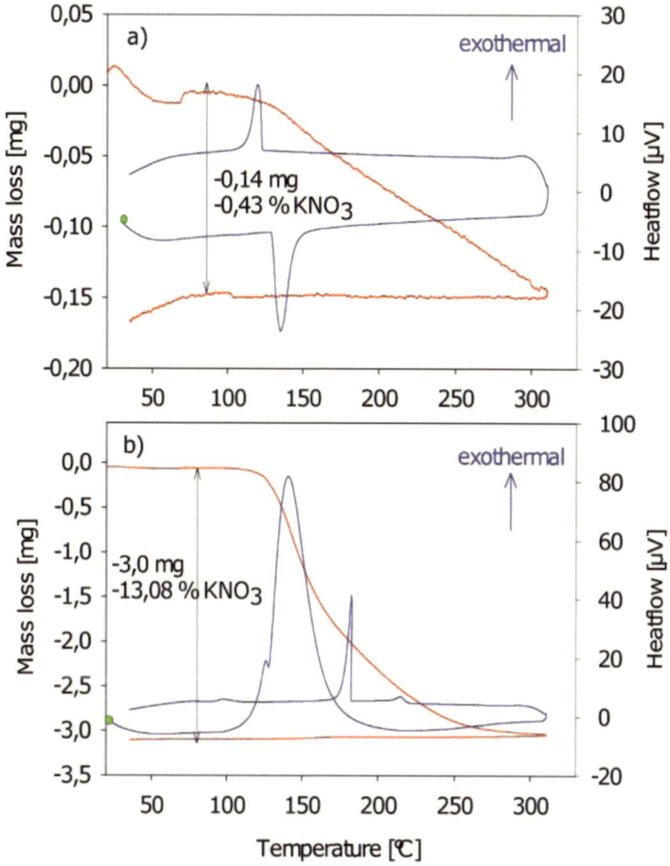

Figure 27. TG/DTA results of heating cooling cycles of KNO_3 powder without (a) and with the addition of Pt powder in flowing forming gas (10/90 H_2/N_2, 50 ml/min, 5°C/min).

First steps of experiments were carried out with the as synthesized BH_4-sodalite in comparison to samples with the addition of about 20 wt% Pt powder using various gases: He (99.9999), synthetic air (80%N_2/20%O_2) and forming gas (10%H_2/90%N_2). The experiments were carried out using always the same conditions with a continuous gas flow of 20 ml/min with heating rate of 4°C/min to 500°C, holding time of 10 min at 500°C, followed by cooling with 4°C/min to room temperature. Spectra of the starting sample and taken after the experiment are compared in Fig. 28. As before the heat treated samples reveal in all cases a reduction in BH_4-absorption intensity and peaks A, B, B′ and C as related to $H_3B(OH)$-, $H_2B(OH)_2$-, $HB(OH)_3$- and BO_2-anion species, respectively. A quantification of the intensity ratio of BH_4^-/BO_2^- and the obtained mass losses with respect to the amount of BH_4-

sodalite is given in Tab. 4. These data sensitively show that the highest reaction rate is achieved under synthetic air conditions – and here to a higher extend in the presence of Pt compared to those in its absence. There is also a slightly lower weight loss in the presence of Pt indicating that here more hydrogen could have been released and less much loss to dehydration occurred. It may, however, also not be ruled out that any uptake of oxygen from the air occurred which could also be indicated by the increased intensity at the B peak position. Moreover the experiments show that the same weight losses and reaction rates BH_4^- to BO_2^- could be seen when He is used with and without Pt addition as well as forming gas only without Pt. If Pt is added the reaction rate becomes smaller and the weight loss increases. This shows that here dehydration takes place with a significantly reduced release of hydrogen. These preliminary experiments show the influence of hydrogen only in the presence of an effective activation of hydrogen which may produce a significant reduction in the concentration gradient for outward diffusion of H-anions.

Sample name	BH_4^-/BO_2^- ratio	Mass loss, Pt-influence subtracted [%]
500°C air, Pt	6,24	1,52
500°C air, no Pt	9,77	1,58
500°C He, Pt	18,35	1,79
500°C He, no Pt	19,81	1,79
500°C H_2/N_2, Pt	18,24	1,76
500°C H_2/N_2, no Pt	10,77	1,99

Table 4. Results of BH_4^-/BO_2^- ratio of integrated intensities from spectra shown in Fig. 28 and mass loss from TG experiments (heating up to 500°C) of a micro-crystalline BH_4-aluminosilicate sodalite sample with and without the addition of Pt powder

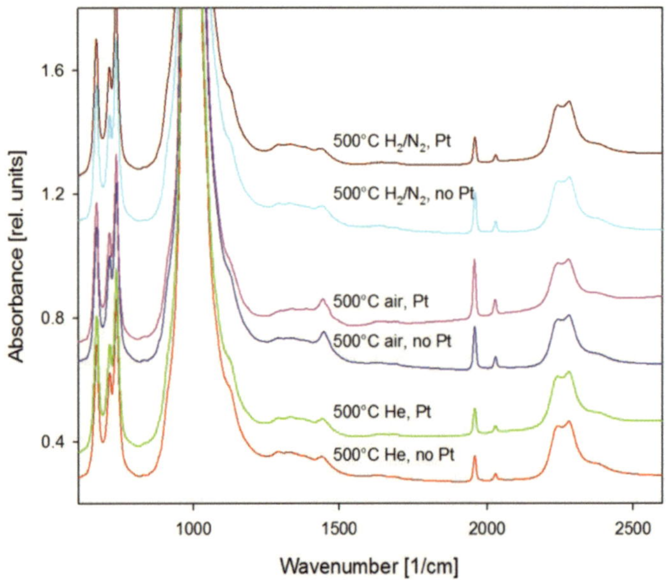

Figure 28. IR absorption spectra of thermally treated micro crystalline BH_4-aluminosilicate sodalite sample with and without the addition of Pt powder under various flowing gas compositions (for details see text).

Further experiments were carried out with BH_4-sodalites with Pt added into the synthesis route. X-ray diffraction pattern could prove the growth of microcrystalline BH_4-sodalite in the presence of nanocrystalline Pt. REM/EDX investigations show a rather homogeneous distribution of Pt. The IR spectra (see below) showed some weak formation of A peaks ($H_3B(OH)$-anion) due to the presence of Pt during synthesis. Typical results of the Pt free and Pt containing as synthesized samples treated with a hydrogen pressure of 162 bar and 3 h holding time at 250°C are shown in Fig. 29. The Pt free sample shows pronounced formation of peaks A and B related to $H_3B(OH)$- and $H_2B(OH)_2$-anions in the sodalite cages, respectively. Contrary to this the sample containing Pt shows almost no change in the infrared absorption beside only a slight increase in A peak intensity. It is interesting to note that there is no peak B′ observed. This can be explained by the too low temperature, but could also be related to the special high pressure conditions. It was observed using closed Au capsules and at 450°C for half an hour pre-reacted samples that the B′ peak completely disappeared at 300°C at autogeneous pressure. This implies a high stability for $H_3B(OH)$- and $H_2B(OH)_2$-anions but less much for $H_3B(OH)$-anions and also for dehydrated BO_2-anions under such conditions. As there is also no indication of $BO(OH)_2$-anions it may be concluded here that the peak at 3620-3640 cm^{-1} could be related to $B(OH)_4$-type species present already in the as synthesized sample.

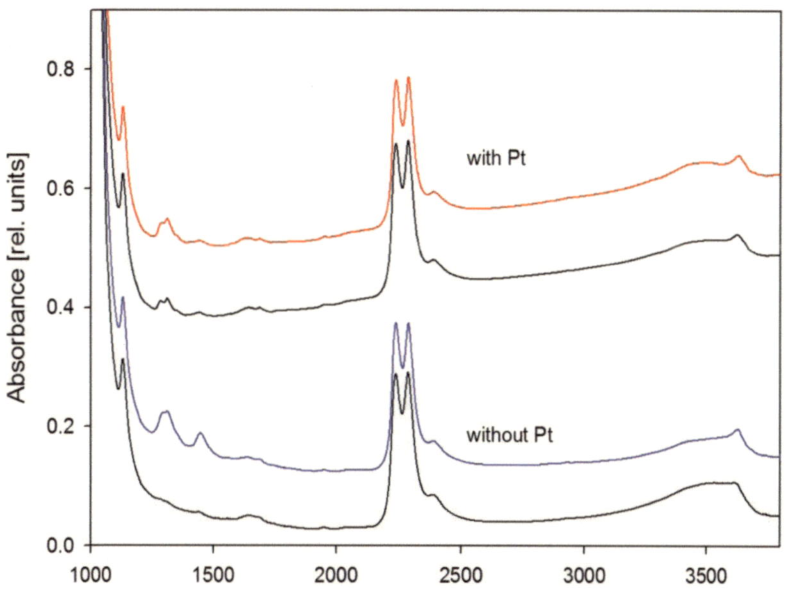

Figure 29. IR absorption spectra of microcrystalline BH_4-aluminosilicate sodalite sample with and without the addition of Pt during synthesis before (lower curve) and after (upper curve) treated for 3 h in 160 bar H_2 at 250°C (compare text).

In Fig. 30 the result of a reinsertion experiment carried out using 160 bar H_2 at 200°C of the Pt containing sample which was pre-reacted at 400°C in air is shown [25]. It can be seen that peak A increases in intensity whereas peak B decreases.

This first step of hydrogen reinsertion concerning the reaction eq. 2 could be supported in further systematic experiments where the influence of reaction temperature and time has been investigated at 55 bar hydrogen pressure on the Pt-containing sample pre-reacted at 400°C for 1 h. The results showed that with increasing temperature at 200, 250 and 300°C the BH_4-anion concentration remains constant and the same as in the pre-reacted sample. The A peak increases in intensity by about 10%, whereas the B peak decreases by about 5%. Moreover the $B(OH)_4^-$ related intensity did not vary within the error of estimation. Thus it is likely that under such conditions reaction eq. 2 could effectively be reversed. The evaluation of peak intensities with increasing reaction time at 250°C and 55 bar H_2 between minutes up to 6 h also showed that the BH_4-anion concentration remains rather constant also reproducing the finding of the former series. The content of $H_3B(OH)$-anions is higher compared to the pre-reacted material in minutes and slightly increases with increasing time. This shows that a fast reversion of reaction eq. 2 can be achieved.

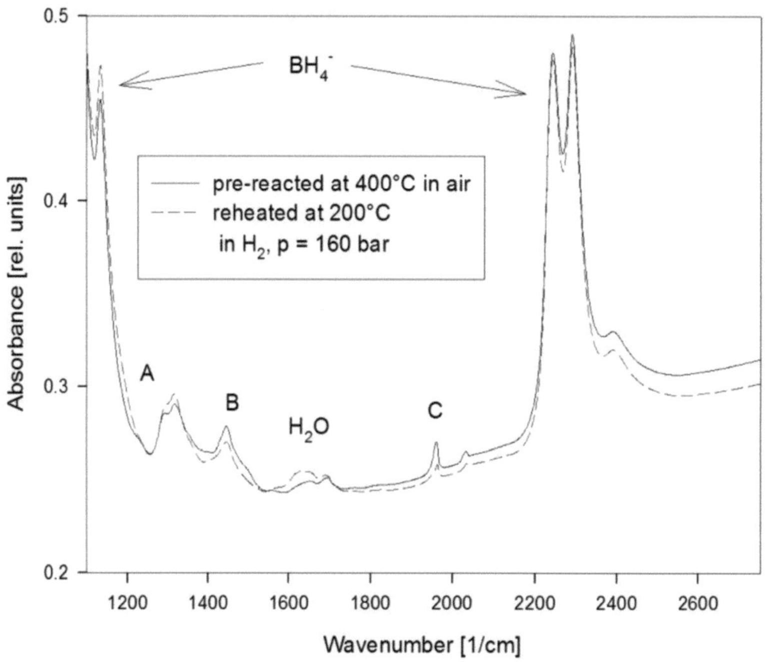

Figure 30. IR absorption spectra of microcrystalline BH_4-aluminosilicate sodalite sample with addition of Pt pre-reacted for 1 h at 400°C in air and reheated for 3 h in 160 bar H2 at 200°C, after [25] (compare text).

The reinsertion experiments have shown that a slight but significant back reaction in the content of $H_2B(OH)_2$-anions to $H_3B(OH)$-anions could be achieved. A higher efficiency might be reached with a further optimisation of the catalyst function. An improved back reaction might also be expected with the help of MgH_2 addition, which could increase the concentration gradient of H^- significantly. In some preliminary experiments commercially available 30 wt% MgH_2 powder (Merck) was simply mixed with two different pre-reacted samples and the spectra were taken of these samples and those after 3 h exposure to 150 bar hydrogen pressure at 200°C. The spectra shown in Fig. 31 reveal first of all the strong contribution of MgH_2 which can be seen by direct comparison with the spectrum of MgH_2 also shown. The special pre-reaction conditions of the Pt-containing BH_4-sodalite are 4 h at 360°C (sample I) and 1 h at 450°C (sample II). The most obvious and important observations here are that in both cases some part of MgH_2 forms $Mg(OH)_2$ as is identified by the new sharp peak at 3699 cm^{-1} for the samples treated in 150 bar hydrogen. The evaluation of the intensity of the triplicate BH_4-peaks reveal an increased intensity by about 4 % and 2 % for sample I and II, respectively. By the same time the A-peak intensity decreased by about 6.2 % in both cases. These results indicate the formation of BH_4-anions in the sodalite cages according to eq. 1.

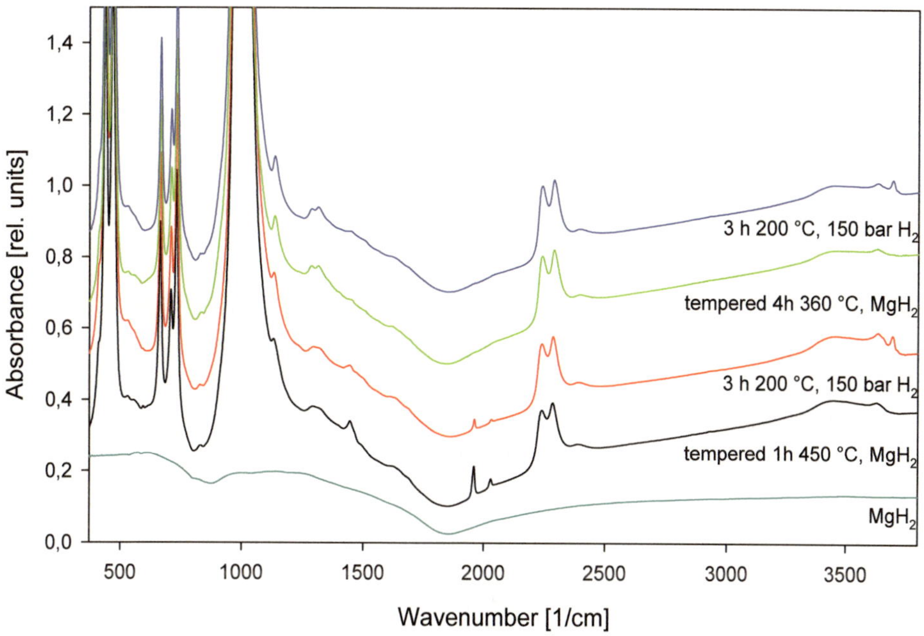

Figure 31. IR absorption spectra of microcrystalline BH_4-aluminosilicate sodalite sample pre-reacted, then mixed with 30 wt% MgH_2 and reheated (conditions as denoted). For comparison the spectrum for MgH_2 is also shown.

5. Summary and conclusion

Mixtures of $NaBH_4$ containing alkaline aluminate and silicate solutions form gels. During hardening at 110°C re-crystallization of $NaBH_4$ occurs together with the appearance of some sodalite type aluminosilicate within a remaining matrix of short range ordered sialate (Si-O-Al) bonds. The new material could easily be handled in water and at elevated temperatures up to about 300°C without significant destruction or release of hydrogen. Series of mixtures show that the aluminosilicate gel could contain up to 72 wt% of $NaBH_4$ content, which could totally be used for hydrogen release at room temperature by the addition of weak acid solutions. A further increase in effective $NaBH_4$ content could be obtained by reducing the temperature for hardening. Therefore, such type of handling of $NaBH_4$ could open new possibilities for future applications for the energy source hydrogen

Another new type of hydrogen storage is the encapsulation of the BH_4-anion in the sodalite cage revealing a whole family of compounds. Here the total content of the hydrogen is 10 times smaller compared to the pure $NaBH_4$-salt. However, it shows the big advantage that all the consecutive reaction steps for hydrogen release via the reaction with water could be discovered and mostly controlled. The hydrogen release reveals stepwise a zoned crystal system of $H_3B(OH)^-$, $H_2B(OH)_2^-$, $HB(OH)_3^-$ and $B(OH)_4$-anions in the sodalite cages. A dehydration of $B(OH)_4$-SOD to BO_2-SOD follows and could be controlled as well. It is very likely that hydrogen release from BH_4-sodalite resembles a diffusion controlled process with H^- and OH^- being the diffusing species. In section 4 experimental observations were given, supporting this interpretation. According to this it can be considered that H_2O becomes first dissociated and OH^- could exchange H^- in the cage, which recombines with H^+ to H_2. Conclusively first steps of a regeneration of the BH_4-anion could be obtained by realizing a diffusion gradient of H^- from outside to inside for pre-reacted BH_4-sodalite samples. It could be concluded that basically the Na-cations effectively close the sixring windows of the sodalite cages. With raising temperature above about 250°C there is an increasing probability of opening this window for an increasing ion exchange. Closing this window effectively protects the BH_4-anion from water attack. The absence of this effect in zeolite LTA for the grc- and toc-units thus could explain the failure of any successful stabilization of BH_4-anion or larger salt-type units in such frameworks. On the other hand for the BH_4-sodalite a further optimization of the reactor regeneration, e.g. by fine tuning of the geometrical parameter concerning the cage sizes as well as the distribution of the appropriate catalyst, for example the development of thin film technique, the BH_4-sodalite could gain some future application as hydrogen storage and hydrogen fuel processing from water.

Author details

Josef Christian Buhl, Lars Schomborg and Claus Henning Rüscher
Institut für Mineralogie, Leibniz Universität Hannover, Hannover

Acknowledgement

Some results were obtained with the financial support for students by the "Land Niedersachsen" and by fund of the Leibniz University of Hannover. Results obtained by LS could be reported prior to publication in his PhD thesis work.

6. References

[1] Bogdanivic B, Schwickardi (1997) M Ti-doped alkali metal aluminium hydrides as potential novel reversible hydrogen storage material. J. Alloys Compd., 253: 1-9.

[2] Schlapbach L, Züttel (2001) A Hydrogen storage materials for mobile applications. Nature 414: 353-358.

[3] Pradhan B K, Harutyunyan A R, Stojkovic D, Grossman J C, Zhang P, Cole M W, Crespi V H, Goto H, Fujiwara J, Eklund P C (2002) Large Cryogenic Storage of Hydrogen in Carbon Nanotubes at Low Pressures. J. Mat. Res. 17: 2209-2216.

[4] Hu Y H, Ruckenstein E (2006) Clathrate hydrogen hydrate – A promising material for hydrogen storage. Angew. Chem. Int. Ed. 45: 2011-2013.

[5] Schüth F (2006) Mobile Wasserstoffspeicher mit Hydriden der leichten Elemente. Nachrichten aus der Chemie 54: 24-28.

[6] Jia C, Yuan X, Ma Z (2009) Metal-organic frameworks (MOFs) as hydrogen storage materials. Prog. in Chem. 21: 1954-1962.

[7] Kuppler R J, Timmons D J, Fang Q-R, Li J-R, Makal T A, Young M D, Yuan D, Zhao D, Zhuang W, Zhou H-C (2009) Potential applications of metal-organic frameworks. Coordination Chemistry Reviews 253: 3042-3066.

[8] Murray L J, Dinca M, Long J R (2009) Hydrogen in metal-organic frameworks. Chem. Soc. Rev. 38: 1294-1314.

[9] Sculley J, Yuan D, Zhou H-C (2011) The current status of hydrogen storage in metal-organic frameworks updated. Energy Environ. Sci. 4: 2721-2735.

[10] Davis R E, Bromels E, Kibby Ch L (1962) Boron hydrides. III. Hydrolysis of sodium borohydride in aqueous solution. J. Am. Chem. Soc. 84: 885-892.

[11] Li Z P, Liu B H, Arai K, Morigazaki N, Suda S (2003) Anodic Oxidation of Alkali Borohydrides Catalyzed by Nickel. J. Alloys Compd. 356-357: 469-474.

[12] Liu B H, Li Z P (2009) Hydrogen generation from borohydride hydrolysis reaction J. Power Sources 187 (2): 527-534.

[13] Cao D, Chen D, Lan J, Wang G (2009) An alkaline $NaBH_4$-H_2O_2 fuel cell with high power density. J. Power Sources 190: 346-350.

[14] Buhl J-Ch (2011) Synthesis and properties of $NaBH_4$-imbibed aluminosilicate gels and its partial crystalline secondary products. Z. Kristallogr. Supplement Issue No. 31: 23-24.

[15] Schomborg L, Rüscher C, Buhl J-C (2012) Thermal Stability and quantification of hydrogen release of $NaBH_4$ enclosed in aluminosilicate gels. Z. Kristallogr. Suppl. Issue No. 32: 62

[16] Barrer R M (1982) Hydrothermal chemistry of zeolites. London: Academic Press. 348 p.

[17] Buhl J-Ch, Gesing T M, Rüscher C H (2005) Synthesis, crystal structure and thermal stability of tetrahydroborat sodalite $Na_8[AlSiO_4]_6(BH_4)_2$. Micropor. Mesopor. Mater. 80: 57-63.

[18] Buhl J-Ch, Gesing T M, Höfs T, Rüscher C H (2006) Synthesis and crystal structure of gallosilicate- and aluminogermanate tetrahydroborate sodalites $Na_8[GaSiO_4]_6(BH_4)_2$ and $Na_8[AlGeO_4]_6(BH_4)_2$. J. Solid State Chem. 179: 3877-3882.

[19] Höfs T K (2009) Synthese und thermisches Reaktionsverhalten $NaBH_4$-haltiger Sodalithe mit alumosilikatischem, gallosilikatischem und alumogermanatischem Strukturgerüst. Theses, Institute of Mineralogy, Leibniz University Hannover, Hannover.

[20] Höfs T K, Buhl J-Ch (2011) Thermal behavior of $NaBH_4$-sodalites with alumosilicate framework: Influence of cage water content and the surrounding conditions. Mat. Res. Bull. 46: 1173-2178.

[21] Poltz I, Robben L, Buhl J-Ch, Gesing T M (2011) Synthesis, crystal structure and high temperature behaviour of gallogermanate tetrahydroborate sodalite $Na_8[GaGeO_4]_6(BH_4)_2$. Z. Kristallogr. Suppl. Issue No. 31: 103.

[22] Buhl J-Ch, Schomborg L, Rüscher C H (2010) Tetrahydroborate sodalite nanocrystals: Low temperature synthesis and thermally controlled intra cage reactions for hydrogen release of nano- and micro crystals. Micro. Meso. Mater. 132: 210-218.
[23] Buhl J-Ch, Rüscher C H Schomborg L, Stemme F (2010) Nanocrystalline $NaBH_4$-enclathrated zeolite SOD: a model for improvement of safeness and reactivity of boron hydride based hydrogen storage systems. Clean Technology, www.ct-si.org, ISBN 978-1-4398-3419-0
[24] Rüscher C H, Stemme F, Schomborg L, Buhl J-Chr (2010) Low temperature hydrogen release from borontetrahydride-sodalite and ist reloading: Observations in in-situ and ex-situ TIR experiments Ceramic Transactions 215: 65-70.
[25] Schomborg L, Rüscher C H, Buhl J-Ch (2011) Hydrogen release and reinsertion reactions in $(BH_x\text{-}(OH)_y\text{-}O_z)$-SOD zoned crystal systems. Z. Kristallogr. Suppl. Issue 31: 24.
[26] Rüscher C H, Mielcarek E, Lutz W, Jirasit F, Wongpa J (2010) New insights on geopolymerisation using molybdate, Raman and infrared spectroscopy. Ceramic Engineering and Science Proceedings 31: 19-35
[27] Hermeler G, Buhl J-Ch, Hoffmann W (1991) The influence of carbonate on the synthesis of an intermediate phase between sodalite and cancrinite. Catalysis Today 8 415-426.
[28] Waddington T C (1958) Thallous borohydride $TlBH_4$. J. Chem. Soc.: 4783-4784.
[29] Goubeau J, Kallfass H (1959) Die Reaktion von Natriumborhydrid und Wasser. Z. Anorg. Allg. Chem. 299: 160-169.
[30] Schutte C J H (1960) The infra-red spectrum of thin films of sodium borohydride. Spectrochim. Acta 16: 1054-1059.
[31] Ketelaar J A A, Schutte C J H (1961) The boronhydride ion (BH_4^-) in a face centered cubic alkali-halide salt. Spectrochim. Acta 17: 1240-1243.
[32] Filinchuk Y, Hagemann H (2008) Structure and Properties of $NaBH4{\cdot}2H_2O$ and $NaBH_4$. Europ. J. Inorganic Chemistry 20: 3127-3133
[33] Hisatsune I C, Suarez N H (1964) Infrared Spectra of Metaborate Monomer and Trimer Ions. Inorg. Chem. 3: 168-174.
[34] Kessler G, Lehmann H A (1965) IR-spektroskopische Untersuchungen an Boraten: I. Natrium(1:1:4)borathydrat. Z. Anorg. Allg Chem. 338: 179-184.
[35] Pietsch H H E, Fechtelkord M, Buhl J-Ch (1997) The formation of unusual twofold coordinated boron in a sodalite matrix, J. Alloys Compd. 257: 168.
[36] Nakamoto K (1978) Infrared and Raman Spectra of Inorganic and Coordination Compounds. New York: John Wiley & Sons. 400 p.
[37] Weidlein J, Müller U, Dehnicke K (1981) Schwingungsfrequenzen I. Stuttgart/New York: Georg Thieme Vlg. 339 p.
[38] Rüscher C H (2005) Chemical reactions and structural phase transitions of sodalites and cancrinites in temperature dependent infrared (TIR) experiments. Microp. Mesop. Materials 86: 58-68.
[39] Davis R E, Swain C G (1960) The general acid catalysis of the hydrolysis of sodium borohydride. J. Am. Chem. Soc. 82: 5949-5950.
[40] Weitkamp J, Fritz M, Ernst S (1995) Zeolites as media for hydrogen storage. J. Hydrogen Energy 20: 967-970.

[41] Baur W H, Fischer R X (2008) A historical note on the sodalite framework: The contribution of Frans Maurits Jaeger. Microp. Mesopor. Mater. 116: 1-3.

[42] Fischer R X, Baur W H (2008) Symmetry relationships of sodalite (SOD) – type crystal structures. Z. Kristallogr. 224: 185-197.

[43] Johnson G M, Mead P J, Weller M T (2000) Synthesis of a range of anion-containing gallium and germanium sodalites. Microp. Mesopor. Mater. 38: 445-460.

[44] Wiebcke M, Sieger P, Felsche J, Engelhardt G, Behrens P, Schefer J (1993) Sodium Aluminogermanate Hydroxosodalite Hydrate $Na_{6+x}[Al_6Ge_6O_{24}](OH)X{\cdot}nH_2O$(X-Approximate-to-1.6, n-Approximate-to-3.0) - Synthesis, Phase Transitions and Dynamical Disorder of the Hydrogen Dihydroxide Anion, $H_3O_2^-$, in the Cubic High-Temperature Form. Z. Anorg. Allg. Chem. 619: 1321-1329.

[45] Gesing T M (2007) Structure and properties of tecto-gallosilicates II. Sodium chloride, bromide, bromide and iodide sodalites. Z. Kristallogr. 222: 289-296.

[46] Liebau F (1983) Zeolites and clathrasils-Two distinct classes of framework silicates. Zeolites 3: 191-193.

[47] Weller M T, Dodd S M, Myron Jiang M R (1991) Synthesis, structure and ionic conductivity of nitrite sodalite. J. Mater. Chem. 1: 11-15.

[48] Buhl J-Ch, Mundus C, Löns J, Hoffmann W (1994) On the enclathration of NaB(OH)4 in the ß-cages of sodalite: crystallization kinetics and crystal structure. Z. Naturforsch. 49a: 1171-1178.

[49] Buhl J-Ch, Gesing T M, Gurris C (2001) Synthesis and crystal structure of rhodanide-enclathrated sodalite $Na_8[GaSiO_4]_6(SCN)_2$. Micropor. Mesopor. Mater. 50: 25-32.

[50] Rüscher C H, Gesing T M, Buhl J-Ch (2003) Anomalous thermal expansion behaviour of $Na_8[AlSiO_4]_6(NO_3)_2$-sodalite: P4-3n to Pm3-n phase transition by untilting and contraction of TO_4 units Z. Kristallogr. 218: 332-344.

[51] Mikheeva V I, Bredtsis V B (1960) Solubility isotherm for sodium boron hydride and sodium hydroxyde in water at zero degrees. Dokl. Akad. Nauk SSSR 131: 1349-1350.

[52] Mesmer R E, Jolly W L (1962) Hydrolysis of aqueous hydroborate. Inorg. Chem. 1 (3): 608-612.

[53] Hadan M, Fischer F (1992) Synthesis of fine grained NaA-type Zeolites from superalkaline solutions. Cryst. Res. Technol. 27: 343-350.

[54] Fischer F, Hadan M, Fiedrich G (1992) Zeolite syntheses from superalkaline reaction mixtures. Collect. Czech. Chem. Commun. 57: 788-793.

[55] Fan W, Morozumi K, Kimura R, Yokoii T, Okubo T (2008) Synthesis of nanometer-sized sodalite without adding organic additives. Langmuir 24: 6952-6958.

[56] Mashal K, Harsh J B, Flury M, Felmy A R (2005) Analysis of precipitates from reactions of hyperalkaline solutions with soluble silica. Appl. Geochem. 20: 1357-1367.

[57] Wang L Q, Mattigod S V, Parker K E, Hobbs D T, McCready D E (2005) Nuclear magnetic resonance studies of aluminosilicate gels prepared in high-alkaline and salt-concentrated solutions. J. Non-Cryst. Solids 351: 3435-3442.

[58] Abts L M, Langland J T, Kreevoy M M (1975) Role of water in hydrolysis of BH_4^-. J. Am. Chem. Soc. 97 (11): 3181-3185.

[59] Fleet M E (1989) Structures of sodium alumino-germanate sodalites [$Na_8(Al_6Ge_6O_{24})A_2$, A = Cl, Br, I]. Acta. Cryst. C45: 843-847.
[60] Flanigen E M, Khatami H, Szymanski H A (1971) Infrared structural studies of zeolite frameworks. Advan. Chem. Ser. 101: 201-209.
[61] Löns J, Schulz H (1967) Strukturverfeinerung von Sodalith $Na_8Si_6Al_6O_{24}Cl_2$. Acta Cryst. 23: 434-436.
[62] Davis R L, Kennard C H L (1985) Structure of sodium tetradeutoroborate, $NaBD_4$. J. Solid State Chem. 59: 393-396.
[63] Izumi F (1993) Rietveld analysis programmes Rietan and Premos and special applications. In: Young R A, editor. The Rietveld Method. Oxford: Oxford University Press. pp. 236-253.
[64] Marcus M, Bredow T, Schomborg L, Rüscher C H, Buhl J-C (2012) Structure and IR spectra of $Na_8[AlSiO_4]_6(BH_4)_2$ sodalite: Comparison between theoretical predictions and experimental data. Z. Kristallogr. Suppl. Issue No. 32: 89-90
[65] Buhl J-Ch, Murshed M-M (2009) $(Na_4BH_4)^{3+}$ guests inside aluminosilicate, gallosilicate and aluminogermanate sodalite host frameworks studied by 1H, 11B, and 23Na MAS NMR spectroscopy. Mat. Res. Bull. 44: 1581-1585.
[66] Wrackmeyer B (1988) NMR Spectroscopy of Boron Compounds Containing Two-, Three- and Four-Coordinate Boron Ann. Rep. NMR Spectrosc. 20: 61-203.
[67] Bray P J (1999) NMR and NQR studies of boron in vitreous and crystalline borates. Inorg. Chim. Acta 289: 158-173.
[68] Hansen M R, Madsen G K H, Jakobsen H J, Skibsted J (2005) Refinement of borate structures from 11B MAS NMR spectroscopy and density functional theory calculations of 11B electric field gradients. J. Phys. Chem. A 109: 1989-1997.
[69] Buhl J-Ch, Engelhardt G, Felsche J (1989) Synthesis, X-ray diffraction and MAS n. m. r. characteristics of tetrahydroxoborate sodalite, $Na_8[AlSiO_4]_6$ $[B(OH)_4]_2$. Zeolites 9: 40-44.
[70] Massiot D, Fayon F, Capron M, King I, Le Calve S, Alonso B, Durand J, Bujola B, Gan Z, Hoatson G (2002) Modelling one- and two-dimensional solid state NMR spectra. Mag. Res. Chem. 40: 70-76.
[71] Engelhardt G, Sieger P, Felsche J (1993) Multinuclear solid state NMR of host-guest systems with TO_2 (T=Si, Al) host-frameworks. A case study on sodalites. Analytica Chimica Acta 283: 967-985.
[72] Pietsch H-H E, Fechtelkord M, Buhl J C (1992) The formation of unusually twofold coordinated boron in sodalite matrix. J. Alloys and Compounds 257: 168-174
[73] Fechtelcord M (2000) Influence of sodium ion dynamics on the ^{23}Na quadrupolar interaction in sodalite: A high temperature ^{23}Na MAS NMR study. Solid State Nucl. Magn. Res 18: 70-88
[74] Rüscher C H, Schomborg L, Buhl J-C (2010) Thermally Controlled Water Injection into BH_4-Sodalie for Hydrogen Formation Investigated by IR Absorption. Diffusion Fundamentals 12: 37-39
[75] Phair J W (2007) Stability of alkali nitrate/Pd composites for hydrogen separation membranes. Energy and Fuels Vol. 21 Issue 6: 3530-3536

The Preparation and Hydrogen Storage Performances of Nanocrystalline and Amorphous Mg_2Ni-Type Alloys

Yanghuan Zhang, Hongwei Shang, Chen Zhao and Dongliang Zhao

Additional information is available at the end of the chapter

http://dx.doi.org/10.5772/50281

1. Introduction

The earth owning better environment, including fresh air, clean rain and healthy sunshine, is always the best wish of human. However, statistics indicate that air pollution and the greenhouse effect are becoming more and more serious caused by exhaust gas of vehicles, which seriously endanger human health. Petroleum fuels are the chief criminal. Furthermore, petroleum fuels belong to non-renewable resources, which need a long time to recycle and the reserves are limited. Especially in China, along with the development of domestic economy, the heavy demand of cars is taking on unceasing climbing in a straight line. The difficulties we met are more serious. Therefore, under this precondition, the need of seeking a kind of convenient, clean efficient sources of energy is a very pressing task for researchers. China provides broad space for clean energy exploitation, including solar, hydrogen, and wind power. Right now, the researches focus on hydrogen due to some advantages. As is known to everybody, hydrogenous sources of energy is widespread, such as water covers most of the earth surface and a wide range of hydride exists in natural gas, coal, plants on earth, providing an inexhaustible resource for our society.

Hydrogen energy is a kind of high efficient, clean secondary energy. However, hydrogen storage is one of the main obstacles in hydrogen utilization. The traditional way such as high pressure tanks and liquid hydrogen tanks are once regarded as candidates for road tests, but it is proved from practice that these ways have certain fatal flaws in some respects. These methods not only occupy a large space of cars but also put in safe hidden trouble. In addition, both high pressure tanks and liquid hydrogen tanks are facing another problem — the lack of hydrogen refueling stations. The cost of hydrogen refueling which is far beyond than that of gasoline filling station, revealing the reasons why most of the major oil companies have been loathed to set up hydrogen refueling stations. In comparison with the above traditional methods, alloy hydrides with lots of advantages such as high hydrogen storage capacity, good reversibility and low cost have been identified as an ideal fuel for many energy converters in recent years [1]. Electrochemical hydrogen storage is a more convenient method, by which atomic hydrogen is adsorbed in hydrogen storage materials during electrochemical decomposition of an aqueous medium [2]. The only product of the whole reaction is just water. So, it is "totally clean" and extremely useful for application to emerging vehicles based on hydrogen fuel cell technologies.

At present, the new energy resources cars mainly involve Hybrid Electrical Vehicle (HEV) and electric vehicles (EV). Substituting petrol-driven car with HEV and EV has been a primary way for solving the above environmental problems for researchers in this area. Compared with the traditional cars, the usage cost of EV and HEV have been reduced to some extent in recent years. So far, the current new energy vehicles in market, primarily uses Ni/MH batteries, lead-acid batteries, and lithium iron phosphate batteries. Lead-acid batteries are usually used in the start-up process of vehicles. The performance of Lithium iron phosphate batteries is more outstanding than Ni/MH batteries. The Chevy Volt, relying on lithium ion batteries, is able to travel 65 km on a single charge. However, it is reported that lithium batteries may be caught fire or exploded after certain overheating, which is a major drawback for its application and development. Among all the batteries, Ni/MH batteries make maximum potential due to it being the most technically matured. And it gets the most extensive of application. Predictably, Ni/MH batteries will still occupy a market share in the first place in the future development of the new energy vehicles. Therefore, in order to occupy the future market, some famous auto manufacture enterprises already have brought some hybrid or electric cars to market. Renault and Nissan are developing electric cars and forecast that plug-in hybrids will become the mainstream products. In the mean time, Toyota plans to start global trials of 500 plug-in hybrids. It is well known that China is always investing heavily in cleaner cars. In the "863" High-tech Plan, "973" Plan and the National Natural Science Foundation, hydrogen storage material is limited as one of the key research areas. China carmakers even have a chance to leapfrog an entire generation of automotive technology and narrowing the gap with developing countries by actively participating in the new energy vehicle development, according to some industry observers. This show the importance and necessity of giving priority to the development of new energy vehicles. Chang'an Jie Xun is the domestic first hybrid cars. Chery Automobile's new A5 also have been available to Chinese consumers whose price is less than 80000 RMB. In addition to the above-mentioned new energy vehicles, Faw-Besturn and Buick Lacrosse are also Eco-friendly cars which using Ni/MH battery as energy supplying device. Yet, barriers against running cars with hydrogen is still quite tough. As an on-board secondary batteries, Ni/MH battery must have higher capacity and good gaseous and electrochemical hydriding and dehydriding kinetics for enhancing the electric power continue voyage.

The negative electrode is a crucial component for hydrogen storage alloy, which is the main bottleneck of Ni/MH batteries for further application. At present, all the Ni/MH batteries sold in the market adopt $LaNi_5$ hydrogen storage alloy as negative electrode, which was obtained by pure chance in 1969. As a typical commercial alloy, the rare earth based AB_5-type alloy has good cycling stability. Prius is regarded as the iconic founder of Ni/MH rechargeable battery which uses $LaNi_5$ hydrogen storage alloys. For the redesigned Prius, Toyota will stick with the current generation's nickel-metal hydride batteries. However, the discharge capacity of the currently advanced AB_5-type materials has reached 320–340mAh/g at 0.2–0.3C rate at room temperature. It seems to be difficult to further improve the capacity of the AB_5-type alloys since the theoretical capacity of $LaNi_5$ is about 372 mA·h/g [3]. Therefore, the investigations of the new type electrode alloys with higher capacity and good gaseous and electrochemical hydriding and dehydriding kinetics are extremely important to exalt the competition ability of the Ni/MH battery in the rechargeable battery field.

Recently, many researchers have been focused on solving the above problems. And several new and good hydrogen storage alloys were reported., among which the most promising candidates are the Mg and Mg-based metallic hydrides in view of their major advantages such as low specific weight, low cost and high hydrogen capacity, e.g. 7.6 wt.% for MgH_2, 3.6 wt.% for Mg_2NiH_4, 4.5 wt.% for Mg_2CoH_5 and 5.4 wt.% for Mg_2FeH_6 [4,5]. However, the commercial application to hydrogen suppliers has been limited mainly due to their sluggish hydriding/dehydriding kinetics as well as high thermodynamic stability of their corresponding hydride. Therefore, the investigation in this area has been focused on finding the ways to substantially ameliorate the hydriding/dehydriding kinetics of Mg and Mg-based alloys. Various attempts, particularly mechanical alloying (MA) [6] and melt spinning [7], have been conducted to ameliorate the kinetic property of Mg-based metallic hydrides. And the worldwide researchers have carried out a lot of investigations and have obtained some extremely important results. It was documented that Mg and Mg-based alloys with a nanocrystalline/amorphous structure exhibit higher H-absorption capacity and faster kinetics of hydriding/dehydriding than crystalline Mg_2Ni, which is ascribed to the enhanced hydrogen diffusivity and solubility in amorphous and nanocrystalline microstructures [8,9]. It was reported that the addition of third element greatly facilitates the glass forming of the Mg–Ni-based alloys[10,11] and that the partial substitution of M (M=Co, Cu) for Ni in Mg_2Ni compound decreases the stability of the hydride and makes the desorption reaction easier [12].

2. Preparation of alloys

It has come to light that high energy ball-milling (HEBM) has been regarded as a very effective method for the preparation of nanocrystalline and amorphous Mg and Mg-based alloys. Particularly, it is suitable to solubilize particular elements into MgH_2 or Mg_2NiH_4 above the thermodynamic equilibrium limit, which is helpful to destabilize MgH_2 or Mg_2NiH_4 [13]. However, the milled Mg and Mg-based alloys exhibit very poor hydrogen absorbing and desorbing stability in the light of the evanishment of the metastable structures formed by ball milling during the multiple hydrogen absorbing and desorbing cycles [14] . Furthermore, the HEBM process has some insurmountable disadvantages such as the necessity of long times to produce an amorphous alloy, difficulty for mass-production, and contamination from the chamber and balls used in ball-milling. Alternatively, the melt-spinning technique is the most useful method to obtain an amorphous and/or nanocrystalline phase in the absence of disadvantages inherent to the HEBM process and is more suitable for mass-production of amorphous alloys. Now, vacuum rapidly quenching technologies are used in the preparation of various metals and alloys, and the nano-crystalline and amorphous structures are obtained. The principle of the melt spinning technology is that the liquid metal and alloy can be solidified at great degree of super cooling by means of vacuum rapidly quenching. It was reported that nanocrystalline alloys produced by melt-spinning could have excellent hydriding characteristics even at room temperature, similar to the alloys produced by the HEBM process[15,16].

In order to improve the hydriding and dehydriding kinetics of the Mg_2Ni-type alloy, Ni in the alloy was partially substituted by element M (M=La, Co, Cu,). We have investigated systematically the effect of M (M=La, Co, Cu,) partial substitution for Ni on the microstructure characteristics, gaseous and electrochemical hydriding and dehydriding kinetics properties. The compositions of the alloys were $Mg_{20-x}La_xNi_{10}$ (x=0-6) and $Mg_{20}Ni_{10-x}M_x$ (M=Co, Cu; x=0, 1, 2, 3, 4). The experimental alloys were prepared by using a vacuum induction furnace in a helium atmosphere at a pressure of 0.04 MPa in order to prevent the volatilization of element Mg during melting. A part of the as-cast alloys was re-melted and spun by melt spinning with a rotating copper roller cooled by water, and the ribbons with width of 5 mm and thickness of 20 to 30 μm were obtained. The spinning rate, a very important parameter, was approximately expressed by the linear velocity of the copper roller because it is too difficult to measure a real quenching rate i.e. cooling rate of the sample during quenching. The spinning rate used in the experiment was 15, 20, 25 and 30 m/s, respectively.

3. Detection method

The phase structures of the as-cast and spun alloys were determined by XRD (D/max/2400). The morphologies of the as-cast alloys were examined by SEM (Philips QUANTA 400). The thin film samples of the as-spun alloys were prepared by ion etching for observing the morphology with HRTEM (JEM-2100F), and for determining the crystalline state of the samples with electron diffraction (ED). The thermal stability of the nanocrystalline and amorphous alloys was determined by Differential Scanning Calorimetry (DSC) instrument (STA449C).

The gaseous hydriding and dehydriding kinetics of the alloys were measured by an automatically controlled Sieverts apparatus. For comparing the hydriding and dehydriding kinetics performance of the alloys, which substituting Ni with Cu, Co, or Mg with La, the alloys were tested in the same temperature and pressure conditions. In our experiments, the hydriding process was carried out under 1.5 MPa hydrogen pressure (which is the initial pressure of hydriding process) at 200 °C, and the dehydriding process was carried out in vacuum (1×10^{-4} MPa) also at 200 °C.

The electrochemical performances of the as-cast and spun alloys were tested by an automatic galvanostatic system. The electrochemical impedance spectrums (EIS), Tafel polarization curves and hydrogen diffusion coefficient (D) were obtained by an electrochemical workstation (PARSTAT 2273). The alloys were pulverized and then mixed with carbonyl nickel powder in a weight ratio of 1:4. The mixture was cold pressed at a pressure of 35 MPa into round electrode pellets of 10 mm in diameter. Electrochemical measurements were performed at 30 °C by using a tri-electrode open cell, consisting of a working electrode (the metal hydride electrode), a sintered $Ni(OH)_2/NiOOH$ counter electrode and a Hg/HgO reference electrode, which were immersed in a 6 M KOH electrolyte. The voltage between the negative electrode and the reference electrode was defined as the discharge voltage. In every cycle, the alloy electrode was firstly charged with a constant current density, and following the rest of 15 min, and then discharged at the same current density to cut-off voltage of –0.5 V.

For the potentiostatic discharge, the test electrodes in the fully charged state were discharged at 500 mV potential steps for 3500s on electrochemical workstation (PARSTAT 2273), using the CorrWare electrochemistry corrosion software.

4. Results and discussion

Element substitution is one of the radical approaches of improving the performance of materials. And based on our former works, we can conclude that the melt spinning can improve the gaseous and electrochemical hydriding and dehydriding kinetics. Therefore, in our studies, we tried to obtain new type hydrogen storage alloys, which have predominant gaseous and electrochemical hydriding and dehydriding kinetics while maintaining good overall performance by the above two methods. It is well known that Mg_2Ni belongs to A_2B-type hydrogen storage alloys, in which Mg represents the A side elements while Ni represents the B side elements. They act differently from each other. Generally, researchers substitute A side element with La, Ce, Pr, Nd, Sm and substitute B side element with Co, Cu, Mn, Al, Zr for improving the gaseous and electrochemical hydriding and dehydriding kinetics. For the sake of simplicity, I will only introduce partial research results of our research team to illustrate the effect of element substitution, such as substituting La for Mg and Co, Cu for Ni, together with the effect of the melt spinning, such as the spun rate being 5, 10, 15, 20, 25m/s, respectively.

4.1. The effect of substituting Mg with La on microstructure characteristics as well as the gaseous and electrochemical hydriding and dehydriding kinetics

4.1.1. Microstructure characteristics

Fig. 1 shows the ribbons which obtained by substituting Mg with La (x=2) and melting spinning as the spun rate being 20m/s. It is about 5 mm wide and 20 to 30 μm thick by measurement.

The XRD patterns of the as-cast and spun $Mg_{20-x}La_xNi_{10}$ (x=0-6) alloys are shown in Fig. 2. It is evident that no amorphous phase is detected in the as-spun La_0 alloy, but all the as-spun alloys substituted by La display an obvious amorphous structure, suggesting that the substitution of La for Mg facilitates the glass forming of the Mg_2Ni alloys.

Figure 1. Actual picture of the as-spun (20 m/s) La_2 alloy

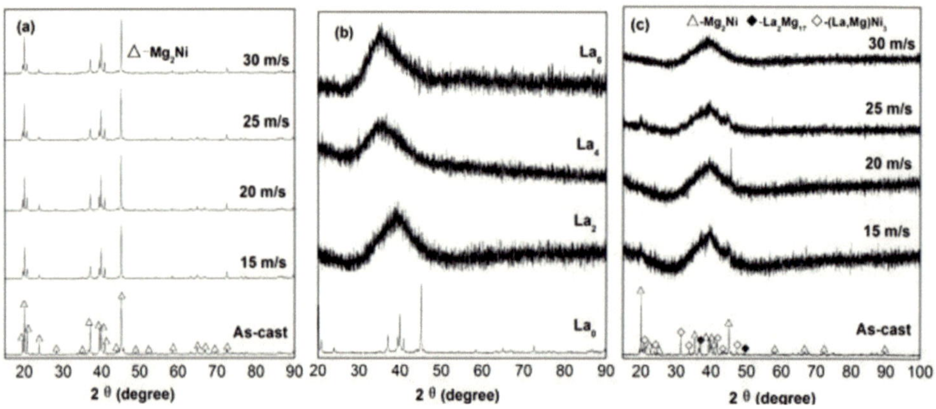

Figure 2. XRD patterns of the as-cast and spun (30 m/s) alloys: (a) La_0 alloy, (b) As-cast and quenched alloys, (c) La_2 alloy

The HRTEM images of the as-spun (30 m/s) alloys are illustrated in Fig. 3. It can be seen that the as-spun La_0 alloy displays a complete nanocrystalline structure, and its electron diffraction (ED) pattern appears sharp multi-haloes corresponding to a crystal structure. The morphologies of the as-spun La_2 and La_4 alloys exhibit a feature of the nanocrystalline embedded in the amorphous matrix. The morphology of the as-spun La_6 alloy shows a nearly complete amorphous character, and its electron diffraction pattern consists of only broad and dull halo, displaying an amorphous structure. It is noteworthy that the amount of the nanocrystalline in the as-spun alloys clearly decreases with increasing La content, meaning that the substitution of La for Mg increases the glass forming ability of the M_2Ni-type alloy, which agrees very well with the results of the XRD observation. Two elucidations can be regarded as the reasons for the above result. On the one hand, the addition of third element to Mg-Ni or Mg-Cu alloys facilitates the glass-formation [17]. On the other hand, the glass forming ability of an alloy is closely associated with the difference of the atomic radii in the alloy. The larger difference between the atom radii suggests the higher glass forming ability [18].

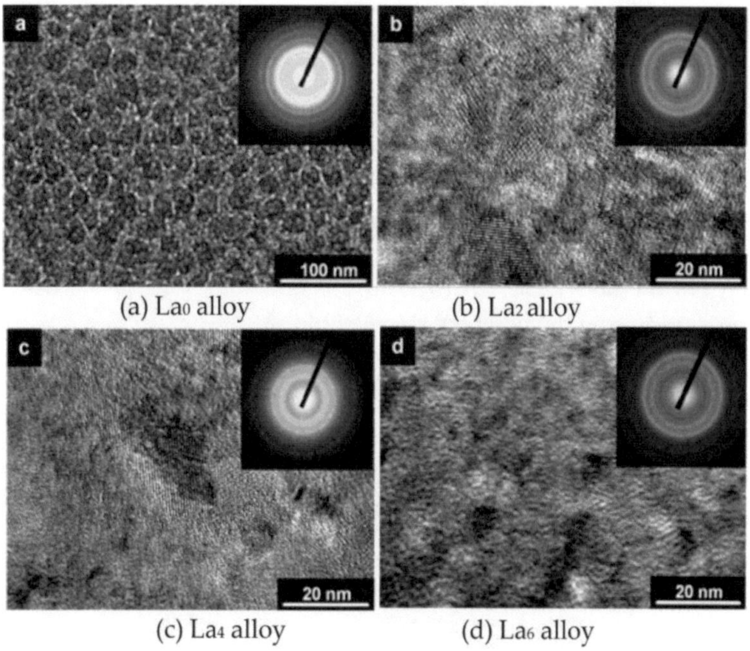

Figure 3. HRTEM micrographs and ED patterns of the as-spun (30 m/s) alloys

4.1.2. Thermal stability and crystallization

In order to examine the thermal stability and the crystallization of the as-quenched amorphous and nanocrystalline/amorphous alloys, DSC analysis was conducted. The resulting profiles shown in Fig. 4 reveals that during heating the alloys crystallized completely, and the crystallization process of La_2 alloy consisted of several steps. The first crystallization reaction at about 210 °C is connected with a sharp exothermic DSC peak, followed by a smaller and wider peak (312 °C) corresponding to a second crystallization reaction. At higher temperatures (at about 396 °C) a third exothermic effect can be detected. It was proved that the first sharper peak corresponds to the crystallization (ordering) of the amorphous into nanocrystalline Mg_2Ni [19]. Based on the results in Fig.2 (c), it was speculated that the second and the third exothermic peaks correspond to the crystallization of the amorphous into nanocrystalline (La, Mg)Ni_3 and La_2Mg_{17}, respectively. It was also seen that the crystallization temperatures of the alloys slightly rose with the increasing quenching rate, which was probably relevant to the influence of the quenching rate on the amorphized degree of the alloy.

4.1.3. Gaseous and electrochemical hydriding and dehydriding kinetics

The hydrogen absorption curves of as-cast and spun La_0 and La_2 alloys are presented in Fig. 5. Upon the first contact with hydrogen at 200 °C and 1.5 MPa, the activated alloys rapidly absorb copious amounts of hydrogen, and the initial hydriding rate increases with the rising of the spinning rate. The hydrogen absorption capacity in 10 min is increased from 1.21 to 3.10 wt% for the La_0 alloys, and from 1.26 to 2.60 wt% for the La_2 alloy by growing spinning rate from 0 to 30 m/s, indicating that the hydrogen absorption capacity and kinetics of all the as-spun alloys studied are superior to those of conventional polycrystalline materials with similar composition. The improved hydrogen absorption characteristics are attributed to the amorphous and nanocrystalline microstructures created by melt spinning in the light of those structure displaying high hydrogen diffusivity and solubility. In order to reveal the mechanism of the melt spinning improving hydrogen absorption kinetics of the alloy, it is evidently necessary to investigate the influences of the melt spinning on the hydrogen diffusion ability in the alloy. The hydrogen diffusion coefficients in the as-cast and spun alloys were measured using the potential step technique.

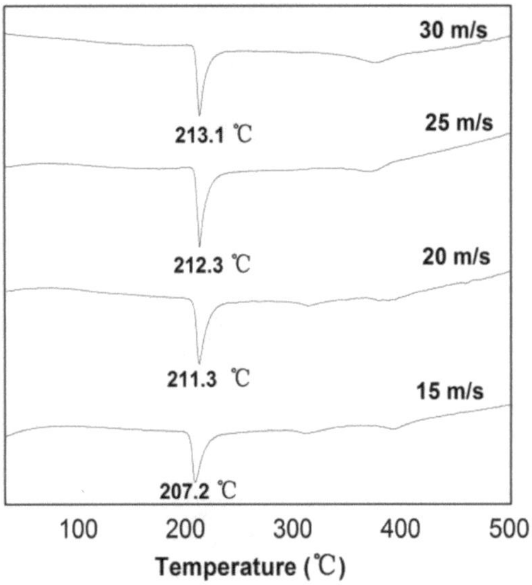

Figure 4. DSC profiles of La_2 alloy quenched at different quenching rates

Fig. 6 shows the semilogarithmic curves of anodic current versus working duration of the as-cast and spun alloys. The diffusion coefficient D of the hydrogen atoms in the bulk of the alloy can be calculated by following formulae [20]:

$$\log i = \log\left(\pm \frac{6FD}{da^2}\left(C_0 - C_s\right)\right) - \frac{\pi^2}{2.303}\frac{D}{a^2}t \tag{1}$$

$$D = -\frac{2.303a^2}{\pi^2}\frac{d\log i}{dt} \tag{2}$$

The D values are also illustrated in Fig. 6, indicating that the melt spinning has a visible effect on hydrogen diffusion in the alloy. As the spinning rate grows from 0 to 30 m/s, the hydrogen diffusion coefficient D always increases from 1.350×10^{-11} to 2.367×10^{-11} cm^2/s for the La_0 alloy, and from 8.122×10^{-12} to 1.7987×10^{-11} cm^2/s for the La_2 alloy. The above results indicate that a higher spinning rate is always beneficial for enhancing the diffusion ability of hydrogen atoms in the alloys, for which the refined grain and the increased internal stress by melt spinning are mainly responsible due to diffusion coefficient being directly proportional to the internal strain [21].

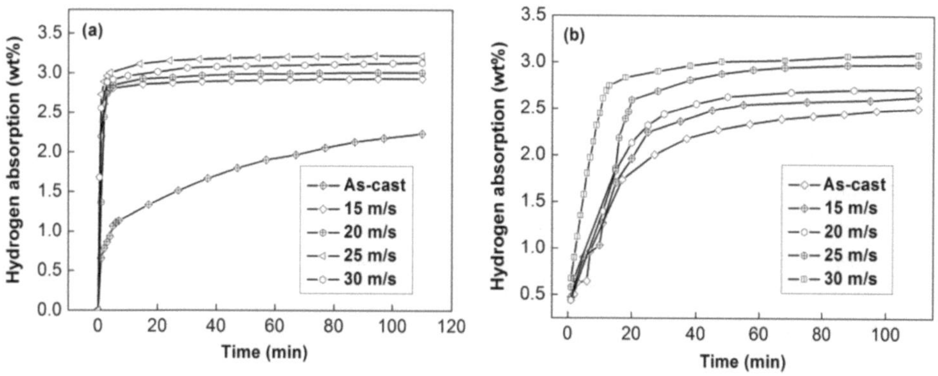

Figure 5. Hydrogen absorption curves of the as-cast and spun alloys: (a) La_0 alloy, (b) La_2 alloy

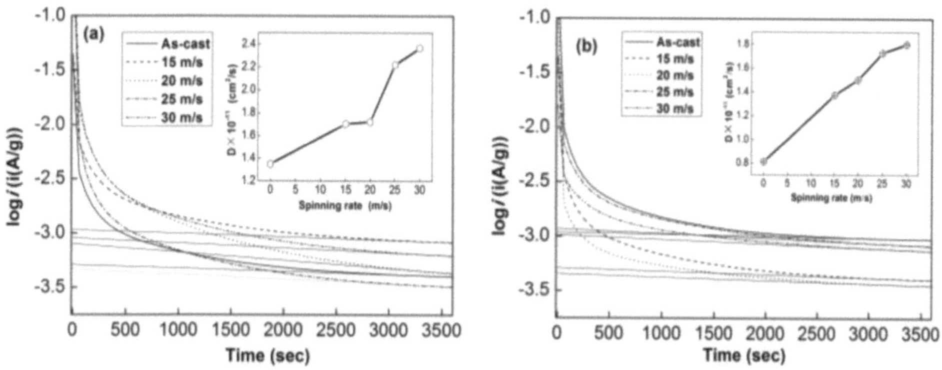

Figure 6. Semilogarithmic curves of anodic current vs. time responses of the as-cast and spun alloy electrodes: (a) La_0 alloy, (b) La_2 alloy

The dehydriding process was carried out in vacuum at 200 °C. The results are shown in Fig. 7. The figures indicate that the hydrogen desorption capacities and kinetics of the alloys are significantly improved by rapid quenching. It can be seen from Fig. 7 that rapid quenching greatly enhanced the hydrogen desorption capacities and improved the dehydrogenation kinetics of the alloys. When the quenching rate rose from 0 to 30 m/s, the hydrogen desorption capacity of the La_0 alloy in 10 min increased from 0.23 to 0.99 wt.%, and from 1.48 to 2.46 wt.% for La_2 alloy, for which the formed nanocrystalline and nanocrystalline/amorphous structures by rapid quenching are mainly responsible.

4.2. The effect of substituting Ni with Co on microstructure characteristics and gaseous as well as electrochemical hydriding and dehydriding kinetics

4.2.1. Microstructure characteristics

The XRD patterns of the as-spun alloys are shown in Fig. 8. It is indicated that the substitution of Co for Ni leads to the formation of secondary phase $MgCo_2$ instead of changing the major phase Mg_2Ni in the alloy. It is evidently visible that the amount of $MgCo_2$ phase grows with the increase in Co content. Fig. 8 displays that no amorphous phase is detectable in the as-spun Co_0 alloy, but the as-spun Co_4 alloy obviously shows the presence of an amorphous phase. Thus, it can be concluded that the substitution of Co for Ni intensifies the glass forming ability of the Mg_2Ni-type alloy. As a third element, Co added to Mg-Ni alloys significantly facilitating the glass formation [17]. Furthermore, the atomic radius of Co is larger than that of Ni, according to the theory of H.S. Chen [18], the glass forming ability of the alloy is enhanced. Table 1 lists the lattice parameters, cell volume as well as the full width at half maximum (FWHM) values of the main diffraction peaks of the as-spun (15 m/s and 30 m/s) alloys, which are calculated by the software of Jade 6.0. It can be derived from Table 1 that the FWHM values of the main diffraction peaks of alloys significantly increase with the increase of the Co content. The substitution of Co for Ni leads to the enlargement of the lattice parameters and the cell volume of alloys. It is also clear in Table 1 that the melt spinning yields the broadened diffusion peaks, indicating the refinement of the average grain size and stored stress in the grains.

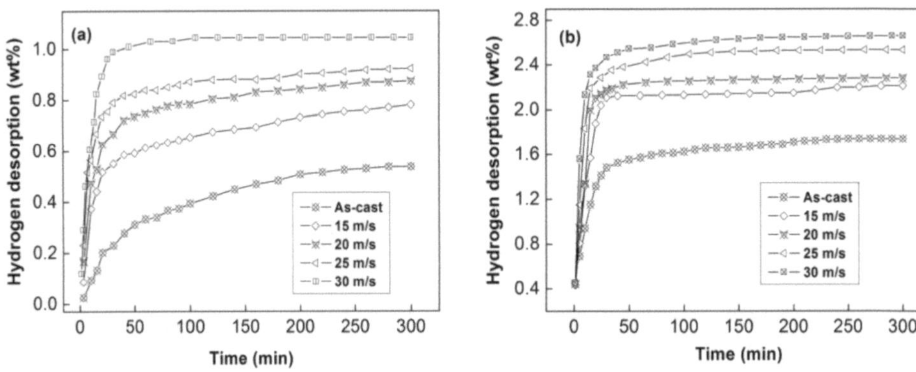

Figure 7. Hydrogen desorption curves of the as-cast and quenched alloys: (a) La_0, (b) La_2

The TEM micrographs of the as-spun (30 m/s) Co_0, Co_1, Co_3 and Co_4 alloys are presented in Fig. 9. It is quite visible that the as-spun Co_0 and Co_1 alloys display a nanocrystalline structure with the grain size of about 20 nm along with the exhibition of their electron diffraction (ED) patterns as sharp multi-haloes. The morphology of the as-spun Co_3 alloy presents a nanocrystalline feature, and a small amount of amorphous phases can be found at its grain boundaries. The Co_4 alloy also exhibits a feature of the nanocrystalline with a grain size of about 20 nm embedded in the amorphous matrix. However, its (Co_4 alloy) electron diffraction pattern consists of broad and dull halo, confirming the presence of an amorphous structure. This result is quite in conformity with the XRD observation shown in Fig. 8.

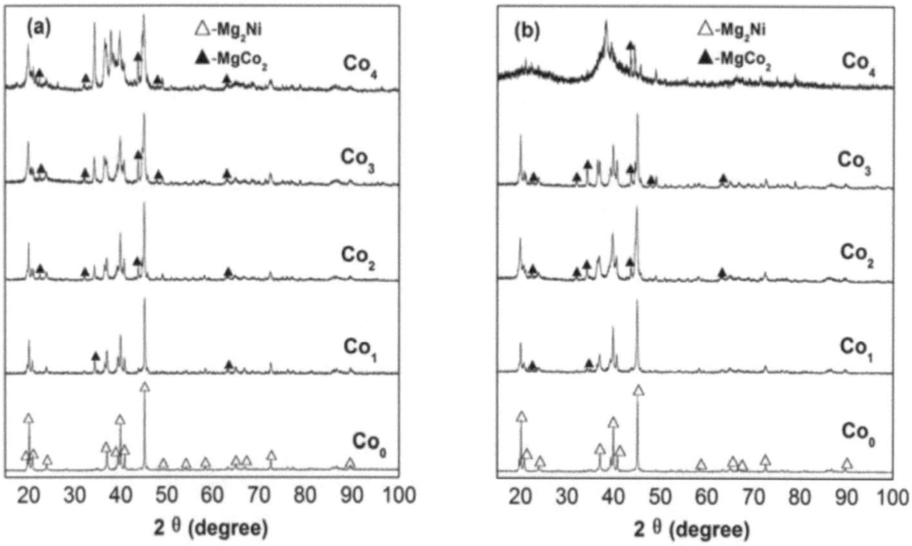

Figure 8. XRD patterns of as-spun alloys with different spinning rates: (a) 15 m/s, (b) 30 m/s

Alloys	FWHM values				Lattice parameters and cell Volume					
	2θ(20.02°)		2θ(45.14°)		a (nm)		c (Å)		V (Å³)	
	15 m/s	30 m/s	15 m/s	30 m/s	15 m/s	30 m/s	15 m/s	30 m/s	15 m/s	30 m/s
Co_0	0.125	0.133	0.171	0.182	0.5210	0.5211	1.3251	1.3287	0.3115	0.3134
Co_1	0.185	0.206	0.191	0.225	0.5213	0.5219	1.3256	1.3323	0.3120	0.3142
Co_2	0.210	0.246	0.265	0.290	0.5217	0.5285	1.3305	1.3336	0.3136	0.3226
Co_3	0.336	0.536	0.357	0.329	0.5224	0.5287	1.3312	1.3412	0.3146	0.3246
Co_4	0.583	—	0.529	—	0.5225	—	1.3318	—	0.3148	—

Table 1. The lattice parameters, cell volume and the FWHM values of the major diffraction peaks of the alloys

4.2.2. Thermal stability and crystallization

In order to examine the thermal stability and crystallization of the as-spun amorphous and nanocrystalline alloys, DSC analysis has been conducted. The resulting profiles of the as-spun (30 m/s) alloys depicted in Fig. 10 reveal that the crystallization of the as-spun alloys completes during the heating process, furthermore, the crystallization process consists of two steps. The first crystallization reaction is associated with a sharp exothermic DSC peak followed by a smaller and wider peak corresponding to the second crystallization reaction. It is also testified that the first sharper peak corresponds to the crystallization (ordering) of the amorphous into nanocrystalline Mg_2Ni [19]. The crystallization temperature of the amorphous phase in the as-spun alloys first increases and then decreases with the increasing of Co content.

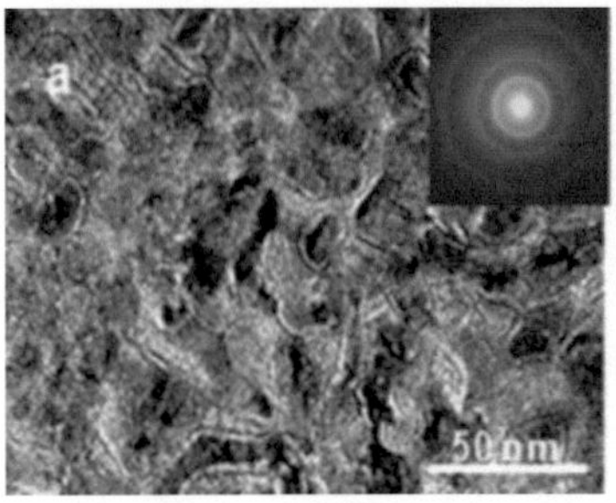

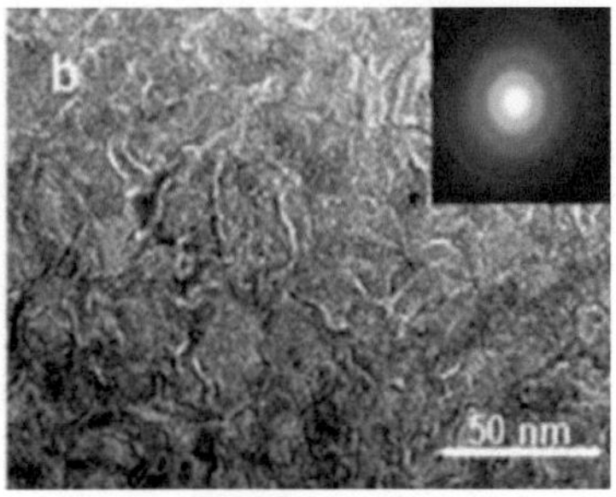

(a) Co_0 (b) Co_1

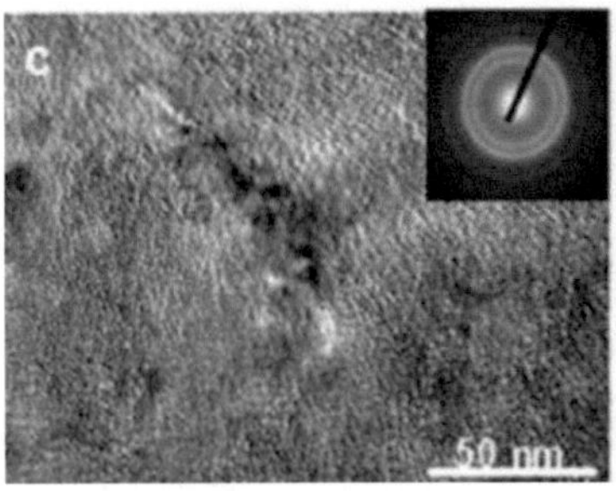

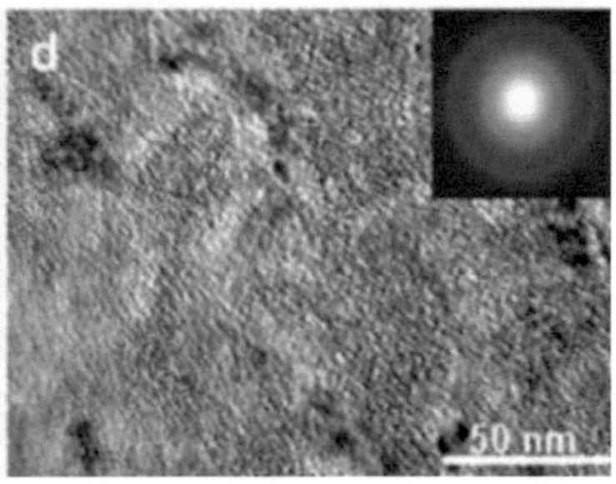

(c) Co_3 (d) Co_4

Figure 9. TEM images and electron diffraction patterns of as-spun (30 m/s) alloys

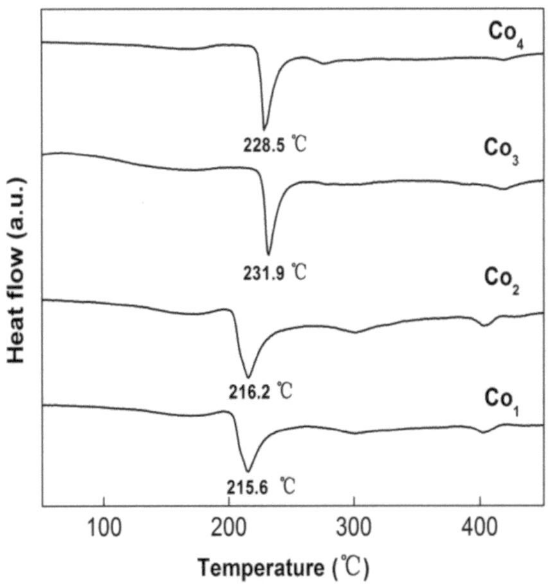

Figure 10. DSC profiles of as-spun (30 m/s) alloys

4.2.3. *Gaseous and electrochemical hydriding and dehydriding kinetics*

The hydrogen absorption kinetic curves of the as-spun alloys are plotted in Fig. 11. It is visible that all the as-spun alloys demonstrate rapid hydrogen absorption rates and nearly reach their saturation capacities in 10 min. The high hydrogen absorption capacities and fast hydrogen absorption rates of the Co_2 and Co_3 alloys (Fig. 11) may be associated with the increased cell volume resulting from the Co substitution. The hydrogen absorption capacity of the $MgCo_2$ phase is very low, which has resulted in the Co_4 alloy displaying a low hydrogen absorption capacity. The enhancement in the hydrogenation characteristics can be associated with the enhanced hydrogen diffusivity in the amorphous and nanocrystalline microstructures because the amorphous phase around the nanocrystalline leads to an easier access of hydrogen to the nano-sized grains, avoiding the long-range diffusion of hydrogen through an already formed hydride, which is often regarded as the slowest stage of absorption. According to the results reported by Orimo and Fujii [22], the distribution of the maximum hydrogen concentrations in three nanometer-scale regions, i.e. grain region, grain boundary region as well as amorphous region, have been experimentally determined to be 0.3% H in the grain region of Mg_2Ni, 4.0% H at the grain boundary and 2.2% H in the amorphous region. This indicates that the hydrides mainly exist in the grain-boundary region and the amorphous phase region.

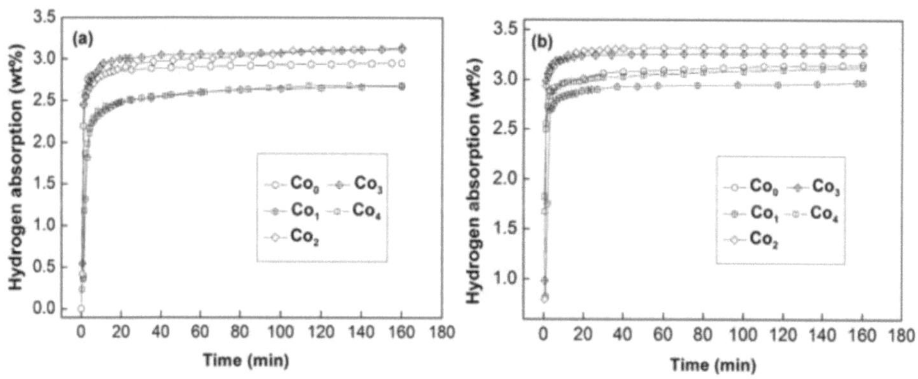

Figure 11. Hydrogen absorption kinetic curves of as-spun alloys: (a) 15 m/s, (b) 30 m/s

The hydrogen absorption kinetic curves of the as-cast and spun Co_1 and Co_4 alloys are plotted in Fig. 12. The results indicate that the hydrogen absorption property of the alloys is significantly enhanced by melt spinning. As the spinning rate grows from 0 to 30 m/s, the hydrogen absorption capacity in 10 min rises from 2.35% to 2.88% for Co_1 alloy, and from 1.91% to 2.96% for Co_4 alloy. These observations indicate that the hydrogenation kinetics and storage capacity of the as-spun nano-crystalline and amorphous Mg_2Ni-type alloys under investigation are superior to those of the conventional polycrystalline materials with similar compositions.

Fig. 13 depicts the hydrogen desorption kinetic curves of the as-spun alloys with spinning rates of 15 m/s and 30 m/s. It is quite visible that the Co substitution significantly ameliorates the hydrogen desorption capacity and kinetics of the as-spun alloys. As Co content (x) increases from 0 to 4, the hydrogen desorption capacity of the as-spun (15 m/s) alloy in 20 min rises from 0.51% to 1.72%, and from 0.89% to 2.15% for the as-spun (30 m/s) alloy. It is interesting to notice that the hydrogen absorption capacity (Fig. 11) of the as-spun Co_0 alloy is similar to that of the Co_4 alloy, but its hydrogen desorption capability is much lower than that of the Co_4 alloy. Therefore, it can be deduced that the substitution of Co for Ni improves the dehydriding performance of the Mg_2Ni-type alloy. The enhanced hydrogenation characteristics may be ascribed to two reasons. First, the Co substitution significantly intensifies the glass forming ability of Mg_2Ni-type alloy and the amorphous Co for Ni in Mg_2Ni compound lowers the stability of hydride and also Mg_2Ni exhibits an excellent hydrogen desorption capability [23]. Besides, the substitution of facilitates the desorption reaction [12,24].

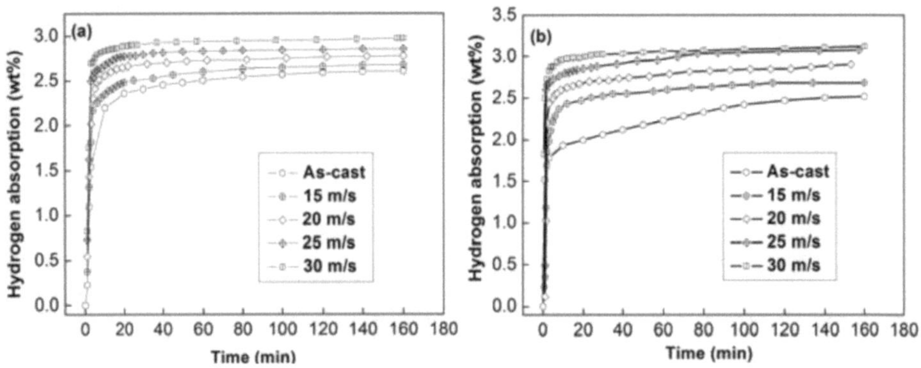

Figure 12. Hydrogen absorption curves of as-cast and spun alloys: (a) Co_1 alloy, (b) Co_4 alloy

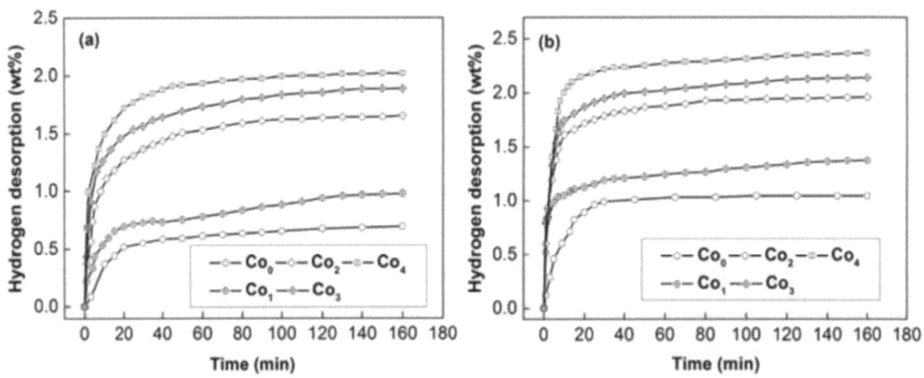

Figure 13. Hydrogen desorption curves of as-spun alloys with different spinning rates: (a) 15 m/s, (b) 30 m/s

Fig. 14 shows the hydrogen desorption kinetic curves of the Co_1 and Co_4 alloys with different spinning rates. The results clearly demonstrate that the dehydriding capability of the alloys increases with the increase of the spinning rate. As the spinning rate grows from 0 to 30 m/s, the hydrogen desorption capacity of the Co_1 alloy in 20 min increases from 0.39%

to 1.13%, and from 1.39% to 2.15% for the Co_4 alloy. In addition, it is clear that the as-cast and spun Co_1 alloy possesses inferior hydrogen desorption capacity and poorer dehydrogenation kinetics comparing to that of the Co_4 alloy. This behavior may be ascribed to the high stability of the crystal Mg-based hydride due to the fact that the melt spinning is unable to change the crystal state of the Co_1 alloy. Hence, the as-spun Co_4 alloy reveals high hydrogen desorption capacity and very fast dehydriding rate. It can be deduced from the aforementioned results that the substitution of Co for Ni significantly ameliorates the hydrogen absorption and desorption capacities and kinetics of the M_2Ni-type alloy. A similar result has already been reported by Khrussanova et al. [25]. The magnesium composites containing 10% and 15% (mass fraction) $Mg_{20}Ni_{10-x}Co_x$ (x=1 and 3) have been prepared by mechanical alloying and it has been found that an intermetallic addition exerts an evident catalytic effect on the hydriding of magnesium. Furthermore, it has been confirmed that the formation of $Mg_{20}Ni_8Co_2$ results in a much stronger catalytic action on the hydrogen absorption kinetics of magnesium than Mg_2Ni alone does [26]. The catalytic effect of the $Mg_{20}Ni_{10-x}Co_x$ additive may be ascribed to the inhomogeneity of the alloys containing a $MgNi_2$ phase along with cobalt and nickel clusters. This inhomogeneity is directly associated with the chemisorption of hydrogen and hence facilitates the hydriding reaction. The catalytic effect of Co has been well elaborated by Bobet et al. [27]. Ni is a largely known catalyst, which is utilized for the hydriding reactions. In this work, the catalytic effect of Co substitution on the hydrogen absorption kinetics of the as-spun Mg_2Ni alloy is ascribed to the enhanced glass forming ability caused by Co substitution. Lee et al. [28] have made efforts to improve the hydrogen sorption properties of Mg by mechanical grinding in H_2 (reactive grinding) with Co, finding that the addition of smaller particles of Co (0.5–1.5 μm) exerts a significant impact on the hydrogen sorption properties of Mg. It has been reckoned that Co with smaller particle sizes may act as a grain refiner for the magnesium. In addition, the expansion and contraction of the particles of Mg during the hydriding/dehydriding cycling also make them finer. Therefore, the addition of Co with smaller particle sizes as well as the hydriding/dehydriding cycling will aid the particles of magnesium to exhibit the higher hydrogen sorption rates. Based on the above mentioned discussion, it can be concluded that the substitution of Co for Ni produces a significant catalytic effect on the hydrogen absorption and desorption capacity and kinetics of Mg and Mg-based alloy. However, it must be pointed out that the action mechanism of Co additive is directly associated with the preparation technology of the alloy.

The electrochemical hydrogen storage kinetics of an alloy is symbolized by its high rate discharge ability (HRD). Fig. 15 describes the evolution of the HRD values of the alloys with the discharge current density. It is fully evident that the HRD values are notably enhanced by both melt spinning and Co substitution. In order to visually describe the influences of the melt spinning and Co substitution on the HRD values of the alloys, The HRD values (i=100 mA/g) as the functions of the spinning rate and the Co content are also plotted in Fig. 15, respectively. It can be derived from Fig. 15 that the HRD value (i=100 mA/g) of the Co_4 alloy is enhanced from 60.3 to 76.0% by increasing the spinning rate from 0 to 30 m/s, and that of the as-spun (20 m/s) alloys is markedly increased from 63.54 to 75.32% by rising Co content from 0 to 4. It has come to light clearly that high rate discharge ability (HRD) basically depends on the charge transfer at the alloy-electrolyte interface and the hydrogen diffusion process from the interior of the bulk to the surface of alloy particle [29]. The substitution of Co for Ni notably enhances the HRD values of the alloys, to be attributed to an impactful positive action of Co substitution on the hydrogen diffusion in the alloy. Furthermore, Co substitution accelerates the formation of a concentrated metallic Ni layer on the surface of the alloy electrode which is highly beneficial to enhance the electrochemical catalytic property and to improve the reaction rate of hydrogen [29]. The hydrogen diffusion coefficients in the as-cast and spun alloys were measured using the potential step technique. Fig. 16 depicts the semilogarithmic curves of anodic current versus working duration of the as-cast and spun alloys. It can be seen from Fig. 16 (a) that the melt spinning has a positive effect on the hydrogen diffusion in the Co_4 alloy. Fig.16 (b) exhibits that an increase in the Co content gives rise to a growth in the D value. The above-mentioned results clarify that hydrogen diffusion ability is a crucial factor of the hydrogen absorption kinetics of the alloy.

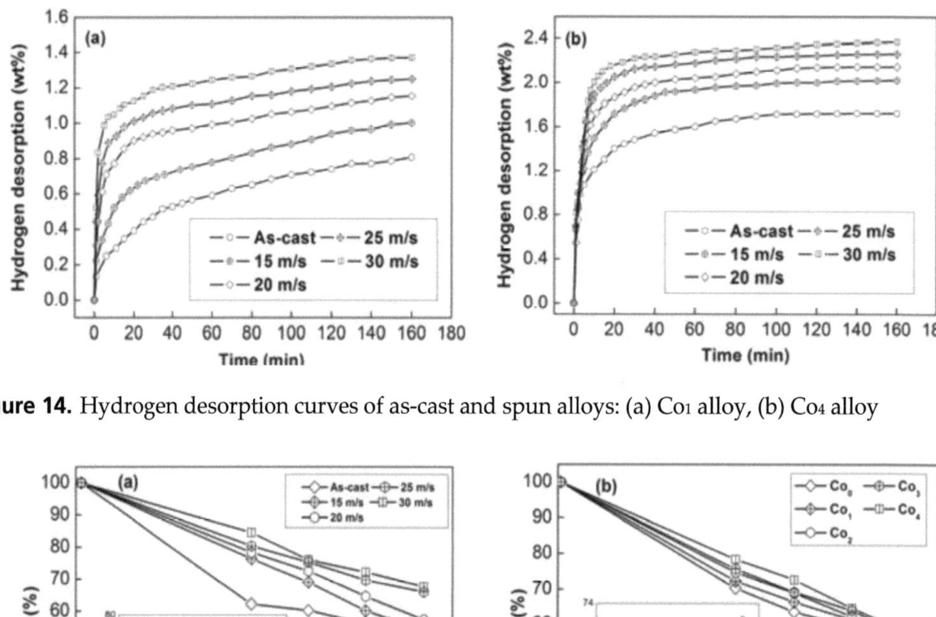

Figure 14. Hydrogen desorption curves of as-cast and spun alloys: (a) Co_1 alloy, (b) Co_4 alloy

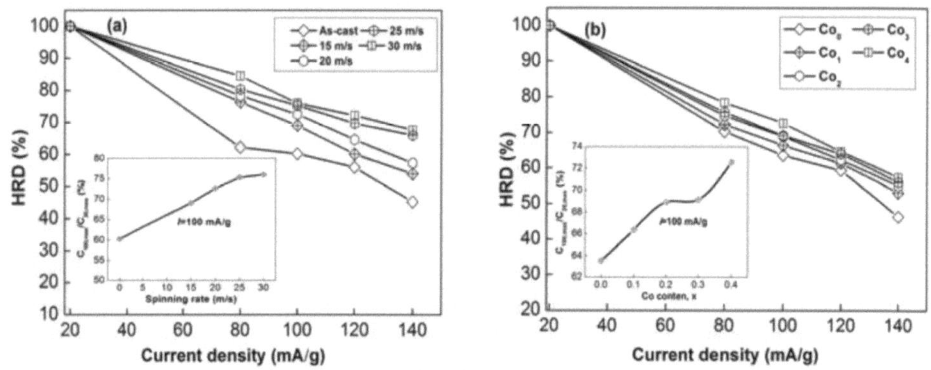

Figure 15. Evolution of the high rate discharge ability (HRD) of the alloys with the discharge current density: (a) Co_4 alloy, (b) As-spun (20 m/s)

Fig. 17 shows the electrochemical impedance spectra (EIS) of the as-spun (15 m/s) alloy electrodes at 50% DOD. It shows that each EIS spectrum contains two semicircles followed by a straight line. According to Kuriyama et al. [30], the smaller semicircle in the high frequency region is attributed to the contact resistance between the alloy powder and the conductive material, while the larger semicircle in the low frequency region is attributed to the charge-transfer resistance on the alloy surface. The linear response at low frequencies is indicative of hydrogen diffusion in the bulk alloy. Hence, the electrode kinetics of the as-spun alloys are dominated a mixed rate-determining process. It can be seen from Fig. 16 that the radius of the large semicircle in the low frequency visibly decreases with the increasing Co content, implying that the refined grain by Co substitution facilitates charge-transfer resistance of the alloy electrode.

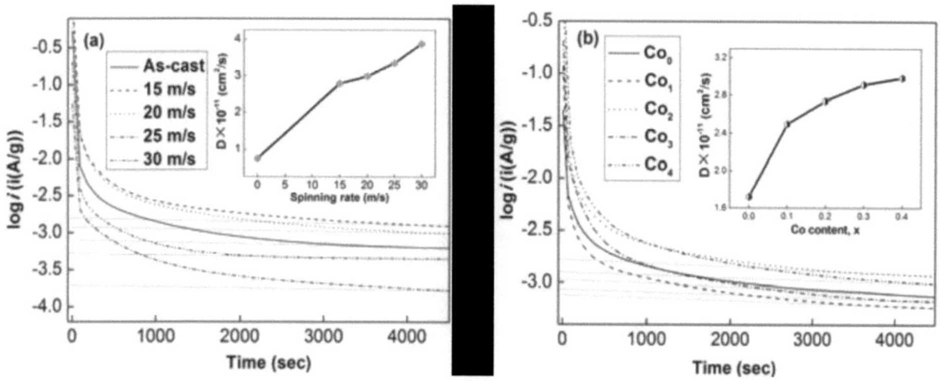

Figure 16. Semilogarithmic curves of anodic current vs. time responses of the alloys: (a) Co_4 alloy, (b) As-spun (20 m/s)

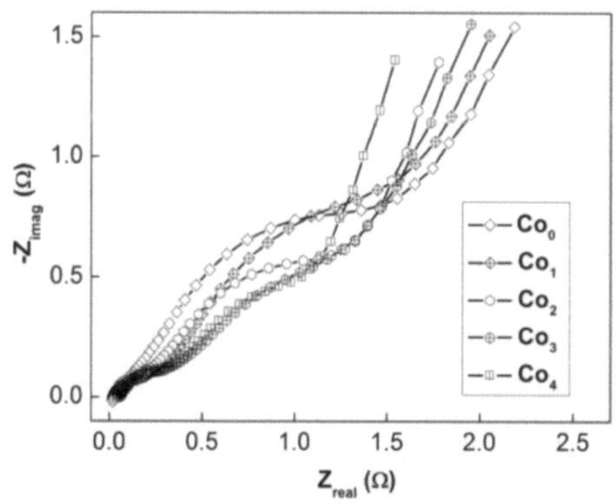

Figure 17. Electrochemical impedance spectra (EIS) of as-spun (15 m/s) alloy electrodes at 50% depth of discharge (DOD)

To determine the kinetics of hydrogen absorption/desorption, Tafel polarization measurements were carried out on the experimental alloy electrodes. Fig. 18 shows the Tafel polarization curves of the as-spun (15 m/s) alloy electrodes at the 50% DOD. It indicates that, in all cases, the anodic current densities increase to a limiting value, then decrease. The existence of a limiting current density, I_L, suggests the forming of an oxidation layer on the surface of the alloy electrode, which resists further penetration of hydrogen atoms [21]. The decrease of the anodic charge current density during cycling implies that charging is becoming more difficult. Hence, the limiting current density, I_L, may be regarded as a critical passivation current density, which limiting current density, which obtained from the

Tafel polarization curves is also presented in Fig. 18. It can be seen from Fig. 18 that I_L values of the alloys notably increase with rising of the Co content. With an increase in Co content from 0 to 4, the I_L value of the as-spun (15 m/s) alloy increases from 46.7 to 191.7 mA/g, indicating a higher rate of hydrogen diffusion caused by substituting Ni with Co. Based on the factors mentioned above, it can be concluded that the substitution of Co for Ni produces a significant improvement on the hydrogen storage kinetics of Mg_2Ni-type alloy. However, it must be pointed out that the action mechanism of Co substitution is directly associated with the preparation technology of the alloy.

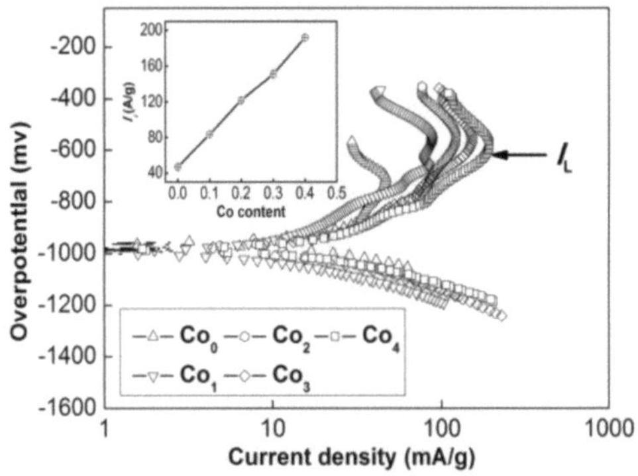

Figure 18. Tafel polarization curves of as-spun (15 m/s) alloy electrodes at 50% DOD and evolution of limiting current density (I_L) with Co content

4.3. The effect of substituting Ni with Cu on microstructure characteristics and gaseous as well as electrochemical hydriding and dehydriding kinetics

4.3.1. Microstructure characteristics

The XRD profiles of the as-cast and spun (25 m/s) alloys are shown in Fig. 19. It is evident that all the as-cast and spun alloys display a single phase structure. Both the substitution of

Cu for Ni and the melt spinning treatment do not change the Mg_2Ni major phase of the alloys. Based on the XRD data, the lattice parameters, cell volume and full width at half maximum (FWHM) values of the main diffraction peaks of the as-cast and spun (25 m/s) alloys were calculated by software of Jade 6.0. The obtained results are listed in Table 2. It is found from Table 2 that the melt spinning renders not only an evident enlargement in the lattice parameters and cell volume, but also a visibly increase in the FWHM values of the main diffraction peaks of the alloys, which is doubtless attributed to the refined average grain size and stored stress in the grains produced by melt spinning. Based on the FWHM values of the broad diffraction peak (203) in Fig. 19, the grain size D_{hkl} (nm) of the as-spun alloy was calculated using Scherrer's equation, which ranging from 2 to 6 nm, consistent with the results reported by Friedlmeier et al. [31].

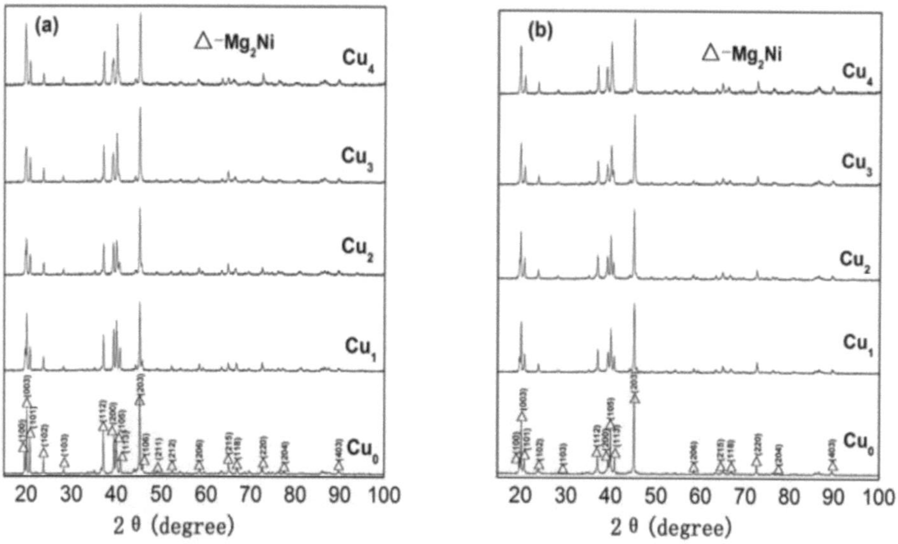

Figure 19. XRD patterns of the as-cast and spun alloys: (a) as-cast, (b) as-spun (25 m/s)

Alloys	FWHM values				Lattice parameters and cell Volume					
	2θ(20.02°)		2θ(45.14°)		a (nm)		c (Å)		V (Å³)	
	As-cast	25 m/s	As-cast	25 m/s	As-cast	25 m/s	As-cast	25 m/s	As-cast	25 m/s
Cu_0	0.122	0.131	0.169	0.179	5.2097	5.2105	13.244	13.265	311.29	311.88
Cu_1	0.133	0.241	0.178	0.237	5.2102	5.2158	13.252	13.277	311.54	312.79
Cu_2	0.148	0.258	0.183	0.242	5.2136	5.2172	13.283	13.311	312.67	313.76
Cu_3	0.151	0.282	0.192	0.259	5.2154	5.2185	13.297	13.319	313.22	314.11
Cu_4	0.165	0.292	0.204	0.273	5.2171	5.2210	13.302	13.323	313.54	314.50

Table 2. The lattice parameters, cell volume and the FWHM values of the major diffraction peaks of the alloys

The SEM images of the as-cast alloy are presented in Fig. 20, displaying a typical dendrite structure. The substitution of Cu for Ni renders an evident refinement of the grains instead of changing the morphology of the alloys. The result obtained by EDS indicates that the major phase of the as-cast alloys is Mg_2Ni phase (denoted as A). Some small massive matters in the alloys substituted by Cu can clearly be seen in Fig. 20, which are determined by EDS to be Mg_2Cu phase (denoted as B), this apparently contrary to the result of XRD observation. It is most probably associated with the fact that the amount of the Mg_2Cu phase is rare so that the XRD observation can not detect it.

The HRTEM micrographs and electron diffraction patterns of the as-spun (30 m/s) Cu_2 and Cu_4 alloys are illustrated in Fig. 21, exhibiting a nanocrystalline microstructure with an average crystal size of about 2 nm to 5 nm. It is found from HRTEM observations that the as-spun alloys are strongly disordered and nano-structured, but no amorphous phase is detected, which consistent with the XRD observation. The crystal defects in the as-spun alloy, stacking faults (denoted as A), twin-grain boundary (denoted as B), dislocations (denoted as C) and sub-grain boundary (denoted as D), are clearly viewable in Fig. 22.

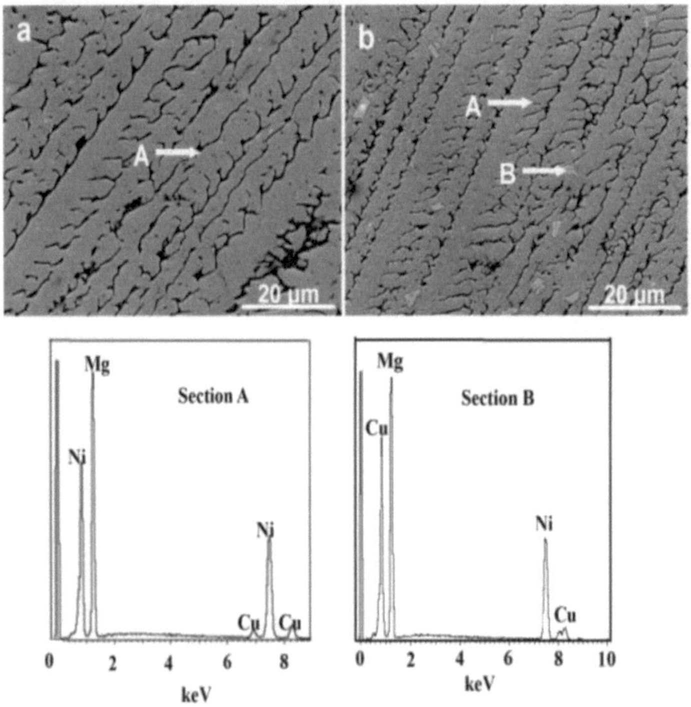

Figure 20. SEM images of the as-cast alloys together with typical EDS spectra of sections A and B in Fig.2b: (a) Cu_0 alloy, (b) Cu_3 alloy

Figure 21. HRTEM images and ED patterns of the as-spun (30 m/s) alloys: (a) Cu_2 alloy, (b) Cu_4 alloy

4.3.2. Gaseous and electrochemical hydriding and dehydriding kinetics

The hydrogen absorption kinetic curves of the as-cast and spun (25 m/s) alloys are plotted in Fig. 23. It is quite evident that the hydrogen absorption capacity of the alloys is visibly enhanced by substituting Ni with Cu. It is noteworthy that, as the amount of Cu substitution increased to 3, such substitution causes a decline of hydrogen absorption capacity, which primarily attributed to the increase of the $MgCu_2$ phase due to the fact the hydrogen absorption capability of the $MgCu_2$ phase is very low. It is viewed from Fig. 23 (a) that the substitution of Cu for Ni markedly ameliorates the hydrogen absorption kinetics of the as-cast alloys, to be ascribed to the increased cell volume and the refined grain caused by Cu substitution in virtue of the grain boundary possessing the largest hydrogen absorption capability [22].

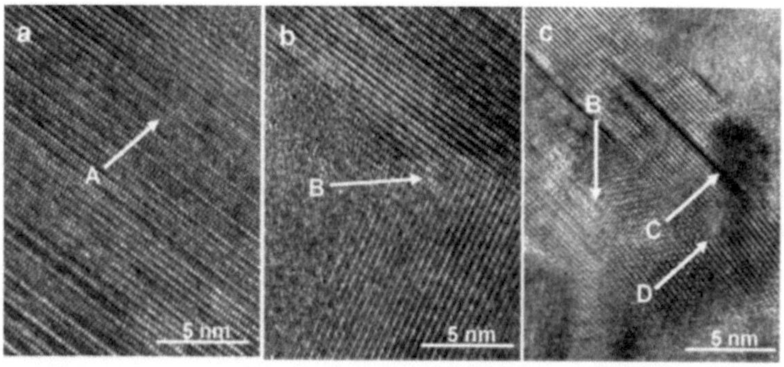

Figure 22. defects in the as-spun (30 m/s) Cu_4 alloy taken by HRTEM: (a) stacking fault, (b) twin-grain boundary, and (c) dislocations and sub-grain boundaries

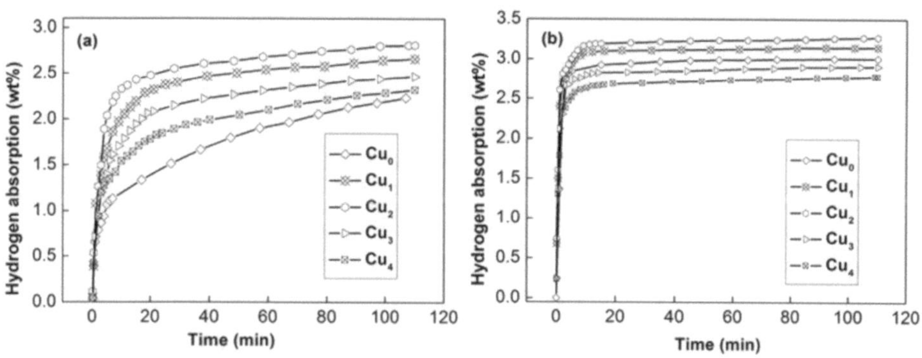

Figure 23. Hydrogen absorption kinetic curves: (a) as-cast, (b) as-spun alloys(25 m/s)

The hydrogen absorption capacity and kinetics of the as-cast and spun Cu_1 and Cu_3 alloys are shown in Fig. 24. It is viewable that the melt spinning evidently improves the hydrogen absorption property of the alloys. The hydrogen absorption capacity in 10 min is increased from 1.99 to 3.12 wt% for Cu_1 alloy, and from 1.74 to 2.88 wt% for the Cu_3 alloy by the increasing spinning rate form 0 (As-cast was defined as spinning rate of 0 m/s) to 30 m/s. It is fully evident that the hydrogen absorption capacity kinetics of all the as-spun nanocrystalline and amorphous Mg_2Ni-type alloys studied are superior to those of conventional polycrystalline materials with the same composition. The improved hydrogenation characteristics are attributed to the enhanced hydrogen diffusivity in the nanocrystalline microstructure as the nanocrystalline leads to an easier access of hydrogen to the nano-sized grains, avoiding the long-range diffusion of hydrogen through an already formed hydride, which is often the slowest stage of absorption [22]. It is known that the nanocrystalline microstructures can accommodate higher amounts of hydrogen than the polycrystalline ones. The large number of interfaces and grain boundaries available in the nanocrystalline materials provide easy pathways for hydrogen diffusion and promote the absorption of hydrogen.

Fig. 25 describes the influence of substituting Ni with Cu on the hydrogen desorption kinetics of the as-cast and spun alloys. It is observable that both the hydrogen desorption capacity and the kinetics of the alloys increase with the growing amount of Cu substitution. With the increase in Cu content from 0 to 4, the hydrogen desorption capacity in 10 min rises from 0.11 to 0.56 wt% for the as-cast alloy, and from 0.56 to 1.31 wt% for the as-spun (25 m/s) alloy. The enhanced hydrogen desorption kinetics can be ascribed to the fact that the substitution of Cu for Ni in the Mg_2Ni compound decreases the stability of the hydride and makes the desorption reaction easier [12,24]. A similar result was reported by Simićić et al. [32].

The hydrogen desorption kinetic curves of the as-cast and spun alloys are shown in Fig. 26. The figures display that the dehydriding capability of the alloys notably meliorated with the rising of the spinning rate. The hydrogen desorption capacity in 10 min is increased from 0.21 to 0.83 wt% for the Cu_1 alloy, and from 0.44 to 1.29 wt% for the Cu_3 alloy by enhancing of spinning rate from 0 to 30 m/s. It is found that the as-spun nanocrystalline Mg_2Ni alloy exhibits a superior hydrogen desorption kinetics comparing to the crystalline Mg_2Ni, which consistent with the result reported by Spassov et al. [15]. The observed essential differences in the hydriding/dehydriding kinetics of the melt-spun nanocrystalline Mg_2Ni type alloys studied are most probably being associated with the composition of the alloys as well as the differences in their microstructure due to the different spinning rates. It was reported that the high surface to volume ratios, i.e. high specific surface area, and the presence of large numbers of grain boundaries in nanocrystalline alloys enhance the hydriding and dehydriding kinetics [24]. Zaluski et al. [33] and Orimo et al. [34] validated that the hydriding/dehydriding kinetics of the milled nanocrystalline Mg_2Ni alloys at low temperatures (lower than 200 °C) can be improved by reducing the grain size, due to the fact that hydrogen atoms mainly occupied in the disordered interface phase and the grain boundary.

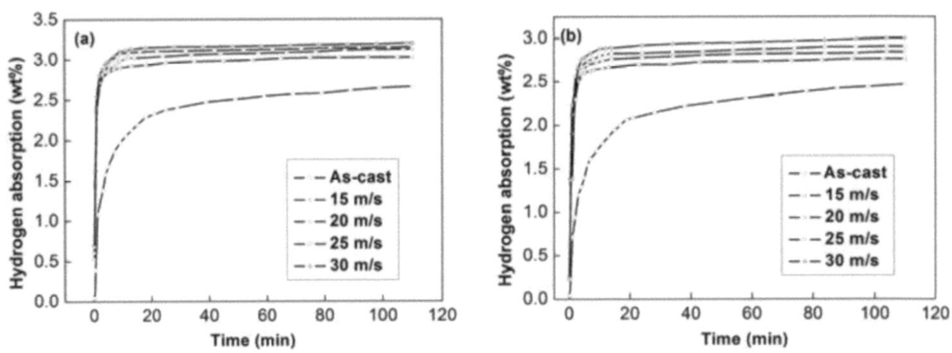

Figure 24. Hydrogen absorption kinetic curves of the as-cast and spun alloys: (a) Cu_1 alloy, (b) Cu_3 alloy

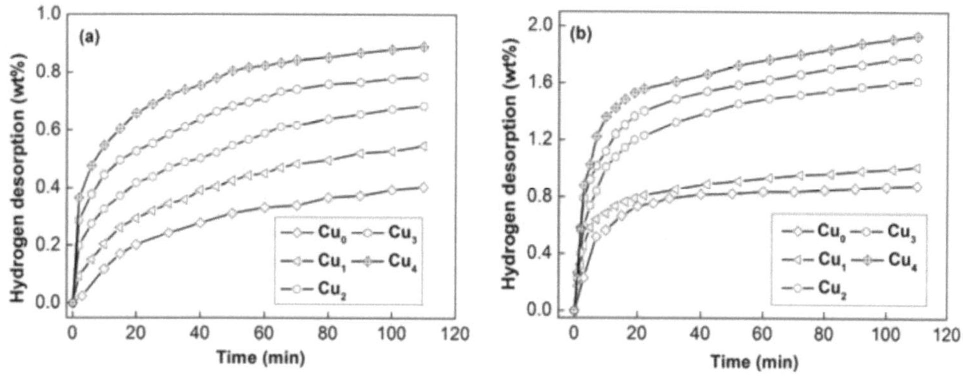

Figure 25. Hydrogen desorption kinetic curves : (a)the as-cast, (b)as-spun alloys (25 m/s)

Fig. 27 describes the evolution of the HRD values of the alloys with the discharge current density. It shows that the HRD values are visibly enhanced by both Cu substitution and melt spinning. In order to visually indicate the influences of the Cu substitution and melt spinning on the HRD values of the alloys, the HRD values (*i*=100 mA/g) as the functions of the Cu content and the spinning rate are also plotted in Fig. 27, respectively. It can be derived from Fig. 27 that the HRD value (*i*=100 mA/g) of the as-spun (30 m/s) alloys is markedly increased from 62.4 to 78.9% by rising the Cu content from 0 to 4, and that of the Cu_4 alloy is enhanced from 44.8 to 78.9% by increasing the spinning rate from 0 to 30 m/s.

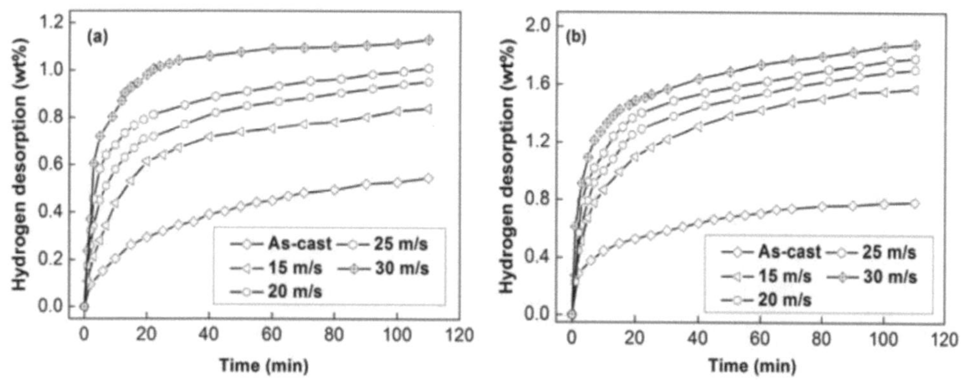

Figure 26. Hydrogen desorption kinetic curves of the as-cast and spun alloys: (a) Cu_1 alloy, (b) Cu_3 alloy

It has come to light clearly that high rate discharge ability (HRD) is basically dominated by the hydrogen diffusion process from the interior of the bulk to the surface of alloy particle and the charge transfer at the alloy-electrolyte interface [35]. The hydrogen diffusion coefficients in the as-cast and spun alloys were measured using the potential step technique. Fig. 28 presents the semilogarithmic curves of anodic current versus working duration of the as-cast and spun alloys. The D values calculated by Eq. (2) are also presented in Fig. 28. It indicates that an increase in the Cu content turns out a growth in the D value. It can be seen in Fig.28 (b) that the melt spinning has a positive effect on the hydrogen diffusion in the Cu4 alloy. The above-mentioned results clarify that hydrogen diffusion ability is a crucial factor of the hydrogen absorption kinetics of the alloy.

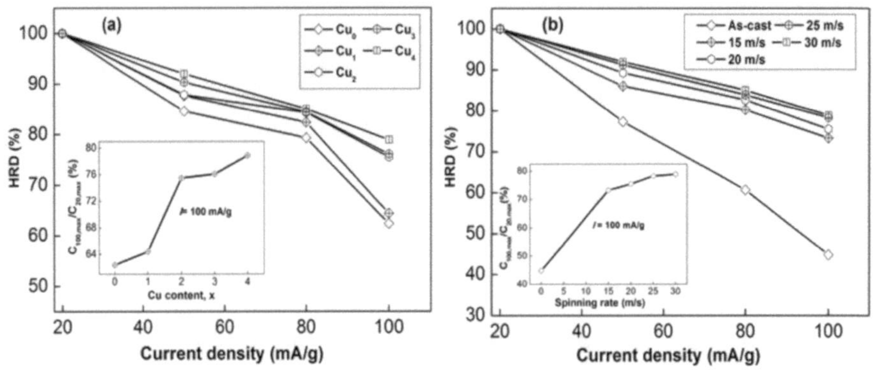

Figure 27. Evolution of the high rate discharge ability (HRD) of the alloys with the discharge current density: (a) As-spun (30 m/s), (b) Cu4 alloy

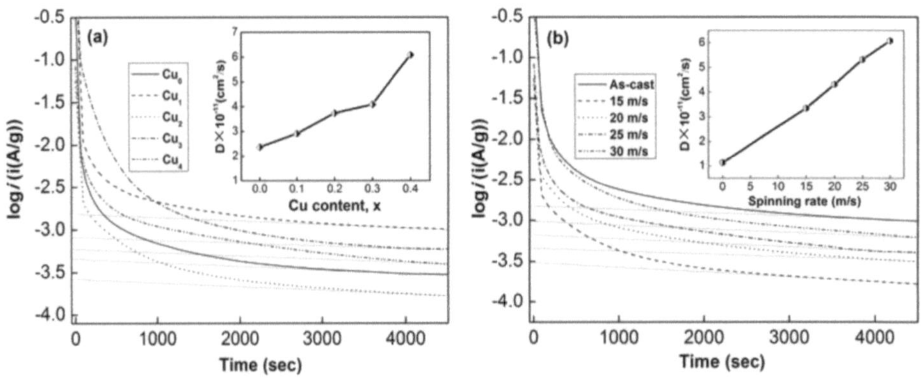

Figure 28. Semilogarithmic curves of anodic current vs. time responses of the as-cast and spun alloys: (a) as-spun (30 m/s), (b) Cu4 alloy

5. Conclusion

The Mg_2Ni-type $Mg_{20-x}La_xNi_{10}$ (x=0-6) and $Mg_{20}Ni_{10-x}M_x$ (M= Co, Cu; x=0~4) alloys with a nanocrystalline and amorphous structure were successfully fabricated by melt spinning technology. We looked carefully at the SEM, XRD, HRTEM images of the as-cast and spun alloys and did find the formation of amorphous structure in alloys is which attributed to the substitute element and the melt spinning. Our investigation indicated that the substitution of La for Mg and Co for Ni notably enhances the glass forming ability of the Mg_2Ni-type. The amorphized degree of the alloys is associated with the spinning rate. The higher the spinning rate is, the larger the amount of the amorphous phase will be. The quenching rate induced a light influence on the crystallization temperature of the amorphous phase, and it significantly improved the initial hydrogenation rate and the hydrogen absorption capacity of the alloys.

1. Rapid quenching caused an insignificant change of the phase structure of the La_0 alloy, but it led to a great variety of the phase structure of the La_2 alloy. There was no amorphous phase in the as-quenched La_0 alloy, whereas the as-quenched La_2 alloy presented a feature of the nanocrystalline embedded in the amorphous matrix, which ascribed to the fact that the substitution of La for Mg significantly enhanced the glass forming ability of the $Mg_{20-x}La_xNi_{10}$ (x = 0, 2) hydrogen storage alloys. Rapid quenching markedly increased the hydrogen absorption and desorption capacities and kinetics of the $Mg_{20-x}La_xNi_{10}$ (x = 0, 2) alloys, which mainly ascribed to the nanocrystalline and amorphous microstructures formed by rapid quenching.
2. The substitution of Co for Ni does not alter the major phase of Mg_2Ni but results in the formation of the secondary phase $MgCo_2$. No amorphous phase is detected in the as-spun Co-free alloy, but a certain amount of amorphous phase is clearly found in the as-spun Co-containing alloys. The substitution of Co for Ni exerts a slight influence on the hydriding kinetics of the as-spun alloy. However, it dramatically enhances the dehydriding kinetics of the as-cast and spun alloys.
3. The substitution of Cu for Ni leads to the formation of the secondary phase Mg_2Cu in the as-cast alloys instead of changing of the Mg_2Ni-type major phase in the alloy. Additionally, Cu substitution visibly refines the grains of the as-cast alloy, whereas it causes an imperceptible impact on the glass forming ability of the alloy. With the increase in the amount of Cu substitution, the hydrogen absorption capacity of the as-cast Mg_2Ni-type alloys first increases and then decreases. But it markedly improves the hydrogen desorption capability of the as-cast and spun alloys. Melt spinning evidently promotes the hydriding and dehydriding performances of the Mg_2Ni-type alloys. Hydriding/dehydriding capacities and rates of the alloys markedly rise with the increasing of the spinning rate.

Author details

Yanghuan Zhang, Hongwei Shang, Chen Zhao and Dongliang Zhao
Department of Functional Material Research, Central Iron and Steel Research Institute, China

Yanghuan Zhang, Hongwei Shang and Chen Zhao
Elected State Key Laboratory, Inner Mongolia University of Science and Technology, China

6. References

[1] Das, D., Veziroglu, TN. Hydrogen production by biological processes: a survey of literature. International Journal of Hydrogen Energy 2001; 26 (1) 13-28.

[2] Frackowiak, E., Be´guin, F. Electrochemical storage of energy in carbon nanotubes and nanostructured carbons. Carbon 2002; 40 (10) 1775-1787.

[3] Reilly, JJ., Adzic, GD., Johnson, JR., Vogt, T., Mukerjee, S., Mcbreen, J. The correlation between composition and electrochemical properties of metal hydride electrodes. Journal of Alloys and Compounds 1999; 293–295 569-582.

[4] Jain, IP., Lal, C., Jain, A. Hydrogen storage in Mg: A most promising material. International Journal of Hydrogen Energy 2010; 35 (10) 5133-5144.

[5] Jain, IP. Hydrogen the fuel for 21st century. International Journal of Hydrogen Energy 2009; 34 (17) 7368-7378.

[6] Kwon, SN., Baek, SH., Mummb, DR., Hong, SH., Song, MY. Enhancement of the hydrogen storage characteristics of Mg by reactive mechanical grinding with Ni, Fe and Ti. International Journal of Hydrogen Energy 2008; 33 (17) 4586-4592.

[7] Palade, P., Sartori, S., Maddalena, A., Principi, G., Lo, Russo S, Lazarescu M, Schinteie G, Kuncser V, Filoti G. Hydrogen storage in Mg–Ni–Fe compounds prepared by melt spinning and ball milling. Journal of Alloys and Compounds 2006; 415 (1-2) 170-176.

[8] Spassov, T., Köster, U. Hydrogenation of amorphous and nanocrystalline Mg-based alloys. Journal of Alloys and Compounds. 1999; 287 (1-2) 243–250.

[9] Orimo, S., Fujii, H. Hydriding properties of the Mg_2Ni-H system synthesized by reactive mechanical grinding. Journal of Alloys and Compounds 1996; 232 (1-2) L16–L19.

[10] Inoue, A., Masumoto, T. Mg-based amorphous alloys. Materials Science and Engineering A. 1993; 173 (1-2) 1–8.

[11] Kim, S.G., Inoue, A., Masumoto, T. High Mechanical Strengths of Mg–Ni–Y and Mg–Cu–Y Amorphous Alloys with Significant Supercooled Liquid Region. Materials Transactions, JIM 1990; 31 (11) 929–934.

[12] Woo, JH., Lee, KS. Electrode characteristics of nanostructured Mg_2Ni-type alloys prepared by mechanical alloying. Journal of the Electrochemical Society 1999; 146 (3) 819-823.

[13] Liang, G. Synthesis and hydrogen storage properties of Mg-based alloys. Journal of Alloys and Compounds 2004; 370 (1-2) 123–128.

[14] Song, MY., Kwon, SN., Bae, JS., Hong, SH. Hydrogen-storage properties of Mg–23.5Ni–(0 and 5)Cu prepared by melt spinning and crystallization heat treatment. International Journal of Hydrogen Energy 2008; 33 (6) 711-1718.

[15] Spassov, T., Köster, U. Thermal stability and hydriding properties of nanocrystalline melt-spun $Mg_{63}Ni_{30}Y_7$ alloy. Journal of Alloys and Compounds 1998; 279 (2) 279–286.

[16] Huang, LJ., Liang, GY., Sun, ZB., Wu, DC. Electrode properties of melt-spun Mg–Ni–Nd amorphous alloys. Journal of Power Sources 2006; 160 (1) 684-687.

[17] Yamaura, SI., Kim, HY., Kimura, H., Inoue, A., Arata, Y. Thermal stabilities and discharge capacities of melt-spun Mg–Ni-based amorphous alloys. Journal of Alloys and Compounds. 2002; 339 (1-2) 230-235.

[18] Chen, H.S. Thermodynamic considerations on the formation and stability of metallic glasses. Acta Metallurgica 1974; 22(12) 1505-1511.

[19] Spassov, T., Solsona, P., Suri˜nach, S., Baró, MD. Optimisation of the ball-milling and heat treatment parameters for synthesis of amorphous and nanocrystalline Mg_2Ni-based alloys. Journal of Alloys and Compounds 2003; 349 (1-2) 242–254.

[20] Zheng, G., Popov, BN., White, RE. Electrochemical Determination of the Diffusion Coefficient of Hydrogen Through an $LaNi_{4.25}Al_{0.75}$ Electrode in Alkaline Aqueous Solution. Journal of the Electrochemical Society 1995; 142 (8) 2695-2698.

[21] Niu, H., Derek, ON. Enhanced electrochemical properties of ball-milled Mg_2Ni electrodes. International Journal of Hydrogen Energy 2002; 27 (1) 69-77.

[22] Orimo, S., Fujii, H. Materials science of Mg-Ni-based new hydrides. Applied Physics A, 2001; 72 (2) 167–186.

[23] Goo, NH., Jeong, WT., Lee, KS. The hydrogen storage properties of new Mg_2Ni alloy. Journal of Power Sources 2000; 87 (1-2) 118-124.

[24] Takahashi, Y., Yukawa, H., Morinaga, M. Alloying effects on the electronic structure of Mg_2Ni intermetallic hydride. Journal of Alloys and Compounds 1996; 242 (1-2) 98–107.

[25] Khrussanova, M., Grigorova, E., Bobet, J.-L., Khristov, M., Peshev, P. Hydrogen sorption properties of the nanocomposites Mg-$Mg_2Ni_{1-x}Co_x$ obtained by mechanical alloying. Journal of Alloys and Compounds 2004; 365 (1-2) 308–313.

[26] Khrussanova, M., Mandzhukova, T., Grigorova, E., Khristov, M., Peshev, P. Hydriding properties of the nanocomposite 85wt.%Mg-15wt.% $Mg_2Ni_{0.8}Co_{0.2}$ obtained by ball milling. Journal of Materials Science, 2007; 42 (10) 3338–3342.

[27] Bobet JL., Chevalier B., Darriet B. Effect of reactive mechanical grinding on chemical and hydrogen sorption properties of the Mg+10wt.%Co mixture. Journal of Alloys and Compounds 2002; 330-320 738-742.

[28] Lee, DS., Kwon, IH., Bobet, JL., Song, MY. Effects on the H_2-sorption properties of Mg of Co (with various sizes) and CoO addition by reactive grinding. Journal of Alloys and Compounds 2004; 366 (1-2) 279–288.

[29] Gasiorowski A., Iwasieczko W., Skoryna D., Drulis H., Jurczyk M. Hydriding properties of nanocrystalline Mg2–xMxNi alloys synthesized by mechanical alloying (M=Mn, Al). Journal of Alloys and Compounds 2004; 364 (1-2) 283–288.

[30] Kuriyama, N., Sakai T., Miyamura H., Uehara I., Ishikawa H., Iwasaki T. Electrochemical impedance and deterioration behavior of metal. Journal of Alloys and Compounds 1993; 202 (1-2) 183–197.

[31] Friedlmeier, G., Arakawa, M., Hiraia, T., Akiba, E. Preparation and structural, thermal and hydriding characteristics of melt-spun Mg-Ni alloys. Journal of Alloys and Compounds 1999; 292 (1) 107–117.

[32] Simićić, MV., Zdujić, M., Dimitrijević, R. Hydrogen absorption and electrochemical properties of Mg_2Ni-type alloys synthesized by mechanical alloying. Journal of Power Sources 2006; 158 (1) 730-734.

[33] Zaluski, L., Zaluska, A., Strön-Olsen, JO. Nanocrystalline metal hydrides. Journal of Alloys and Compounds 1997; 253-254 70-79.

[34] Orimo, S., Fujii, H., Ikeda, K. Notable hydriding properties of a nanostructured composite material of the Mg_2Ni-H system synthesized by reactive mechanical grinding. Acta Materialia 1997; 45 (1) 331-341.

[35] Wu, Y., Han, W., Zhou, SX., Lototsky, MV., Solberg, JK., Yartys, VA. Microstructure and hydrogenation behavior of ball-milled and melt-spun Mg-10Ni-2Mm alloys. Journal of Alloys and Compounds 2008; 466 (1-2) 176–181.

Chapter 5

Hydrogen Storage Properties and Structure of Magnesium-Based Alloys Prepared with Melt-Spinning Technique

Kazuhide Tanaka

Additional information is available at the end of the chapter

http://dx.doi.org/10.5772/50170

1. Introduction

Magnesium is known to have a relatively high hydrogen storage capacity of 7.6 wt. % by forming a non-metallic hydride MgH_2, and has long been a target of many research works for developing excellent hydrogen storage materials. However, owing to very small hydrogen solubility and diffusivity in metallic Mg, together with very low chemical reactivity of its surface with hydrogen gas, the hydrogenation (and dehydrogenation) rate of Mg is generally quite slow, and temperatures above ~400°C and gas pressures exceeding ~10 MPa are required to cause the reaction. Several techniques have been developed to overcome this drawback of Mg without significantly reducing its hydrogen capacity. Among them, ball-milling and melt-spinning techniques appear to be most important. In the former, commercially available MgH_2 powder is mixed with a small amount of transition metal powder (Ti, V, Mn, Fe, Ni, Nb, Pd, etc., and their oxides or fluorides) and ball-milled for appropriate time periods. The mechanically milled powder exhibits an excellent reactivity with hydrogen, absorbing H_2 gas even at ~100°C or lower temperatures, and desorbing it above ~200°C. Here, the transition metal additives play crucially important roles by working as chemical catalysts for the abs/des reactions on the surface of micron-size Mg powder (Barkhordarian et al., 2004; Liang et al., 1999; Zaluska et al., 1999). On the other hand, in the latter, Mg alloyed with small amounts of transition metal (Ni, Pd, etc.) and/or rare earth (La, Ce, Nd, etc.) is melt-spun to form a thin metallic ribbon of normally amorphous structure. Upon crystallization, it changes into a stable multi-phase nanostructure where nano-sized grains of the alloying elements or their compounds are homogeneously precipitated in a nano-crystalline Mg matrix. The ribbon thus produced also exhibits an excellent reactivity with hydrogen. Here, nano-sized precipitates promote the chemisorption of H_2 molecules on the ribbon surface and enhance the flow of H atoms into the interior to form alloy hydrides (Spassov & Köster, 1999; Tanaka et al., 1999).

We have studied the hydrogen storage properties and structures of melt-spun and crystallized Mg-(Ni, Pd)-(La, Nd) alloys using several experimental and analytical techniques, and have made clear the absorption and desorption processes of hydrogen in the alloys (Tanaka, 2008; Tanaka et al., 2009; Yamada et al., 2001; Yin & Tanaka, 2002). In this chapter, some of our results specifically on Mg-Ni-La alloy, in comparison with Mg-Ni alloy, are reviewed and desorption mechanisms of hydrogen in these alloys are discussed. This chapter is composed of the following sections. In section 2, experimental procedures for the sample preparation with melt-spinning technique and for hydriding/dehydriding (H/D) measurements are described, and phase structures of hydrogenated/dehydrogenated samples characterized by X-ray diffraction (XRD) are shown. In section 3, equilibrium properties and kinetic behavior of hydrogen in the alloys as revealed by pressure-composition isotherms (PCT) and H/D rates, respectively, are described. In section 4, thermal desorption spectra (TDS) of samples subjected to various H/D treatments are provided, where component TDS peaks are related with existing hydride phases in the alloy. Section 5 is devoted to transmission electron microscope (TEM) observation of nanostructures of hydrogenated and dehydrogenated samples. High-resolution TEM (HRTEM) combined with electron energy-loss spectroscopy (EELS) is used to identify the metallic and hydride phases in the samples. An EELS plasmon peak (H-plasmon) of MgH_2 is successfully used to visualize the formation and destruction of the hydride phase in the sample. In section 6, the H-plasmon peak is applied to an in situ observation of the desorption process of the MgH_2 phase during heating. Finally, in section 7, concluding remarks are given, where the importance of the role of nano-grain boundaries as pathways of hydrogen flowing into and out of the nanostructured alloy is emphasized.

2. Sample preparation and phase-structure characterization

Magnesium-rich alloys with nominal compositions, $Mg_{85}Ni_{10}La_5$ and $Mg_{90}Ni_{10}$ (in at.%), have been prepared in this study. Appropriate amounts of raw materials (99.9% pure Mg and 99.9% pure La) and a mother alloy ($Mg_{70}Ni_{30}$) are mixed together, melted at 900°C in sealed steel crucibles under high purity Ar atmosphere and slowly cooled. The chemical compositions of the ingots obtained by SEM-EDX analyses are $Mg_{89.2}Ni_{7.6}La_{3.2}$ and $Mg_{89.5}Ni_{10.5}$. Amorphous ribbons (~20 μm thick and ~1 mm wide) are prepared from these ingots with a melt-spinning technique using a single Cu roll under Ar atmosphere. They are cut into small pieces, slightly ground and subjected to H/D measurements using a Sieverts-type PCT apparatus. The crystallization temperature T_x for these amorphous alloys as determined by DSC measurements is 162±5°C, which is well below the initial activation temperature 300°C adopted for hydrogenation in this study, hence all the H/D measurements have been performed for the crystallized alloys.

The SEM photographs for the as-cast $Mg_{90}Ni_{10}$ and $Mg_{85}Ni_{10}La_5$ alloys shown in Fig. 1 indicate that a primary dendritic phase of Mg (α) is coexistent with Mg (α)/Mg_2Ni (γ) eutectic phases in the former, whereas a primary equiaxial phase of Mg_2Ni (γ) is coexistent with both the Mg/Mg_2Ni eutectic phases and a $Mg_{17}La_2$ compound phase in the latter. These phase structures are consistent with the equilibrium phase diagrams of Mg-Ni (Massalski,

1986) and Mg-Ni-La (De Negri at al.., 2005) systems. Since the grains of all the constituent phases are so coarse that the initial activation and subsequent H/D processes of the as-cast alloys inevitably become sluggish. However, this difficulty is overcome substantially by refining the microstructure employing higher cooling rates of casting (Yamada et al., 2001; Yin et al., 2002). The melt-spinning technique combined with the crystallization treatment employed here is a quite efficient method for attaining the grain refining. An equal channel angular pressing (ECAP) treatment has also been applied for the grain refining in a Mg-Ni-Mm alloy (Løken et al., 2006), and a high hydrogen absorbency exceeding 5 wt%H with moderate H/D kinetics has been attained.

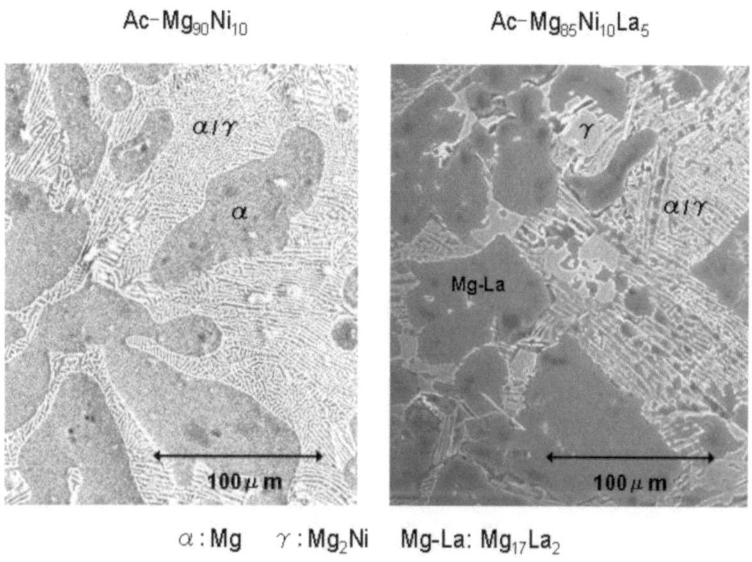

Figure 1. SEM photographs of as-cast $Mg_{90}Ni_{10}$ and $Mg_{85}Ni_{10}La_5$ alloys.

The phase structures of melt-spun $Mg_{90}Ni_{10}$ and $Mg_{85}Ni_{10}La_5$ alloys have been examined for the samples subjected to crystallization annealing, hydriding and dehydriding treatments at 300°C using X-ray diffraction with Cu Kα radiation, as shown in Fig. 2. After crystallization, the $Mg_{90}Ni_{10}$ alloy is composed of Mg + Mg_2Ni phases similarly to the as-cast alloy. After hydriding to 4.8 wt%H, β-MgH_2 + Mg_2NiH_4 phases appear, with a part of the Mg_2Ni phase left unchanged. The Mg_2NiH_4 phase is separated to the low-temperature (LT) and high-temperature (HT) phases. The γ-MgH_2 phase, which commonly exists in ball-milled MgH_2 (Liang et al., 1999), is not formed in the melt-spun alloy. After dehydriding, the alloy recovers to the original Mg + Mg_2Ni phases. In the subsequent H/D processes, the alloy phases cyclically change as:

$$Mg + Mg_2Ni + H_2 \leftrightarrow MgH_2 + Mg_2NiH_4 \quad (Mg_{90}Ni_{10} \text{ alloy})$$

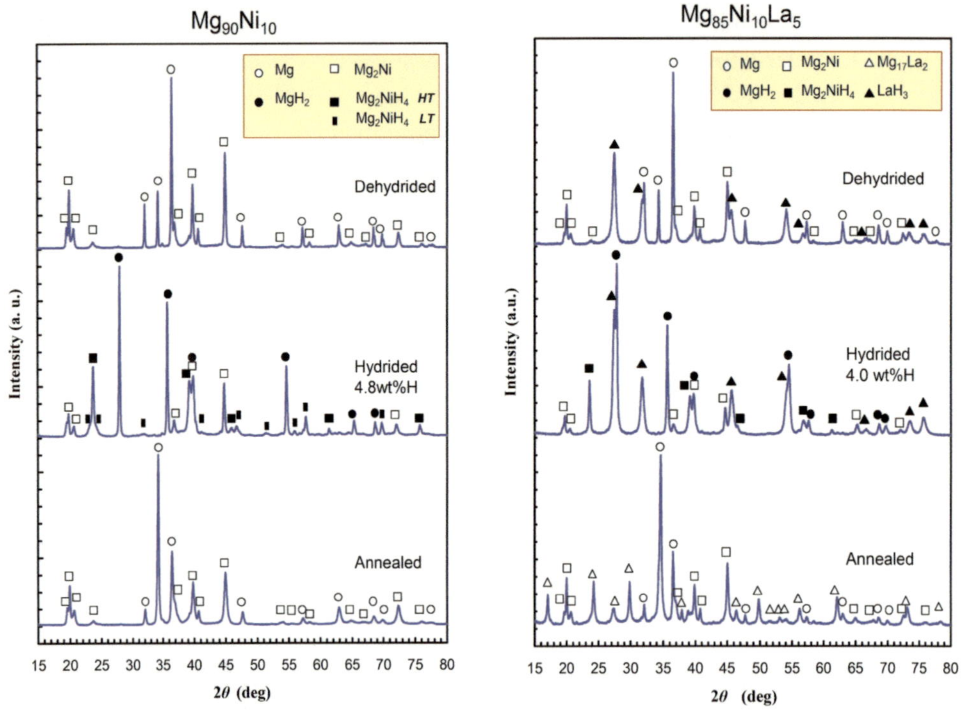

Figure 2. X-ray diffraction patterns (Cu Kα) of melt-spun $Mg_{90}Ni_{10}$ and $Mg_{85}Ni_{10}La_5$ alloys subjected to crystallization annealing, hydriding and dehydriding at 300°C.

On the other hand, the $Mg_{85}Ni_{10}La_5$ alloy is composed of Mg + Mg_2Ni + $Mg_{17}La_2$ phases after crystallization similarly to the as-cast alloy. Upon hydriding to 4.0 wt%H, these metallic phases transform into β-MgH_2 + Mg_2NiH_4 (HT) + LaH_3 hydride phases. In this process, the $Mg_{17}La_2$ phase is decomposed to MgH_2 + LaH_3 through a disproportionation reaction. A part of the Mg_2Ni phase is remaining. After dehydriding, Mg + Mg_2Ni metallic phases recover whereas the thermally stable LaH_3 hydride remains unchanged. Thus, the H/D processes for this alloy are simply written as:

$$Mg + Mg_2Ni + Mg_{17}La_2 + H_2 \rightarrow Mg + Mg_2Ni + LaH_3 + H_2 \leftrightarrow$$

$$MgH_2 + Mg_2NiH_4 + LaH_3 \text{ } (Mg_{85}Ni_{10}La_5 \text{ alloy})$$

The metallic and hydride phases described above provide stable nano-grain structures in these alloys, as shown later (Section 5 and 6).

3. Pressure-composition isotherms and H/D kinetics

Pressure (p) – composition (c) isotherms are measured for the crystallized and activated samples (~1.0 g) with a volumetric method using a Sieverts–type apparatus. Figure 3 shows a result on $Mg_{85}Ni_{10}La_5$ alloy for temperatures (T) between 240° and 330°C, where the absorption and desorption isotherms are displayed with dashed and solid lines, respectively. Here, the H-content c is given by c (wt%) = (weight of H)/(weight of alloy + weight of H) × 100. All the isotherms clearly show two-stage plateaus. The first (lower pressure) stage corresponds to Mg – MgH_2, and the second (higher pressure) one to Mg_2Ni – Mg_2NiH_4 equilibria. The H-capacity ($c \sim 4.6$ wt%) and the ratio of the widths of the two stages, ~ 3.8/0.8, are consistent with those expected for the alloy assuming a complete hydride formation. It can be seen from the figure that the first stage is flat and exhibits only a small hysteresis except for the initial parts of the isotherms, whereas the second one is slightly inclined and exhibits a definite hysteresis which is more emphasized at lower temperatures.

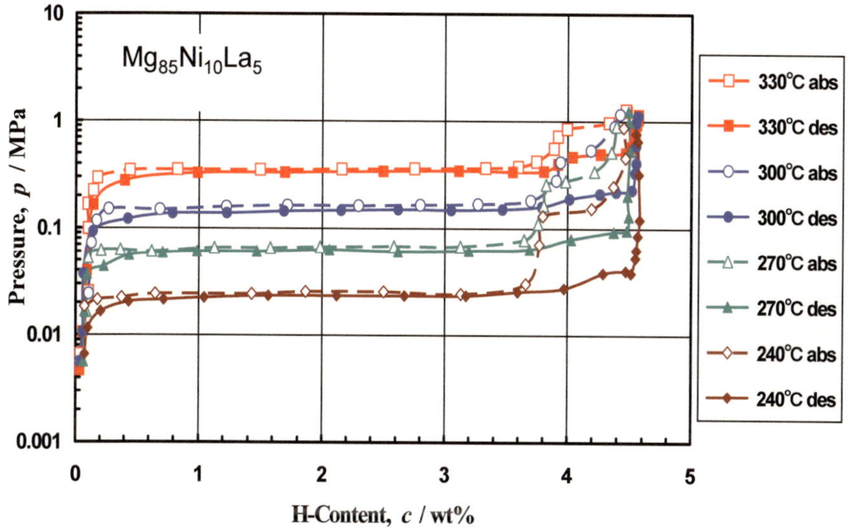

Figure 3. Pressure – composition isotherms for the melt-spun $Mg_{85}Ni_{10}La_5$ alloy. (Tanaka, 2008)

Relationships between the pressure and inverse temperature for the first and second plateaus in the absorption and desorption processes are shown in Fig. 4. The formation enthalpy ΔH_f and standard formation entropy ΔS_f^0 for MgH_2 and Mg_2NiH_4 are related with the plateau pressure p and temperature T as

$$ln(\frac{p}{p^0}) = \frac{\Delta H_f}{RT} - \frac{\Delta S_f^0}{R}, \tag{1}$$

where $p^0 = 0.1$ MPa and R is the gas constant. They can be evaluated from the linear relationships shown in the figure. Values of ΔH_f and ΔS_f^0 are listed in Table 1 and compared with those for an ordinary coarse-grained material (Wiswall, 1978). The values for MgH_2 are comparable between the melt-spun and ordinary alloys. However, those for Mg_2NiH_4 are definitely different between the two. Namely, the formation enthalpy and entropy for Mg_2NiH_4 are closer to those for MgH_2 in the melt-spun alloy than in the ordinary alloys.

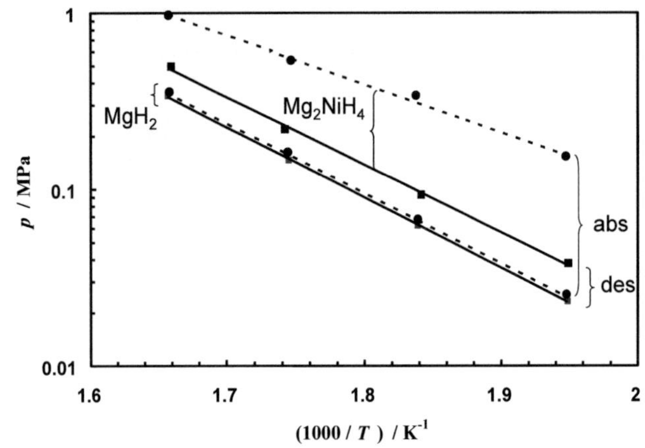

Figure 4. Relationships between plateau pressures p and inverse temperatures T^{-1} for hydrogen absorption (broken lines) and desorption (solid lines) in $Mg_{85}Ni_{10}La_5$ alloy. (Tanaka, 2008)

Hydrides	Process	ΔH_f /(kJ/molH_2)	ΔS_f^0 /(J/K/molH_2)
MgH_2	abs	-75.9	-136
	des	-76.2 (-74.8)	-136 (-136)
Mg_2NiH_4	abs	-52.0	-105
	des	-73.3 (-64.7)	-135 (-122)

Table 1. Thermodynamic parameters ΔH_f and ΔS_f^0 for the formation of MgH_2 and Mg_2NiH_4 in $Mg_{85}Ni_{10}La_5$ alloy. The values in the parentheses are for an ordinary alloy.

Figure 5 shows PCT curves for the as-cast $Mg_{85}Ni_{10}La_5$ alloy. Although the gross features of the isotherms are similar, the as-cast alloy provides more inclined plateaus than the melt-spun alloy. Furthermore, the boundary between the first and second plateaus is less clearly defined in the former than the latter. These unfavorable features of the PCT characteristics of the as-cast alloy are possibly due to the presence of structural and compositional inhomogeneities such as coarse grains and Mg/Mg_2Ni eutectics. The PCT Measurements for the melt-spun $Mg_{90}Ni_{10}$ alloy have also been carried out, and isotherms qualitatively similar to those of melt-spun $Mg_{85}Ni_{10}La_5$ alloy (Fig. 3) have been obtained. However, the measured H-capacity (c ~ 4.6 wt%) is substantially smaller than that expected for the alloy (~ 6.1 wt%). The isotherms show that the magnitude of the second plateau (formation of Mg_2NiH_4) is far smaller than that expected. A much slower reaction rate for the Mg_2NiH_4 formation in the $Mg_{90}Ni_{10}$ alloy than in the $Mg_{85}Ni_{10}La_5$ alloy might have caused this apparent loss of H-capacity in the binary $Mg_{90}Ni_{10}$ alloy.

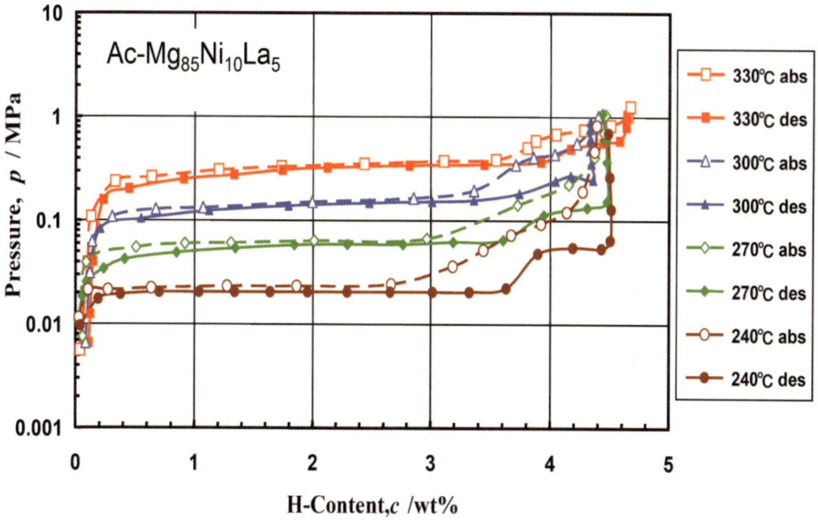

Figure 5. Pressure-composition isotherms for the as-cast $Mg_{85}Ni_{10}La_5$ alloy. (Tanaka, 2008)

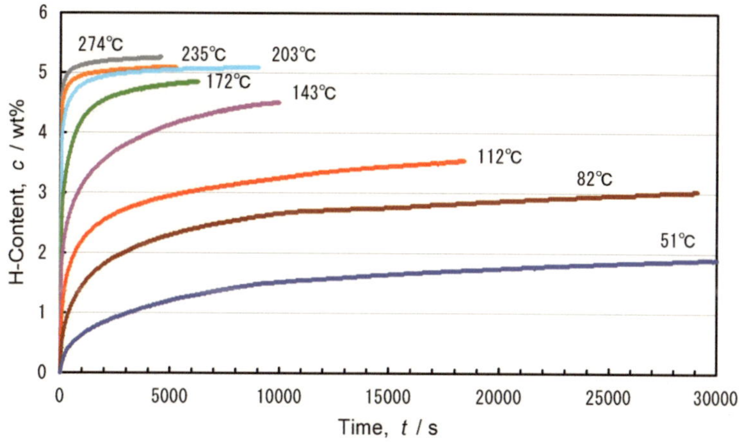

Figure 6. Hydrogen absorption characteristics of melt-spun $Mg_{85}Ni_{10}La_5$ alloy. (Tanaka et al., 2009)

Figure 6 shows hydrogen absorption rates of a melt-spun $Mg_{85}Ni_{10}La_5$ alloy (~150 mg) at temperatures between 51° and 274°C under H_2 pressures well above the corresponding plateau pressures of Mg_2NiH_4 and MgH_2. In the measurements of H/D kinetics, the H-content is defined, for convenience, as c (wt%) = (weight of H)/(weight of alloy) × 100. At temperatures above 200°C, a maximum H-content of c ~ 5 wt% is readily reached within ~0.5 h, although the absorption rate is drastically reduced at lower temperatures. This behavior is qualitatively similar to those of ball-milled MgH_2 added with transition metals (Liang et al., 1999) or transition metal-oxides (Hanada et al. 2006; Oelerich, 2001). These absorption curves reflect complex underlying processes depending on T and p, for which simple analytical expressions are unavailable at present except for the initial parts of absorption (Hanada et al., 2007).

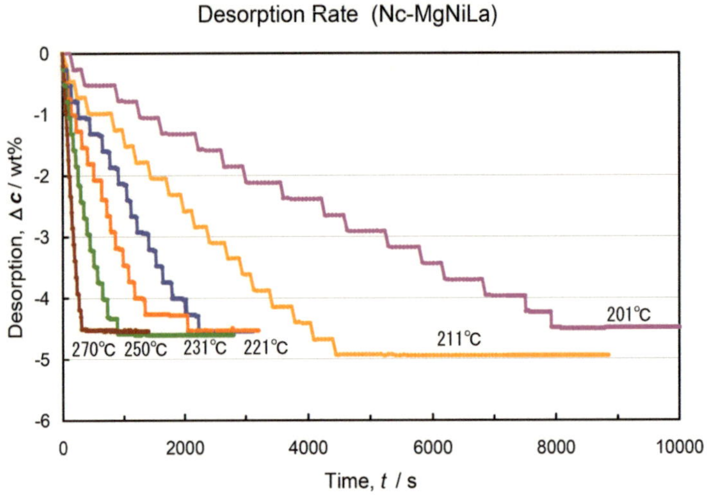

Figure 7. Hydrogen desorption characteristics of melt-spun $Mg_{85}Ni_{10}La_5$ alloy. (Tanaka et al., 2009)

Figure 7 shows hydrogen desorption rates of the same sample as above at temperatures between 201° and 270°C. For these measurements, the sample is initially saturated with hydrogen (~ 5 wt%H) by equilibrating with H_2 gas above the plateau pressure of Mg_2NiH_4 at the measuring temperature. The chamber is then instantly evacuated to lower than 0.001 MPa, well below the plateau pressure of MgH_2, and the subsequent increment in pressure due to desorption from the sample is recorded. It can be seen from this figure that the absorbed hydrogen is almost wholly desorbed in a reasonable time period in the measured temperature range. The desorption appears to take place according to an equation

$$c = c_0 - rt \,, \tag{2}$$

where c_0 is the initial H-content, t the desorbing time and r is a rate constant that depends on the temperature. Assuming that r is given by an Arrhenius-type equation

$$r = r_0 \, exp\left(-\frac{E_d}{RT} \right), \tag{3}$$

E_d (apparent activation energy for desorption) is calculated as 98.1 kJ/mol H_2, as shown in Fig. 8. This value is definitely larger than the enthalpies of decomposition of MgH_2 (76.2 kJ/mol H_2) and Mg_2NiH_4 (73.3 kJ/mol H_2). The linear time dependence given by eq. (2) may reflect a rate-controlling process at the powder surface of the alloy. Namely, hydrogen is desorbed from the surface at a constasnt rate irrespective of the H-content. The activation energy E_d for this alloy is intermediate between those for undoped MgH_2 (~120 kJ/mol H_2) and for Nb_2O_5-doped MgH_2 prepared by ball milling (62 kJ/mol H_2) (Barkhordarian, 2004). The desorption process of hydrogen will be discussed in conjunction with TDS measurements later (Section 4).

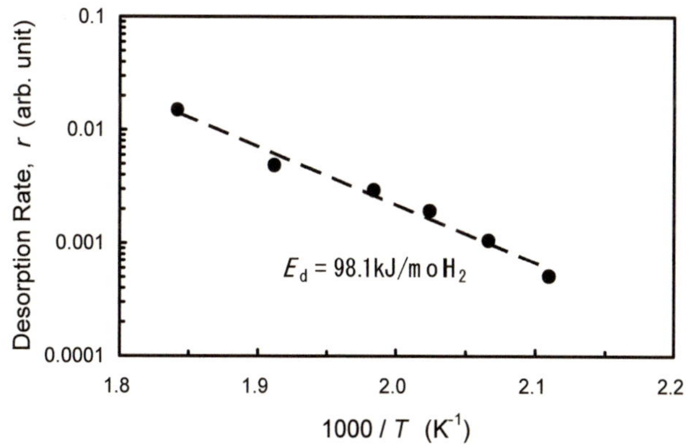

Figure 8. The desorption rate r for melt-spun $Mg_{85}Ni_{10}La_5$ plotted against T^{-1}. (Tanaka et al., 2009)

4. Thermal desorption spectroscopy for hydrogen

The thermal desorption spectroscopy (TDS) is a technique in which a small piece of hydrogenated sample is heated at a constant rate in high vacuum, and the H-desorption rate is measured by monitoring the partial gas pressure of hydrogen as a function of temperature. In the present experiment, a powder specimen (~10 mg) is measured at a heating rate between 0.5° and 4°C/min in a vacuum of ~10^{-6} Pa using an apparatus equipped with a turbo-molecular pumping system and a quadrupole mass spectrometer, as shown in Fig. 9. A sample cell compatible with the PCT measurement is used. With this cell, a hydrogenated sample is transferred from the PCT to TDS apparatus for measuring a desorption spectrum, or vice versa for loading the specimen with hydrogen, without exposing to the air. This process is repeated several times for a single specimen in a series of measurements to examine the heating-rate dependence of a TDS spectrum or to examine the effect of thermal cycles on the spectrum.

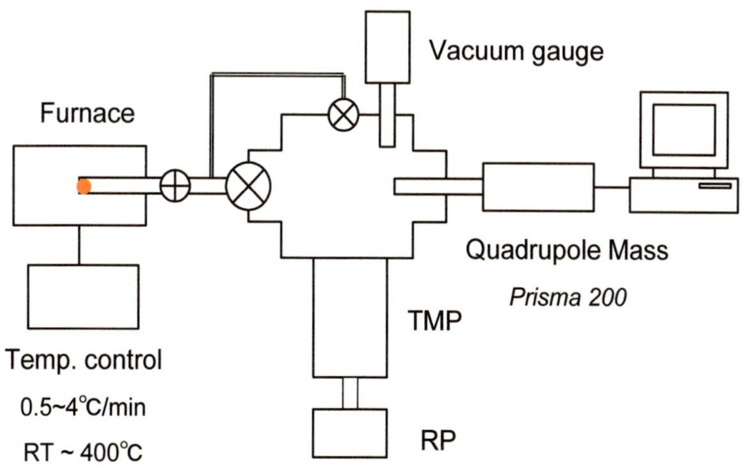

Figure 9. Apparatus used for TDS measurements.

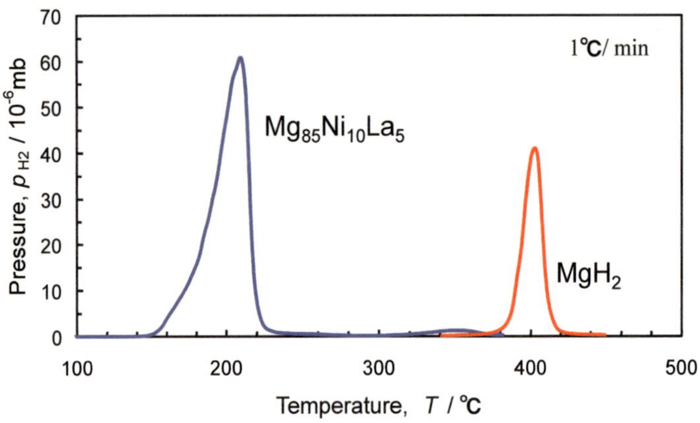

Figure 10. TDS spectrum of a hydrogenated melt-spun $Mg_{85}Ni_{10}La_5$ alloy compared with that of commercial MgH_2 powder.

Figure 10 shows an example of TDS spectrum of a hydrogenated $Mg_{85}Ni_{10}La_5$ alloy stabilized by exposing to a few hydrogenation/thermal desorption cycles at 300°C, compared with that of commercial MgH_2 powder. In this figure, a hydrogen partial pressure derived from a measured ion current of the mass spectrometer is plotted against a specimen temperature. The partial pressure can be used as a measure of hydrogen desorption rate if the pumping speed of the apparatus is assumed to be constant irrespective of the pressure. Clearly, a prominent peak with a few substructures occurs around 200°C in the $Mg_{85}Ni_{10}La_5$ alloy, while a single peak is caused around 400°C in MgH_2, indicating that the desorption temperature is significantly reduced in the melt-spun alloy. Although the nature of the TDS spectrum of the alloy is not clear solely from this figure, it will be shown in the following that the main peak as well as the substructures is related with thermal decomposition of hydride phases present in the alloy.

Figure 11 shows TDS spectra, measured at 1°C/min (left) and 2°C/min (right), of melt-spun $Mg_{85}Ni_{10}La_5$ alloy subjected to hydrogenation/thermal desorption cycles (1st to 4th). Prior to each desorption measurement, the sample has been fully hydrogenated at ~300°C under ~1 $MPaH_2$. Each spectrum is optionally shifted vertically for the sake of clarity. Except for apparent upward shifts of peak temperatures for the higher heating rate, the two sets of spectra provide common features with increasing cycles. In the 1st cycle, a broad peak *c* is dominantly seen together with a shoulder associated with peak *b*. In the 2nd cycle, peak *b* develops while peak *c* diminishes without altering the total peak area appreciably. In the 3rd and 4th cycles, peak *b* further develops and becomes narrower while peak *c* changes less. Throughout these processes, peak *d* changes little. Beyond the 4th cycle, the spectral change almost ceases and all the peaks remain stable if the measuring temperature is not increased beyond ~400°C. A substructure, *a*, is occasionally observable as a shoulder on the lower-temperature tail of peak *b*, but it is not always reproducible in these measurements.

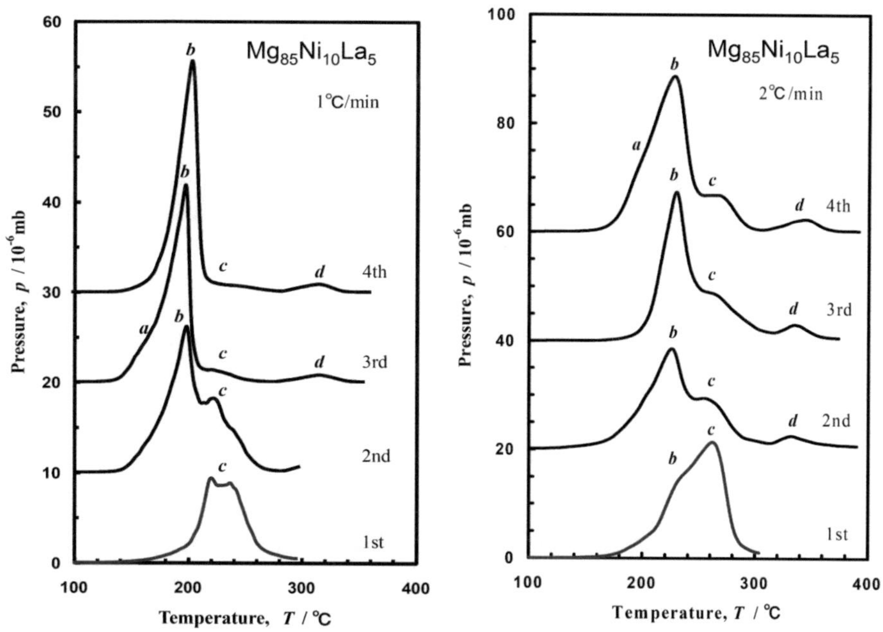

Figure 11. Evolution of TDS peaks of melt-spun $Mg_{85}Ni_{10}La_5$ alloys subjected to 1st to 4th hydrogenation/thermal desorption cycles measured with heating rates of 1°C/min (left) and 2°C/min (right). (Tanaka, 2008)

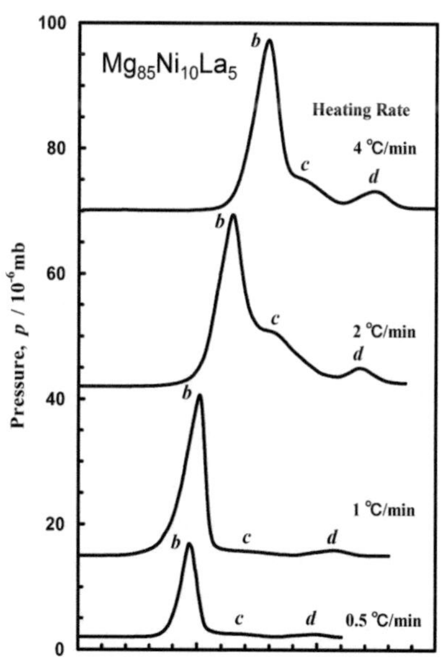

Figure 12. Heating-rate dependence of the TDS profiles of $Mg_{85}Ni_{10}La_5$ subjected to 4th hydrogenation/desorption cycle. (Tanaka, 2008)

In Fig. 12, TDS spectra of $Mg_{85}Ni_{10}La_5$ measured at different heating rates (0.5, 1, 2, and 4°C/min) are compared, where a spectrum stabilized by subjecting at least four hydrogenation/thermal desorption cycles is chosen as a representative for each heating rate. It can be seen from the figure that, with increasing heating rate, peaks *b* to *d* systematically shift toward higher temperatures, although peak *a* is not well defined in these spectra. Furthermore, the integrated strength of each peak tends to increase with increasing heating rate. The latter feature is a natural effect of the TDS technique. The observed peak shifts are analyzed according to Kissinger's method below.

The relation between the heating rate β and peak temperature T_m is shown in Fig. 13, where β/T_m^2 is plotted against T_m^{-1} in a semi-log scale for the peaks *b* to *d*. Nearly linear relations can be seen, indicating that Kissinger's equation,

$$\ln(\beta/T_m^2) = -E_{des}/RT_m + \ln k_0, \tag{4}$$

holds for these peaks. Here, E_{des} denotes an apparent activation energy for desorption and k_0 is a reaction constant. Values of E_{des} obtained in this way are also given in the figure.

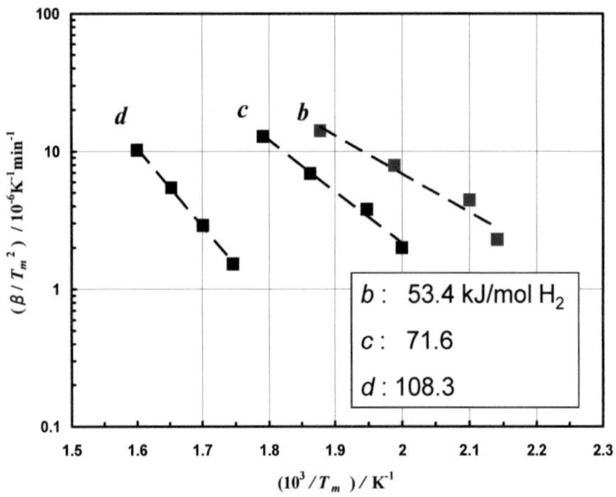

Figure 13. Kissinger's plots for the TDS peaks *b* ~ *d* of Fig.12. Values of E_{des} are given. (Tanaka, 2008)

To examine further the evolution of TDS spectra due to heating up to higher temperatures (~380°C), measurements have been done for a single specimen of melt-spun $Mg_{85}Ni_{10}La_5$ alloy using D_2 gas, instead of H_2 gas, as shown in Fig. 14. The specimen is fully pre-charged at 300°C under 1 MPa of D_2 pressure in every cyclic measurement, and heated from room temperature up to 300°C for the 1st to 5th cycles, and up to ~380℃ for the 6th to 10th cycles at a rate of 1℃/min. The spectral feature and its variation up to the 6th cycle are similar to those shown in Fig. 11. A sharp prominent peak *b* at ~210°C is stabilized if the specimen temperature is kept below 300°C. However, if the temperature is raised to ~380℃ in the 7th to 10th cycles, a drastic change takes place; the main peak *b* undergoes significant broadening and shifts toward higher temperatures, causing a single peak at ~280℃ in the 10th cycle. In this cycle, both the amount of pre-charged hydrogen and its absorption rate are markedly reduced. This result shows that the internal structure of the specimen is altered to some extent by heating up to ~380°C repeatedly.

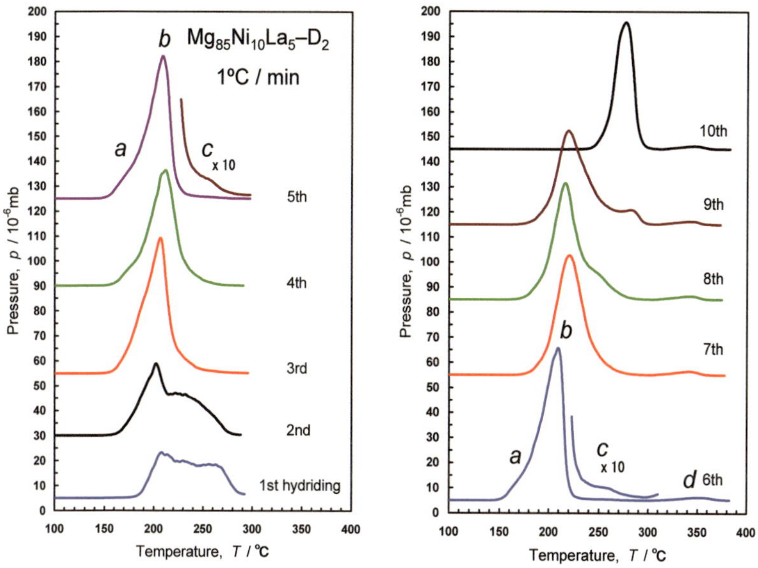

(a) (b)

Figure 14. Evolution of TDS spectra of melt-spun $Mg_{85}Ni_{10}La_5$-D_2 system from 1st to 5th cycles (a) and from 6th to 10th cycles (b). (Tanaka et al., 2009)

Figure 15 shows TDS spectra, measured at 1°C/min, of a melt-spun $Mg_{90}Ni_{10}$ alloy subjected to the hydrogenation/thermal desorption cycle as mentioned above. We find that, in contrast to $Mg_{85}Ni_{10}La_5$, peak *d* is absent in $Mg_{80}Ni_{10}$, indicating that this peak is exclusively derived from LaH_3 hydride present only in the $Mg_{85}Ni_{10}La_5$ alloy. We also find that peak *b* and peak *c* really exist in the 1st cycle, but peak *c* is much smaller than peak *b* even in the 1st cycle. The peak *c* quickly decays with increasing number of cycle, leaving only the peak *b* beyond the 4th cycle. These spectral features of $Mg_{90}Ni_{10}$ are somewhat different from those of $Mg_{85}Ni_{10}La_5$ shown in Fig. 11. This difference of TDS spectra is possibly associated with the difference in the thermal stability of the structures formed in these alloys. The behavior of peak *b* of $Mg_{90}Ni_{10}$ at different heating rates has also been examined and a result analogous to that of $Mg_{85}Ni_{10}La_5$ (Fig. 12) has been obtained.

In the following, we discuss the origins of TDS peaks *a* ~ *d* of melt-spun $Mg_{85}Ni_{10}La_5$ alloy and their behavior with the thermal cycles. It seems to be reasonable to assume that when a hydrogenated sample is slowly heated in vacuum, the thermal desorption, or the decomposition of hydrides, takes place according to a reaction process as manifested by the PCT desorption isotherms of the sample. Referring to Fig. 3, the main peak *b* at ~200°C and its lower-temperature shoulder *a* in Figs. 11 and 14 are attributable to the decompositions of MgH_2 and Mg_2NiH_4, respectively. The plateau pressure for the decomposition of Mg_2NiH_4 becomes closer to that of MgH_2 as the temperature decreases. This might be a reason why the shoulder *a* is not clearly defined as a single peak in the TDS spectra. Now, peak *c* at ~250°C, which is broad but well defined in the 1st cycle, progressively shrinks with increase in the number of cycle; peak *b* develops at the expense of peak *c* instead. This peak might be attributable to desorption of hydrogen dissolved in unstable grain boundaries (nano-boundaries) existing in high density after the crystallization. Most of these boundary regions are expected to transform to nano-crystalline Mg grains to form MgH_2 in the following 2nd to 6th cycles. The fact that the main peak *b* is broadened and shifts toward higher temperatures when the alloy is heated up to ~380℃ (7th to 10th cycles) suggests that the nanostructure is altered to some extent allowing redistribution and coarsening of the nano-grains. In this temperature range, the stable LaH_3 hydrides are also decomposed to form coarse grains of metallic La, resulting in destabilizing the nano-grain structure of the alloy. This structural change will necessarily lead to retarding the decomposition reaction of hydrides and the subsequent hydrogen transport toward the alloy surfaces. Indeed, a much longer time is required for hydrogenating the sample at 300℃ in the 10th cycle in comparison with those in the 2nd to 6th ones. The structural change suggested above has been really confirmed by TEM observation, as shown later (Section 5 and 6).

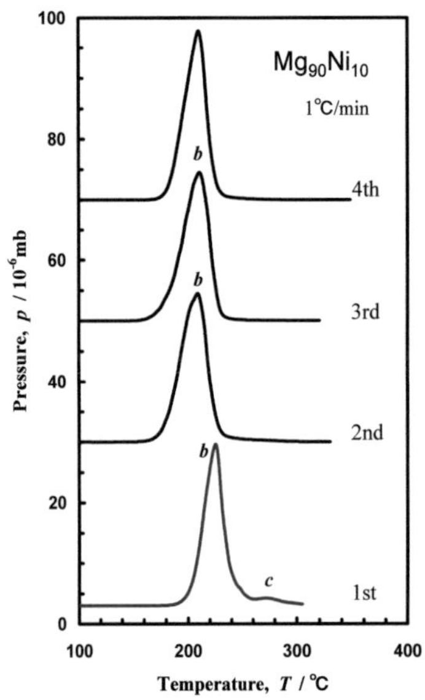

Figure 15. Evolution of TDS peaks of a melt-spun $Mg_{90}Ni_{10}$ alloy subjected to 1st to 4th hydrogenation/thermal desorption cycles. (Tanaka, 2008)

We next discuss the hydrogen desorption process in melt-spun $Mg_{85}Ni_{10}La_5$ alloy. As mentioned in connection with eq. (2), the linear time dependence of H-desorption at a fixed temperature suggests that the surface reaction is a rate-controlling step for the process. Namely, the decomposition of hydrides and subsequent diffusion of hydrogen toward external surfaces of powder grains occur quickly enough to establish a homogeneous hydrogen distribution within the grains. Thus, hydrogen in a grain may be assumed to be in a quasi-equilibrium state during the desorption. Although data for H-diffusion in Mg-rich nano-crystalline alloys are unavailable at present, the H-diffusivity along grain boundaries may be higher than those in Mg and Mg_2Ni grains, or in MgH_2 and Mg_2NiH_4 hydrides (Spassov & Köster, 1999). Hydrogen atoms released from the hydrides inside a grain (diameter < ~5μm) might quickly diffuse through nano-boundaries toward the surface of the grain and be chemisorbed on some surface sites such as Mg_2Ni nano-grains (Yamaga et al., 2008). Hydrogen molecules are desorbed from the chemisorbed sites into vacuum via a recombination reaction.

Now, we consider a thermal desorption spectrum for a simple isolated system of hydrogen atoms adsorbed on an Mg_2Ni grain surface. The peak temperature of the TDS spectrum is just given by eq. (4) with $E_{des} = E_{ad}$, adsorption energy of a hydrogen molecule. In a real system, where hydrogen is continuously supplied from an interior hydride to the surface sites, a modified relation, $E_{des} = E_{ad} + E_{bind}$, may hold, where E_{bind} is an effective binding energy of hydrogen of the hydride relative to the surface sites. The effective binding energy may be evaluated from the formation enthalpy ΔH_f of the hydride, as shown below.

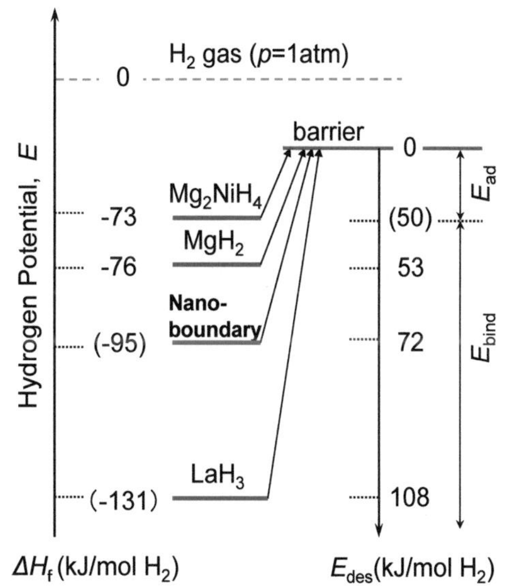

Figure 16. Hydrogen potential energy diagram in melt-spun $Mg_{85}Ni_{10}La_5$ alloy, showing interrelations between the formation enthalpies of hydrides, ΔH_f, and the activation energy of desorption, E_{des}. (Tanaka, 2008)

Figure 15 displays a potential energy diagram of hydrogen existing in the hydrides or nano-boundaries in melt-spun $Mg_{85}Ni_{10}La_5$ relative to that of the gas phase at standard state. Here, the formation enthalpy ΔH_f is taken as the potential energy E in a hydride, and values given in Table 1 (desorption process) are used for the MgH_2 and Mg_2NiH_4 hydrides. The potential energies for hydrogen in the nano-boundaries and LaH_3 are estimated and given in the parentheses. Now, we set an energy barrier of E_{ad} = 50 kJ/mol H_2 above the potential of

Mg_2NiH_4 hydride for the desorption of hydrogen. (In the real system, desorption from Mg_2Ni grain surface will occur, for which an alternate value of E_{ad} must be used. In the present model, direct desorption from Mg_2NiH_4 hydride surface is tentatively assumed.) Then the net activation energy for desorption is given by $E_{des} = E_{ad} + E_{bind}$, where E_{bind} means the potential depth referred to the Mg_2NiH_4 hydride. This value can be obtained from E_{des} measured by TDS (Fig. 13), as shown in the figure. In this way, ΔH_f of nano-crystalline LaH_3 is estimated at –131 kJ/mol H_2, for which a value of –188 kJ/mol H_2 is given in conventional material (Fukai, 1993). Furthermore, the potential energy of hydrogen dissolved in nano-boundaries (mean value) is estimated at –95 kJ/mol H_2, indicating that hydrogen in the nano-boundary is more stable than in MgH_2 hydride by about 20 kJ/mol H_2. It should be argued here that the apparent energy for desorption, E_d = 98.1 kJ/mol H_2, obtained by analyzing the desorption rates, r, at fixed temperatures for melt-spun $Mg_{85}Ni_{10}La_5$ (eq. (3) and Fig. 8) is much larger than E_{des} = 53.1 kJ/mol H_2 obtained from TDS for the decomposition of MgH_2 in the same alloy. Reasons to explain this difference may not be simple. A big difference in the experimental conditions between the two measurements is that, while the TDS is always measured in a high vacuum of 10^{-6} to 10^{-5} Pa, the measurement of the desorption-rate is performed under $p_{H2} \sim 10^3$ Pa. In the latter case, a driving force correction is needed because the plateau pressure for the decomposition of the hydride becomes closer to p_{H2} especially at lower temperatures (Fernandez & Sanchez, 2002). This correction, however, has not been performed in the present study. Therefore, the present E_d value is expected to be lowered toward that of E_{des} if the correction is correctly applied.

5. High-resolution TEM and EELS studies of nanostructures

Nano-sized structures of melt-spun $Mg_{90}Ni_{10}$ and $Mg_{85}Ni_{10}La_5$ alloy subjected to crystallization, hydrogenation and dehydrogenation have been investigated using a Hitachi H-9000 electron microscope operated at 300 kV. High-resolution TEM (HRTEM) and electron energy-loss spectroscopy (EELS) imaging techniques have been employed to reveal the structures in detail. Each sample has been ground in acetone into a fine powder with a diameter of ~1 to ~2 μm, dispersed on a carbon micro-grid, and inserted into the TEM sample chamber for observation. In the following, some results mainly on the $Mg_{85}Ni_{10}La_5$ alloy are presented.

Figure 17 shows a TEM bright-field (BF) image and its electron diffraction (ED) pattern of melt-spun and crystallized $Mg_{85}Ni_{10}La_5$ alloy. The ED pattern exhibits broad diffraction rings of Mg, Mg_2Ni and $Mg_{17}La_2$ phases, consistently with the XRD pattern of the same alloy shown in Fig. 2. The alloy is composed of homogeneous distribution of nano-sized grains of these phases. Figure 18 shows a HRTEM image of the encircled region of Fig. 17. Lattice fringes of Mg, Mg_2Ni and $Mg_{17}La_2$ nano-grains are evidently seen. Their sizes range between 5 and 10 nm. The outer regions of these grains are also composed of nano-grains of the same phases, although their lattice fringes are not observable in this figure. No evidence of the formation of Mg/Mg_2Ni eutectics has been found in the crystallized alloy.

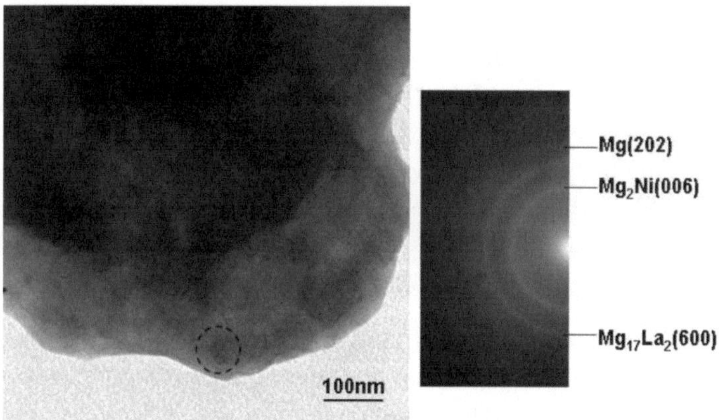

Figure 17. TEM bright-field image and its electron diffraction pattern of melt-spun and crystallized $Mg_{85}Ni_{10}La_5$ alloy.

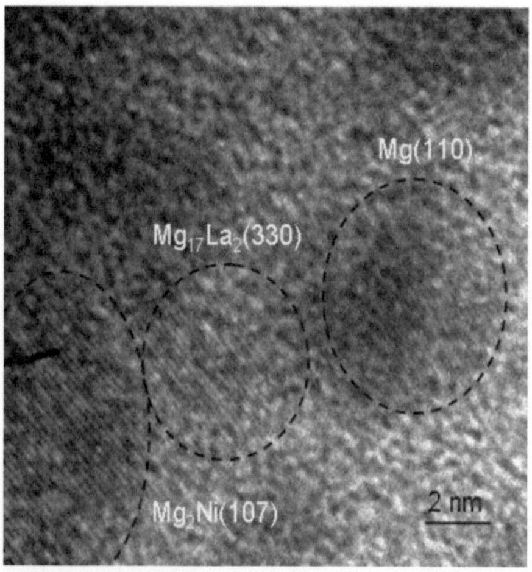

Figure 18. HRTEM image of the encircled region of the BF image shown in Fig. 17.

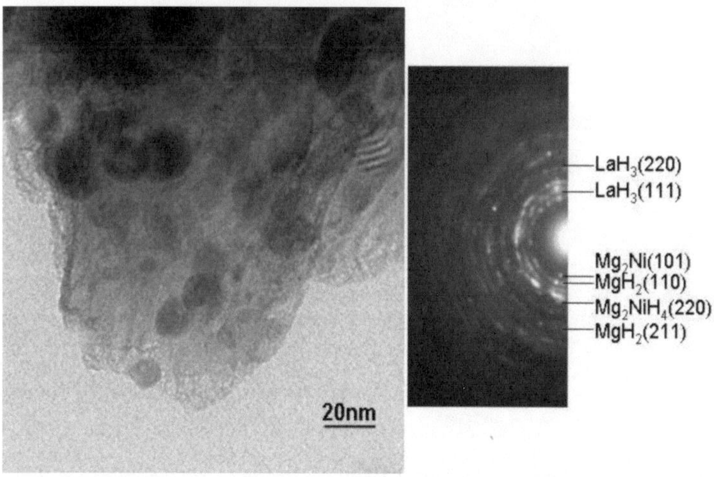

Figure 19. BF image of hydrogenated $Mg_{85}Ni_{10}La_5$ alloy and its ED pattern.

Fig. 19 shows a BF image and its ED pattern of hydrogenated $Mg_{85}Ni_{10}La_5$ alloy. The ED pattern shows that MgH_2, Mg_2NiH_4 and LaH_3 hydrides are present together with remained Mg_2Ni phase, in agreement with the XRD result shown in Fig. 2. The diffraction rings are not continuous but spotty, indicating that some grains of the hydride phases have undergone grain growth during the hydrogenation at 300°C. In the BF image are seen spherical particles with dark-grey contrast dispersed in a matrix of light-grey contrast. The former are identified as nano-grains of Mg_2NiH_4 and LaH_3 or their clusters and the latter as aggregates of MgH_2 nano-grains by EELS and HRTEM micrographs shown below.

Figure 20 shows EELS mapping (Egerton, 1996) for the same sample area as in Fig. 19, where a plasmon loss (~10 eV), Mg-K (1305 eV), Ni-$L_{2,3}$ (872, 855 eV) and La-$M_{4,5}$ (849, 832 eV) absorption edges are employed. The plasmon-loss image, denoted as H-plasmon here, represents the distribution of MgH_2 phase (bright region) present in the sample, as explained in detail later (Section 6).

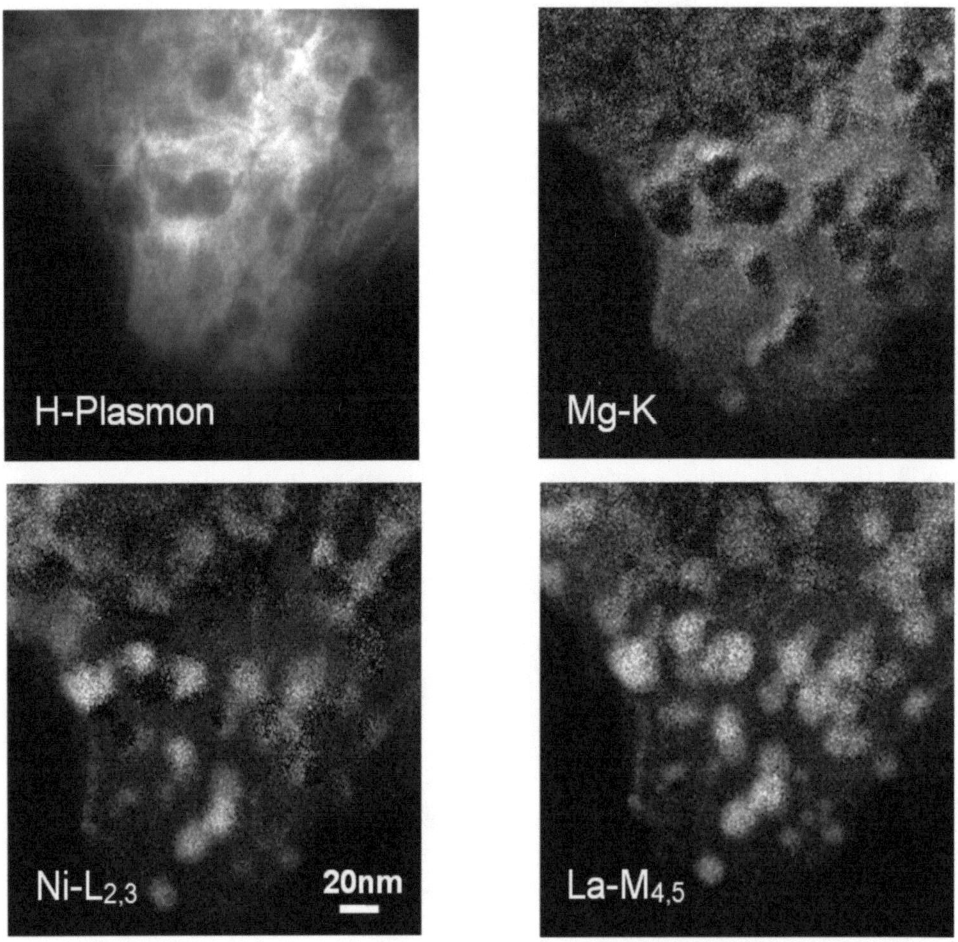

Figure 20. EELS mapping for the sample shown in Fig. 19, using H-plasmon, Mg-K, Ni-$L_{2,3}$ and La-$M_{4,5}$ absorption edges.

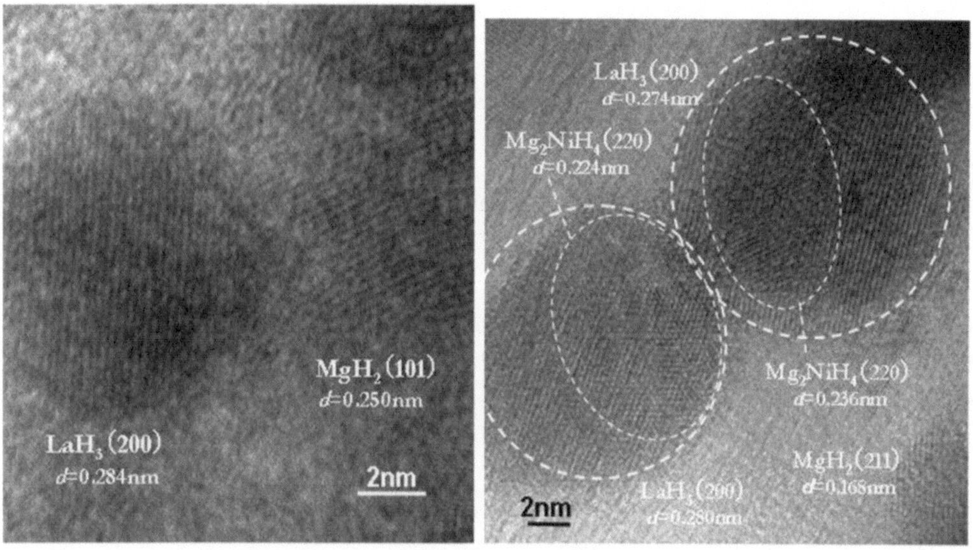

Figure 21. HRTEM images of areas at the lower part of the sample shown in Fig. 19. (Tanaka et al., 2009)

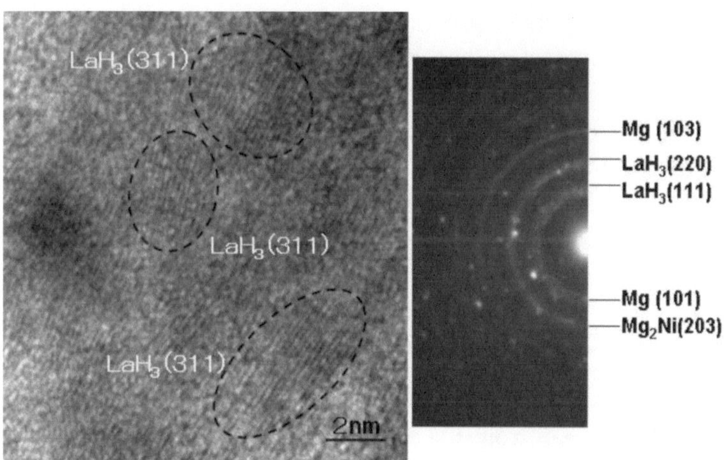

Figure 22. HRTEM image and its ED pattern of dehydrogenated $Mg_{85}Ni_{10}La_5$ alloy.

The Mg-K edge image likewise represents the MgH_2 phase, and it is clearly recognized that the MgH_2 phase provides bright regions common to both the H-plasmon and the Mg-K edge in these pictures. Likewise, the Ni-$L_{2,3}$ and La-$M_{4,5}$ images should represent Mg_2NiH_4 and LaH_3 phases, respectively. Both these phases, observed as bright particles of roughly ~10 nm in size, give rise to dark-grey or light-grey spots in the H-plasmon and Mg-K images. However, owing to a rather poor energy resolution employed in the present EELS mapping, the Ni-$L_{2,3}$ and La-$M_{4,5}$ absorption edges are not necessarily well discriminated from each other and certain overlapping generally occurs, resulting in providing nearly similar patterns of phase distribution in the present pictures. To avoid the confusion involved in specifying the Mg_2NiH_4 and LaH_3 hydride nano-grains, a combined use of EELS and HRTEM techniques appears to be important.

Figure 21 shows HRTEM images of the same sample as shown in Fig. 19. The hydride phase of each nano-grain has been identified in such a way that a measured *d*-spacing of lattice planes is consistent with a calculated one of the hydride. In these figures, nano-boundaries formed between MgH_2 and LaH_3 nano-grains (left) and a small cluster composed of LaH_3 and Mg_2NiH_4 nano-grains embedded in a matrix of MgH_2 nano-grains (right) can be seen.

Figure 22 shows a HRTEM image and its ED pattern of a dehydrogenated $Mg_{85}Ni_{10}La_5$ alloy. After the dehydrogenation at 300°C, the alloy transforms into a structure composed of Mg and Mg_2Ni metallic phases together with the LaH_3 hydride phase remaining unchanged. This phase structure is consistent with the XRD result shown in Fig. 2. Although the grain growth of Mg nano-grains is suppressed during the dehydrogenation, coarsening of Mg_2Ni nano-grains is often observed by HRTEM. It is reflected on the ED pattern as a continuous diffraction ring of Mg and a spotty one of Mg_2Ni.

6. *In situ* TEM-EELS studies of dehydrogenation process

It is interesting to observe the hydrogen desorption process at elevated temperatures directly in the electron microscope. For this purpose, a hydrogenated sample is deposited on a tungsten wire of 25 μm in diameter and heated by passing an electric current under a vacuum of 10^{-5} Pa in the microscope. The temperature of the sample is determined from the current using a calibrated curve with an accuracy of about ± 20°C. Since the sample has been exposed to the air for a few minutes before inserting into the microscope, it inevitably suffers from surface contamination. This may cause a substantial delay of the onset of hydrogen desorption in comparison with those found in the TDS measurements. The desorption process of a hydrogenated sample has been continuously video-recorded for TEM BF and EELS H-plasmon images at a fixed temperature of ~350° and ~450°C.

Figure 23 shows a time evolution of a TEM BF image of a hydrogenated $Mg_{85}Ni_{10}La_5$ alloy held at ~350°C. It can be seen from the figure that, a few minutes later from the start, a bright spot appears at a central part of the sample, goes on extending outwards with the holding time, and covers almost the whole area at the final stage of the heating. This change of the BF image just corresponds to the thermal desorption of hydrogen from the observed area. In fact, the ED patterns taken before and after the heating indicate that, while the sample is composed of nano-grains of MgH_2, Mg_2NiH_4 and LaH_3 hydrides before heating, it changes into the structure consisting of metallic Mg and Mg_2Ni nano-grains in addition to the LaH_3 hydrides, as shown in Fig. 24. Although some grains appear to undergo coarsening during the heating, the others stay mostly in the nano-size range at this temperature.

The desorbing process has also been pursued by observing an EELS H-plasmon image at ~450°C for the same alloy as above. Its change with the holding time is shown in Fig. 25. It can clearly be seen from the figure that the hydrogen desorption which is initiated on an edge side of the sample soon after the heating proceeds into the interior region, and finally covers the whole area of the sample. A bright region corresponding to an MgH_2-enriched zone initially extends throughout the sample, but it tends to shrink with the progress of desorption, and is finally replaced by a dark region completely. This change indicates that the MgH_2 matrix has been altered to a metallic Mg matrix by desorbing hydrogen.

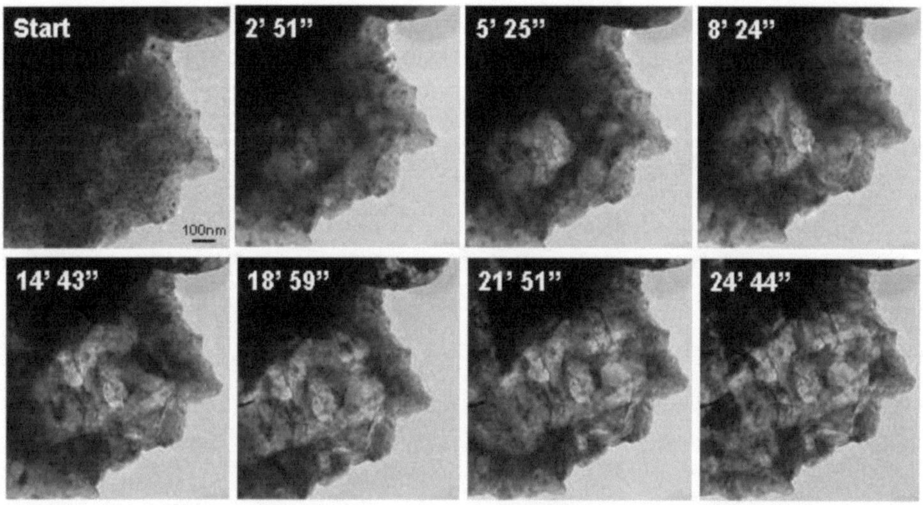

Figure 23. Time evolution of a BF image of a hydrogenated $Mg_{85}Ni_{10}La_5$ alloy during heating at ~350°C.

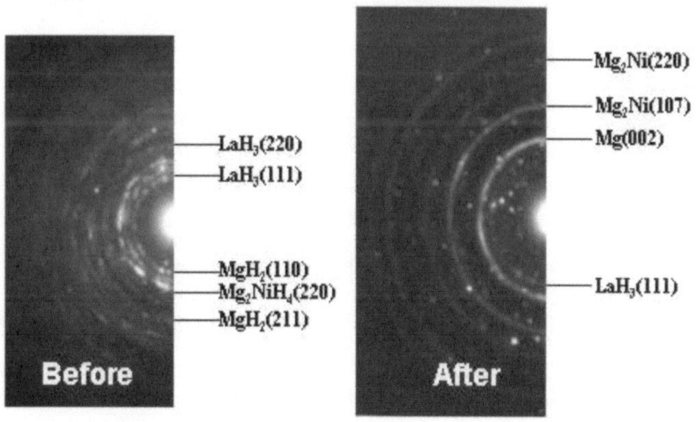

Figure 24. ED patterns of the same sample as in Fig. 23 taken before and after heating at ~350°C.

Figure 26 shows BF images and the corresponding ED patterns of the same sample as above before and after the heating at ~450°C. Before heating, fine dark-grey particles corresponding to LaH_3 and Mg_2NiH_4 hydrides are uniformly distributed in the matrix of MgH_2 throughout the sample. This structure has changed drastically after heating. These fine particles have undergone redistribution and reconstruction forming coarse aggregates or precipitates of metallic Mg_2Ni and La in the metallic Mg matrix. The ED pattern taken after heating exhibits strong diffraction spots lying along weak continuous rings of Mg, Mg_2Ni and La, indicating that a majority of these metallic nano-grains has undergone the grain growth at the end of the heating. All these features have also been well recognized in the corresponding EELS images (Tanaka et al., 2009).

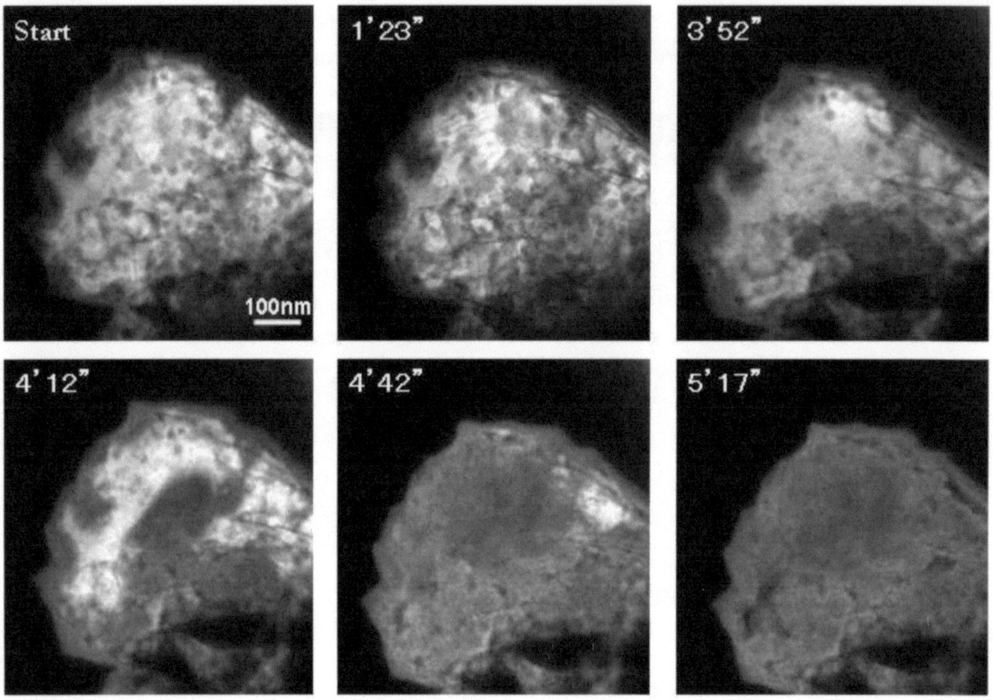

Figure 25. Evolution of EELS H-plasmon image of hydrogenated $Mg_{85}Ni_{10}La_5$ alloy with holding time at ~450°C. (Tanaka et al., 2009)

Figure 27 shows a HRTEM image of the sample subjected to the heating at ~450°C, showing a spheroidal Mg grain grown to ~20 nm in size together with smaller Mg_2Ni and La grains in the close vicinity of the Mg grain. A HRTEM image of a rod-shape La grain grown to ~50 nm in length has also been observed (Tanaka et al., 2009).

The TEM-EELS results presented above clearly show that the nano-crystalline hydrides are really formed by hydrogenation and are completely decomposed by heating up to ~450°C in the melt-spun $Mg_{85}Ni_{10}La_5$ alloy. These results are, however, in contradiction to recent TEM studies (Hanada et al., 2008; Porcu et al., 2008), which claim that MgH_2 is readily decomposed to Mg even at room temperature by the electron beam irradiation during observation. In these studies, ball-milled MgH_2 sample doped with Nb_2O_5, which exhibits excellent hydrogen reactivity, is used in contrast to our melt-spun ternary alloy. A difference in the hydride stability between the two samples may have caused these contradictive results. Another aspect of discriminating the ball-milled sample from the melt-spun alloy is that a thermally unstable hydride, γ-MgH_2, is amply formed and mixed with the stable hydride, β-MgH_2, by ball milling (Danaie, 2010). Such an unstable hydride is not formed in the melt-spun alloy. It must also be taken into account that our powder sample used for the TEM-EELS studies is inevitably contaminated with oxygen on the surface before it is inserted into the electron microscope. This may have prevented the desorption of hydrogen from the sample and hence the decomposition of hydrides during the observation. Further studies are necessary to clarify these points.

Figure 26. BF images and ED patterns of the same sample as in Fig. 25 taken before heating (upper) and after heating (lower) at ~450°C. (Tanaka et al., 2009)

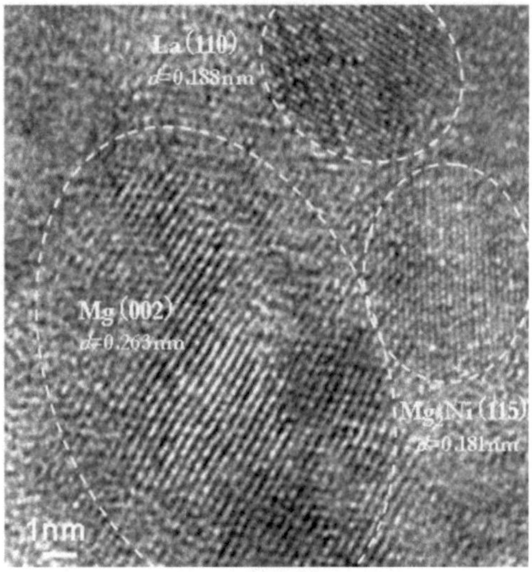

Figure 27. Fig. 26. HRTEM image of the same sample as in Figs. 25 and 26 observed after heating at ~450°C. (Tanaka et al., 2009)

Finally, we discuss the nature of an EELS plasmon peak in a hydrogenated sample. The EELS plasmon (or H-plasmon) peak observed in this study is considered to be associated with an excitation of bulk plasmons by an incident electron beam in the MgH_2 nano-grains. This is justified from the similarity between the EELS H-plasmon and Mg-K images of the hydrogenated $Mg_{85}Ni_{10}La_5$ sample shown in Fig. 20. The bright region of the Mg-K image reflects the MgH_2 phase, which just corresponds to the bright region of the H-plasmon image. This bright region of H-plasmon image has been confirmed to disappear completely after the in situ hydrogen desorption, as shown in Fig. 25, although the Mg-K image still provides a bright region for the metallic Mg phase formed (Tanaka et al., 2009).

To prove the interrelation between the EELS plasmon spectrum and the H-plasmon image more closely, we reexamine

the EELS spectrum and the corresponding image of MgH_2 using a commercially pure sample, as shown in Fig. 28, where the measured EELS spectra and the corresponding images are compared between the as-received sample (MgH_2) and the dehydrogenated one (Mg) prepared by in situ heating at ~450°C. The H-plasmon peak (indicated by arrow) can be clearly seen at an energy-loss E~11eV in MgH_2, whereas it is almost diminished in Mg. The H-plasmon peak causes a bright image of the MgH_2 phase throughout the sample in the former, whereas a dark-grey image of metallic Mg is left in the latter. Another small peak at E~22eV, which is detected in every sample studied, is attributable to a plasmon excitation in a thin MgO layer which covers the sample surface. In the free-electron model, the plasmon energy E_P is given by (Egerton, 1996),

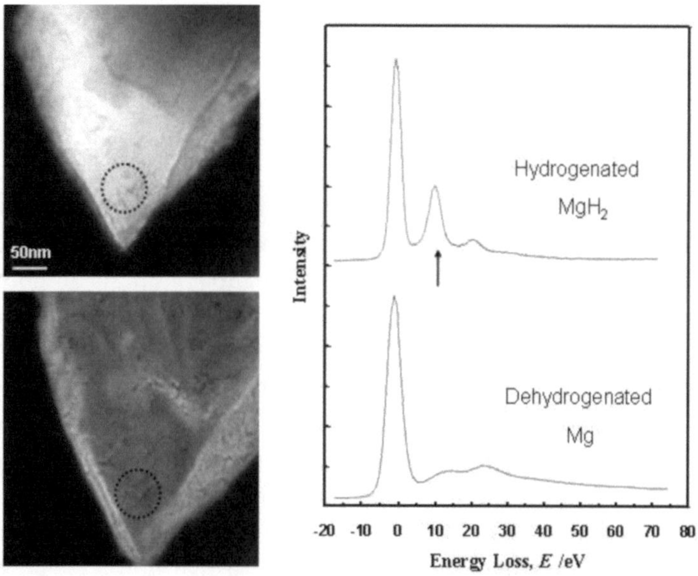

Figure 28. EELS spectra in the low energy-loss region and the corresponding H-plasmon images for MgH_2 (upper) and Mg (lower) for a sample prepared from a commercially pure MgH_2 powder. The H-plasmon peak is indicated by an arrow. The encircled area of the sample shows a spot where the EELS spectra have been taken. (Tanaka et al., 2009)

$$E_p = \frac{h}{2\pi}\omega_p = \frac{h}{2\pi}\left(\frac{ne^2}{\varepsilon_0 m}\right)^{\frac{1}{2}}, \tag{5}$$

where ω_p is the plasmon angular frequency, n is the electron density, and other quantities have their usual meaning. Taking n = 129.8 nm^{-3} (MgH_2) and 428.4 nm^{-3} (MgO), E_p is calculated as 13.3 eV and 24.2 eV, respectively. These calculated energies approximately agree with the observed ones of the EELS spectra shown in Fig. 28. Now, the plasmon energy for metallic Mg is calculated as E_p = 10.8 eV taking n = 86.1 nm^{-3}, which falls in the close vicinity of MgH_2. However, owing to the lack of energy resolution of the present EELS experiment and/or to the lack of the spectral intensity of the plasmon peak of metallic Mg in comparison with that of MgH_2, this peak has not been clearly detected in the present study. In this respect, it is worthwhile to note that both peaks have been detected in a partially dehydrogenated MgH_2 sample subjected to ball milling (Danaie, 2010). An H-plasmon peak similar to our study has also been observed in a ball-milled NbB_5-doped MgH_2 sample (Kim et al., 2010).

7. Summary and conclusions

1. A magnesium-based Mg-Ni-La alloy, specifically $Mg_{85}Ni_{10}La_5$ alloy, prepared by melt-spinning and subsequent crystallization annealing exhibits favorable reaction kinetics and PCT characteristics for hydrogen absorption and desorption. The *p-c* isotherms at 240-330°C manifest a flat two-stage plateau with small hysteresis and a maximum H-

capacity of ~4.6 wt.%. A temperature as low as ~200°C is enough for hydriding and dehydridng the alloy entirely in a moderate time period. XRD results show that a mixed phase structure of Mg and Mg_2Ni before hydrogenation, or MgH_2 and Mg_2NiH_4 after hydrogenation, together with a thermally stable hydride LaH_3, is the main source of these excellent hydrogen storage characteristics.

2. The fully hydrogenated alloy exhibits, in general, four TDS peaks (*a*~*d*). The magnitude and temperature of each peak depend on the specimen history and the heating rate (0.5~4°C/min) for the TDS measurement. Assuming that a quasi-equilibrium condition is established in the specimen, these peaks are attributed to decomposition of Mg_2NiH_4 (peak *a*), decomposition of MgH_2 (peak *b*), release from nano-grain boundaries (peak *c*), and decomposition of LaH_3 (peak *d*). The heating-rate dependence of the TDS peak temperature is successfully analyzed on the basis of Kissinger's equation, from which the activation energy for desorption E_{des} is obtained for each peak. Assuming that the desorption is a surface-controlled process and that E_{des} is the sum of an adsoption energy E_{ad} and an effective binding energy E_{bind} of hydrogen in the alloy, a relationship between E_{bind} and the formation enthalpy ΔH_f of a hydride is proposed. From this relationship, the hydrogen dissolved in the nano-boundaries appears to be ~20 kJ/mol H_2 more stable than in MgH_2.
3. The role of Ni and La for facilitating the hydrogen storage properties of the Mg-based alloy is multifold. Both elements act to promote the amorphization of the alloy by melt-spinning and to produce a homogeneous nano-grain structure by crystallization. Nickel forms stable Mg_2Ni nano-grains, which act as dissociation and recombination catalysts for hydrogen molecules on the surface of the alloy and contribute to enhance the absorption and desorption kinetics of hydrogen. On the other hand, La forms thermally stable LaH_3 nano-hydrides, which work to stabilize the whole nanostructure during the hydriding and dehydriding cycles at temperatures below ~350°C.
4. The improved kinetics of hydrogen absorption and desorption of the alloy is partly due to faster diffusion of hydrogen along the nano-boundaries than through inside the Mg/MgH_2 or Mg_2Ni/Mg_2NiH_4 grains. The relatively fast diffusivity and high solubility of hydrogen in the nano-boundaries will facilitate the hydrogen transport inside the alloy and hence enhance its absorption and desorption rates.
5. It has been confirmed by TEM and HRTEM studies that, after hydrogenation at 300°C, this alloy exhibits a nanostructure consisting of Mg_2NiH_4 and LaH_3 nano-particles (~10 nm) imbedded uniformly in MgH_2 nano-grain (3~5 nm) matrices. After dehydrogenation at 300°C, this alloy exhibits Mg_2Ni and LaH_3 nano-particles imbedded uniformly in the Mg nano-grain matrix. This nanostructure is almost maintained if the hydrogenation and dehydrogenation treatments are performed at temperatures below ~350°C.
6. If the temperature is raised beyond ~400°C, the desorbing as well as the absorbing characteristics of the alloy are significantly degraded. The in situ TEM observation during heating at ~450°C indicates that the original nanostructure undergoes remarkable redistribution and reconstruction of the alloy phases, in parallel with the decomposition of the hydrides. Instead of a uniform distribution of the hydride

particles, segregations of coarse precipitates of Mg_2Ni and La, together with the grain growth of Mg matrix, take place. These coarsening of the alloy phases and their inhomogeneous redistribution must have retarded the reaction kinetics for hydrogen.

7. We have succeeded in clearly visualizing a global hydrogen distribution in the alloy sample by imaging the EELS plasmon peak (H-plasmon) arising from the MgH_2 phase. By using this technique, the thermal desorption process has been directly observed in the microscope. Melt-spun and nano-crystallized Mg-based alloys are quite favorable materials for this kind of studies.

Author details

Kazuhide Tanaka
Nagoya Institute of Technology, Toyota Physical and Chemical Research Institute, Japan

Acknowledgement

The author is grateful to Prof. H. Inokuchi, Prof. A. Ikushima and Prof. Y. Ishibashi of Toyota Physical & Chemical Research Institute for stimulating discussions and comments. He is also grateful to Dr. S. Towata of Toyota Central R/D Laboratories for the collaboration in the PCT measurements and useful advices, to Mr. T. Miwa, Dr. K. Sasaki and Prof. K. Kuroda of Nagoya University for the cooperation in the TEM-EELS studies, and to Dr. M. Yamada and Mr. K. Hibino of Nagoya Institute of Technology for the collaboration in the sample preparation.

8. References

Barkhordarian, G., Klassen, T., & Bormann, R. Effect of Nb_2O_5 content on hydrogen reaction kinetics of Mg. *Journal of Alloys and Compounds*, Vol. 364, (2004), pp. (242-246)

Danaie, M., Tao, S.X., Kalisvaart, P., & Mitlin, D. Analysis of deformation twins and the partially dehydrogenated microstructure in nanocrystalline magnesium hydride (MgH_2) powder. *Acta Materialia*, Vol. 58, (2010), pp. (3162-3172)

De Negri, S., Giovannini, M., & Saccone, A. Phase relationships of the La-Ni-Mg system at 500°C from 0 to 66.7 at.% Ni. *Journal of Alloys and Compounds*, Vol. 397, (2005), pp. (126-134)

Egerton, R.F. (1966). *Electron Energy-Loss Spectroscopy in the Electron Microscope* (2nd edition), Plenum, ISBN 0-306-45223-5, New York

Fernandez, J.F., & Sanchez, C.R. Rate determining step in the absorption and desorption of hydrogen by magnesium. *Journal of Alloys and Compounds*, Vol. 340, (2002), pp. (189-198)

Fukai, Y. (1993). *The Metal-Hydrogen System*, Springer, ISBN 3-540-55637-0, Berlin, Heidelberg, New York

Hanada, N., Ichikawa, T., Hino, S., & Fujii, H. Remarkable improvement of hydrogen sorption kinetics in magnesium catalyzed with Nb_2O_5. *Journal of Alloys and Compounds*, Vol. 420, (2006), pp. (46-49)

Hanada, N., Ichikawa, T., & Fujii, H. Hydrogen absorption kinetics of the catalyzed MgH_2 by niobium oxide. *Journal of Alloys and Compounds*, Vol. 446-447, (2007), pp. (67-71)

Hanada, N., Hirotoshi, E., Ichikawa, T., Akiba, E., & Fujii, H. SEM and TEM characterization of magnesium hydride catalyzed with Ni nano-particle or Nb_2O_5. *Journal of Alloys and Compounds*, Vol. 450, (2008), pp. (395-399)

Kim, J.W., Ahn, J.P., Kim, D.H., Chung, H.S., Shim, J.H., Cho, Y.W., & Oh, K.H. In-situ electron microscopy study on microstructural changes in NbF_5-doped MgH_2 during dehydrogenation. *Scripta Materialia*, Vol. 62, (2010), pp. (701-704)

Liang, G., Huot, J., Boily, S., Van Neste, A., & Schulz, R. Catalytic effect of transition metals on hydrogen sorption in nanocrystalline ball-milled MgH_2-Tm (Tm=Ti, V, Mn, Fe and Ni) systems. *Journal of Alloys and Compounds*, Vol. 292, (1999), pp. (247-252)

Løken, S., Solberg, J.K., Maehlen, J.P., Denys, R.V., Lototsky, M.V., Tarasov, B.P., & Yartys, R.V. Nanostructured Mg-Mm-Ni hydrogen storage alloy: Structure-properties relationship. Journal of Alloys and Compounds, Vol. 446-447, (2007), pp. (114-120)

Massalski, T.B. (Ed.). (1986). *Binary Alloy Phase Diagrams*, American Society of Metals, Matals Park, Ohio

Oelerich, W., Klassen, T., & Bormann, R. Metal oxides as catalysts for improved hydrogen sorption in nanocrystalline Mg-based materials. *Journal of Alloys and Compounds*, Vol. 315, (2001), pp. (237-242)

Porcu, M., Petford-Long, A.K., & Sykes, J.M. TEM studies of Nb_2O_5 catalyst in ball-milled MgH_2 for hydrogen storage. *Journal of Alloys and Compounds*, Vol. 453, (2008), pp. (341-346)

Spassov, T., & Köster, U. Hydrogenation of amorphous and nanocrystalline Mg-based alloys. *Journal of Alloys and Compounds*, Vol. 287, (1999), pp. (243-250)

Tanaka, K., Kanda, Y., Furuhashi, M., Saito, K., Kuroda, K., & Saka, H. Improvement of hydrogen storage properties of melt-spun Mg-Ni-RE alloys by nanocrystallization. *Journal of Alloys and Compounds*, Vol. 293-295, (1999), pp. (521-525)

Tanaka, K. Hydride stability and hydrogen desorption characteristics in melt-spun and nanocrystallized Mg-Ni-La alloy. *Journal of Alloys and Compounds*, Vol. 450, (2008), pp. (432-439)

Tanaka, K., Miwa, T., Sasaki, K., & Kuroda, K. TEM studies of nanostructure in melt-spun Mg-Ni-La alloy manifesting enhanced hydrogen desorbing kinetics. *Journal of Alloys and Compounds*, Vol. 478, (2009), pp. (308-316)

Yamada, T., Yin, J., & Tanaka, K. Hydrogen storage properties and phase structures of Mg-rich Mg-Pd, Mg-Nd and Mg-Pd-Nd alloys. *Materials Transactions*, Vol. 42, No. 11, (2001), pp. (2415-2421)

Yamaga, A., Hibino, K., Suzuki, M., Yamada, M., Tanaka, K., & Ueda, K. Analysis of hydrogen distribution on Mg-Ni alloy surface by scanning electron-stimulated desorption ion microscope (SESDIM). *Journal of Alloys and Compounds*, Vol. 460, (2008), pp. (432-439)

Yin, J. & Tanaka, K. Hydriding-dehydriding properties of Mg-rich Mg-Ni-Nd alloys with refined microstructures. *Materials Transactions*, Vol. 43, No. 7, (2002), pp. (1732-1736)

Wiswall, R. (1978) Hydrogen Storage in Metals, In: *Hydrogen in Metals II*, Alefeld, G. & Völkl, J (Eds.), pp. (201-240), Springer, ISBN 3-540-08883-0, Berlin, Heidelberg, New York

Zaluska, A., Zaluski, L., & Ström-Olsen, J.O. Nanocrystalline magnesium for hydrogen storage. *Journal of Alloys and Compounds*, Vol. 288, (1999), pp. (217-225)

Chapter 6

Application of Ionic Liquids in Hydrogen Storage Systems

Sebastian Sahler and Martin H.G. Prechtl

Additional information is available at the end of the chapter

http://dx.doi.org/10.5772/50154

1. Introduction

In the present world, individual mobility is mainly based on fossil hydrocarbons as transportable energy medium. Since there is a wide agreement on the future depletion of fossil resources, an alternative energy carrier, that can be transported efficiently, has to be found. One material, which is in many considerations suitable to fill this gap, is hydrogen.(Hamilton et al., 2009; Schlapbach & Zuttel, 2001) The major advantages of hydrogen for these purposes are the low weight, the high abundance of the oxide, i.e. water, and the environmental benignity of the waste oxidation product, also water. The major hindrance for a widespread application of hydrogen in mobile applications is its low density in the gaseous state. Since the compression or liquefaction of hydrogen respectively the transportation of the denser forms of hydrogen is accompanied by a manifold of problems, i.e. severe disadvantages in especially gravimetric efficiency, alternatives are sought after.(Eberle et al., 2009; Felderhoff et al., 2007; Hamilton et al., 2009; Marder, 2007; Staubitz et al., 2010)

Since there are some materials that consist of an amount of hydrogen, that is high enough to compete with the aforementioned physical storage solutions, the storage of hydrogen in a chemical compound is discussed intensively lately. One important compound with one of the highest hydrogen contents possible is ammonia borane. The application of ammonia borane and its derivatives for hydrogen storage is matter of research in the last decade as well as a compound as simple as formic acid, which is in this sense the hydrogenation-product of carbon dioxide.(Fellay et al., 2008; Loges et al., 2008; Scholten et al., 2010; Stephens et al., 2007)

The application of ammonia borane itself is more efficient than elemental hydrogen, but nevertheless accompanied by some problems as well. Since there is a worldwide infrastructure suitable for the deployment of liquid fuels, a solid fuel bears certain disadvantages competing with a liquid fuel. Another important aspect of fuels is the applicability, and here liquid fuels are also advantageous in comparison to solid fuels.

In conclusion it seems reasonable to search for a liquid fuel (-system). Regarding this assumption ionic liquids (ILs) promise advantages owing to their tunable physico-chemical properties.(Dupont & Suarez, 2006) Hydrogen-enriched materials or blends are as well in the focus of current research as catalytic decomposition.(Groshens & Hollins, 2009; Hugle et al., 2009; Jaska et al., 2003; Mal et al., 2011; Wright et al., 2011) ILs are investigated for their ability of supporting decomposition as well as solution of certain materials, including in some cases spent fuel products.(Bluhm et al., 2006; Wright et al., 2011) In some extent these approaches can be applied in combination.(Wright et al., 2011)

1.1. Ionic liquids

In the last two decades ionic liquids (ILs) have become very popular in various fields in chemistry.(Dupont et al., 2002; Dupont & Scholten, 2010; Dupont et al., 2011; Migowski & Dupont, 2007; Prechtl et al., 2010; Prechtl et al., 2011; Scholten et al., 2012) Their physico-chemical properties, i. e. non-flammable, non-volatile, highly solvating, in general weakly-coordinating, tunable polarity and good thermal stability make them highly attractive for various applications.(Dupont, 2004; Dupont & Suarez, 2006) Low melting (organic) salts are defined as IL if the melting point is below 100 °C.(Dupont et al., 2002; Hallett & Welton, 2011; Welton, 1999)

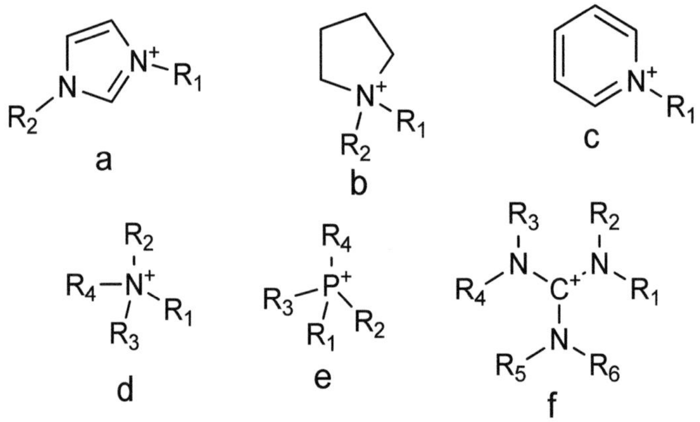

Figure 1. Selected examples for IL-cations: (a) imidazolium, (b) pyrrolidinium, (c) pyridinium, (d) ammonium, (e) phosphonium and (f) guanidinium. The side-chains are alkyl groups, which can also carry functionalities like alcohols, ethers, nitriles, etc. Usually weakly coordinating anions are used as counter-parts such as BF_4^-, PF_6^-, bistrifluoromethanesulfonylimide (NTf_2^-) etc.

The low melting points are a result of the reduced lattice energy originating from large ion-pairs and low symmetry of the cations (Figure 1).(Dupont, 2004; Dupont & Suarez, 2006; Krossing et al., 2006) Ionic liquids have been known since the early 20th century, though their predicted potential and wider application developed rapidly in the last two decades. For a more detailed historic introduction see the article by Wilkes.(Wilkes, 2002)

For different applications the physico-chemical properties of ILs can be modified by tailoring the cation and/or anion structures.(Chiappe et al., 2006; Cui et al., 2010; Fei et al., 2007; Yang et al., 2008; Zhao et al., 2004) It is possible to design the hydrophilicity or hydrophobicity, the viscosity as well as the coordinating properties, especially those implied by the anion. Another important aspect one has to keep in mind is the pK_a value, since especially in imidazolium-ILs the C_2-proton is fairly acidic(Dupont, 2011; Dupont et al., 2002; Prechtl et al., 2010; Welton, 1999) and metal ions form carbene complexes in some cases.(Xu et al., 2000) This can be avoided by protecting the C_2-position with a methyl group. Some ILs are able to stabilize certain homogenous and heterogeneous catalysts against decomposition or agglomeration, which can be of great benefit.(Prechtl et al., 2010)

The outcome of a reaction is strongly influenced by the solubility of the reactants, intermediates and products. In the particular case of dehydrogenation reactions, the reaction is influenced by the rather low solubility of hydrogen gas in ILs driving the equilibrium to the desired side.(Anthony et al., 2002; Jacquemin et al., 2006) In recent research ionic liquids have been employed not only as solvent in dehydrogenation reactions, but also as storage materials (vide infra).

1.2. Hydrogen-rich molecules as storage media

Suitable molecules for the storage of hydrogen have to fulfill several requirements. Since hydrogen is storable in elemental form as a gas or liquid, these molecular systems must bear certain advantages over storage as elemental hydrogen. The most impractical properties of elemental hydrogen are the low density of gaseous hydrogen and the high requirements to the storage system in case of compressed or liquid hydrogen. The amount of hydrogen (atmospheric pressure) required for a 500 km travel (6 kg) would require a tank volume of nearly 67 m^3! To overcome this problem one is able to compress hydrogen into a reasonable volume. A 700 bar hydrogen storage tank is able to store the same amount of hydrogen in the volume of only 260 liters.(Eberle et al., 2009) Due to the severe pressure the tank has to withstand, the fuel including the tank weights 125 kg. This corresponds to a gravimetric efficiency of 4.8 %. Concerning liquid storage the losses due to boil off are considered too high for practical application, since they add up to the problems of a strongly isolated tank, i.e. gravimetric inefficiency. Since molecular hydrogen has these severe disadvantages, it seems reasonable to store hydrogen in a chemical way, where higher volumetric hydrogen density in the condensed phase is an obtainable target.

1.2.1. Ammonia Borane (AB)

One important molecule in the focus of research is the simple adduct of ammonia and borane (AB). (Ahluwalia et al., 2011; Al-Kukhun et al., 2011; Basu et al., 2011; Bluhm et al., 2006; Staubitz et al., 2010; Stephens et al., 2007) This solid Lewis pair is stable towards air and water and consists of 19.6 % hydrogen. When heating this compound to 130 °C it loses 14 % of its weight as hydrogen.(Eberle et al., 2009) Undesired side-products in the decomposition-reaction can be ammonia, diborane and borazine, where the last is a standard decomposition product, that is only critical due to its high volatility contaminating the gas stream and lowering efficiency due to incomplete hydrogen release. For the complete dehydrogenation of AB temperatures of over 500 °C are required. The dehydrogenation of this compound leads via oligoamino boranes and polyamino boranes to insoluble polyborazylene (Figure 2), which is significantly stable in mechanical, chemical and thermal means.

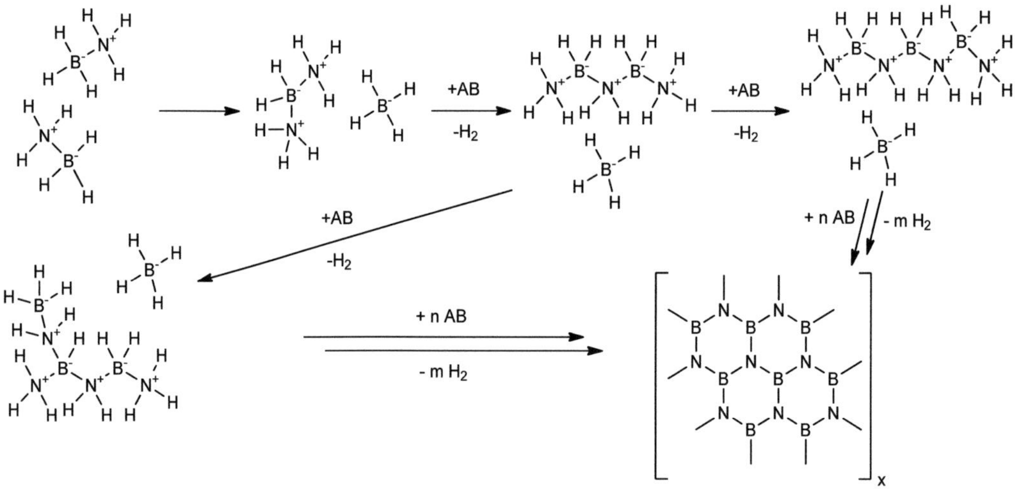

Figure 2. Decomposition Pathway of Ammonia Borane.

Since the temperature to decompose ammonia borane completely is so high and the according product is remarkably stable, complete dehydrogenation is undesired. As above mentioned, partial dehydrogenation leads to a capacity of around 14 wt% which is still enough to be considered as efficient hydrogen storage material.

Regarding the regeneration of spent fuel Sutton and co-workers recently presented meaningful results.(Sutton et al., 2011) The solution of AB spent fuel in liquid ammonia could be quantitatively reduced by hydrazine to yield AB, with gaseous nitrogen as the only by-product.

1.2.2. Ammonia Borane derivatives

Several derivatives of AB have been studied recently. The most important are: hydrazine borane (HB),(Hugle et al., 2009) guanidinium borohydride (GBH),(Groshens & Hollins, 2009) ethylenediamine bisborane (EDB),(Neiner et al., 2011) methylguanidinium borohydride (Me-GBH)(Doroodian et al., 2010) and different alkyl amine boranes (Figure 3).(Bowden et al., 2008; Mal et al., 2011) On the frontier between molecular and metal hydride storage materials are the metal amido borane compounds, which are not in the scope of this chapter.(Chua et al., 2011)

A derivative of AB with a comparable weight efficiency has been studied by Lentz and co-workers recently: hydrazine borane (HB).(Hugle et al., 2009) Much of the hydrogen content of pure HB is thermally available, but the efficiency of release could be significantly improved by combining HB with the hydride-donor LiH: Though blending with LiH in 1:1 molar ratio lowers the theoretical gravimetric hydrogen density from 15.4 wt% to 14.8 wt% the actual release of hydrogen reached nearly 12 wt% at 150 °C in 4.5 h. The idea of combining HB with a hydride-donor arose from the fact that HB has an excess acidic hydrogen atom, since it consists of four acidic but only 3 hydridic hydrogen atoms. The released gas stream showed to consist of mainly hydrogen with an impurity of ammonia present in the range of <1%. The solid residue showed to be reactive towards water and completely insoluble in organic solvents. Hydrazine bisborane, a molecule very similar to HB, with a hydrogen content a little higher than HB, is not considered due to stability problems. The authors reported that rapid heating or temperatures above 160 °C lead to explosive decomposition.

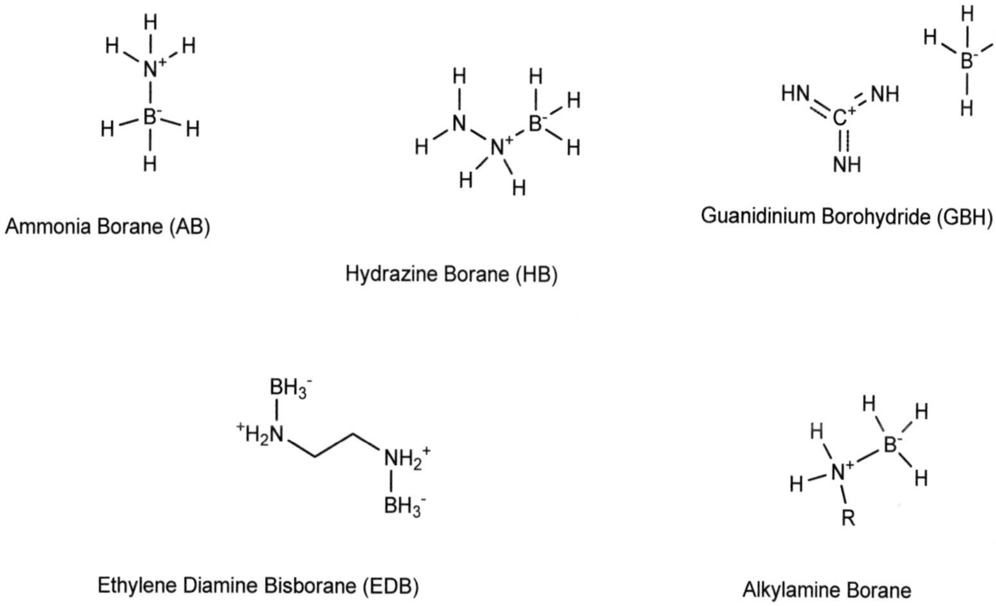

Figure 3. Structures of selected molecular hydrogen storage materials.

Guanidinium borohydride (GBH) and blends containing GBH were investigated by Groshens and co-workers showing remarkable results.(Groshens & Hollins, 2009) The material consists of the guanidinium ion contributing acidic protons and borohydride for hydridic hydrogen atoms. As this material consists only of light weight atoms and hydrogen, the theoretical capacity is 10.8 wt%. The exothermic reaction results in a self-sustaining decomposition when initiated by a heat source. Nearly quantitative release of hydrogen was observed with a distinct impurity of ammonia of about 5 %. The imbalance of acidic and hydridic hydrogen atoms is a factor susceptible to improvement by blending. The material of choice for blending was ethylene diamine bisborane, which contributes two excess hydridic hydrogen atoms to the mixture. As shown in figure 4 the blending results in the maintenance of hydrogen yield, while suppressing the formation of undesirable ammonia. This effect is exceptionally strong in the mixtures containing approximately equimolar amounts of both materials.

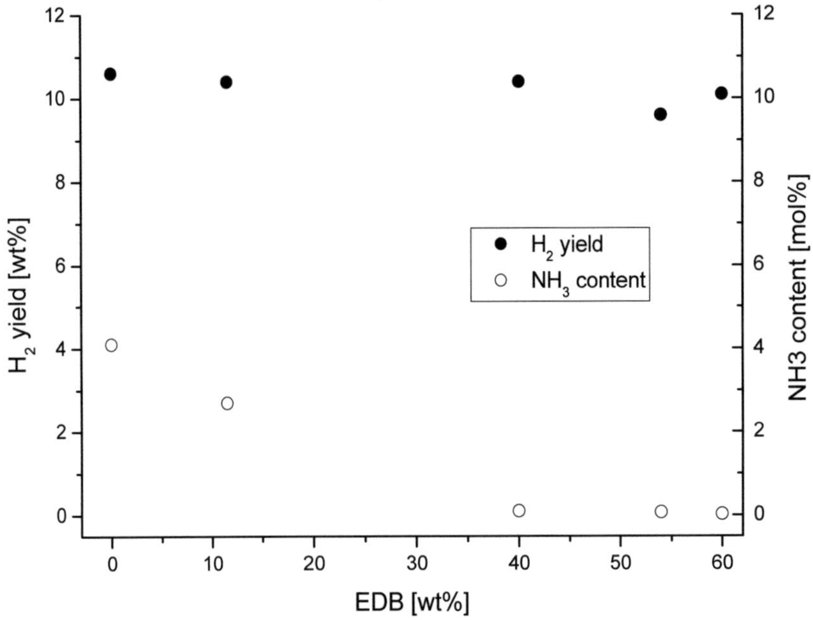

Figure 4. GBH-EDB Self-Sustaining-Thermal-Decomposition. Mixture containing 60 wt% EDB is not self-sustaining.(Groshens & Hollins, 2009)

In comparison to pure GBH, that decomposes with an impurity of ammonia of 4.1 mol% in the gas stream, resulting in a hydrogen yield of 10.6 wt%, the 40:60 wt% mixture of GBH and EDB releases 10.1 wt% of hydrogen with an impurity of ammonia as low as 0.026 mol%. The solid state of this material(-blend) is one major drawback.

The decomposition of pure EDB is comparably efficient to that of the blend mentioned above.(Neiner et al., 2011) The characteristics in decomposition are especially interesting due to the purity of released hydrogen without additives. No volatile by-products are released up to a temperature of 200 °C. In addition the decomposition achieves release rates comparable to AB.

1.2.3. Formic acid

In recent years catalytic splitting of formic acid into hydrogen and carbon dioxide has drawn some interest (Figure 4). (Boddien et al., 2011; Enthaler et al., 2010; Johnson et al., 2010; Scholten et al., 2010; Tedsree et al., 2011; Yasaka et al., 2010; Zhao et al., 2011)

Figure 5. Reversible splitting of formic acid.

Though the hydrogen storage capacity of 4.4 wt% stays behind that of several other materials, the simplicity and wide availability of carbon dioxide as educt make formic acid attractive for hydrogen storage. In chemical means formic acid can be utilized as hydrogen source for catalytic hydrogenation reactions. (Blum et al., 1972; Bulushev & Ross, 2011; Kawasaki et al., 2005)

2. Ionic liquids in hydrogen storage systems

The unique physico-chemical properties of ILs make it possible to use them for different applications in hydrogen storage systems. The most simple one is to use a low molecular weight, hydrogen rich IL as hydrogen storage material. This approach has been realized in two different ways, which are presented below. Other possible applications are as inert solvent, as a catalytically active solvent, as a co-catalyst or as a stabilizing agent e.g. for nanoparticles, as presented afterwards.

2.1. ILs as hydrogen storage materials

Introducing one methyl group into a GBH molecule results in the first reported IL that can be efficiently dehydrogenated (Figure 6).(Doroodian et al., 2010)

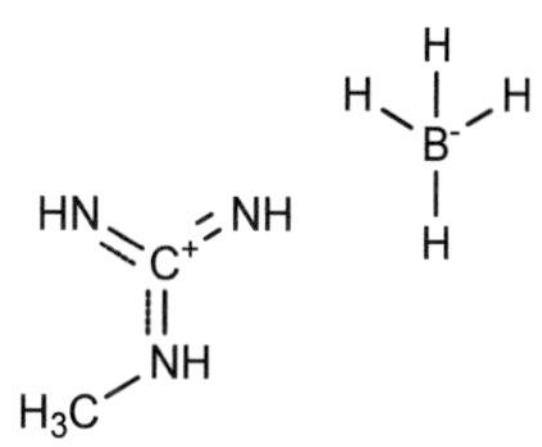

Methylguanidinium Borohydride (Me-GBH)

Figure 6. Structure of methyl guanidinium borohydride.

Me-GBH has a storage capacity slightly reduced in compraison to GBH. Rieger and co-workers modified GBH by introducing a methyl group. As the methyl group lowers the cation's symmetry and, consequently, lowers the melting point,(Mateus et al., 2003) this salt is an IL, i.e. liquid below 100 °C. Furthermore with a melting point of -5 °C Me-GBH is the first room temperature IL that can be readily dehydrogenated with a reasonable gravimetric efficiency having a theoretical capacity of 9.0 wt%. Kinetic investigations by Rieger and co-workers about the thermal decomposition (75 °C) showed the decomposition of this material to be insufficiently slow and too inefficient for application. In the course of the decomposition some mass loss exceeding the theoretical capacity of the substrate is observed and consequently cannot be assigned to hydrogen evolution. Temperatures above 120 °C result in a detectable ammonia amount in the gas stream. The authors reported around 9 wt% hydrogen yield determined by TGA and volumetric quantification. The dehydrogenation product is solid and insoluble in the IL. To improve the rate and extent of hydrogen generation Wilkinson's catalyst ($(PPh_3)_3RhCl$) and $FeCl_2$ were employed with moderate success.

Initial attempts using an imidazolium IL containing a cyclohexyl moiety as hydrogen carrier were reported by Dupont and co-workers.(Stracke et al., 2007) This group used commercial palladium on charcoal catalyst and rather harsh conditions in the dehydrogenation/hydrogenation cycles. Despite the theoretical storage capacity of this system being insignificant, this was the first report of an IL being employed as hydrogen storage material. Chemically this procedure is based on the reversible dehydrogenation of a cyclohexyl group (Figure 7).

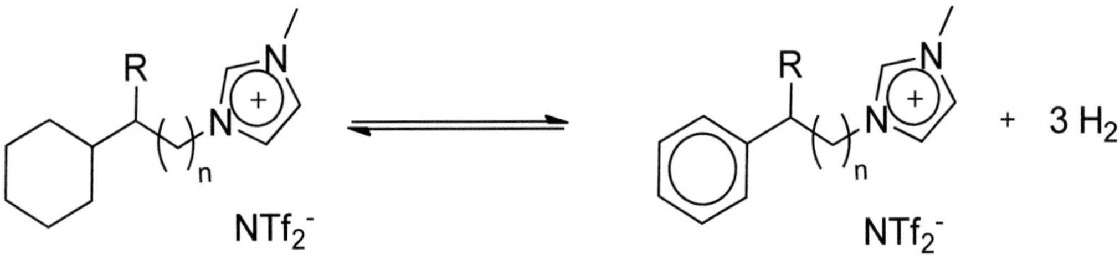

n=0-2 R=H,Cy,Ph

Figure 7. Structure of cyclohexyl-ILs for hydrogen storage by Dupont and co-workers.(Stracke et al., 2007)

The strong thermal stress (300 °C), required for fast dehydrogenation, and low weight efficiency are the most severe drawbacks of this system. The thermal stability of some of the employed ILs is remarkable.

2.2. ILs as stabilizing agents for nanoparticles for catalytic dehydrogenation

Dimethyl amine borane (DMAB) is another derivative of AB able to decompose releasing hydrogen. It has been studied for mechanistic investigations, since the two additional methyl groups inhibit further dehydrogenation after the first equivalent of hydrogen. The group of Manners investigated catalytic dehydrocoupling of different disubstituted amine boranes for the selective formation of cyclic dimers and trimers at ambient temperatures.(Jaska et al., 2003) These investigations were found to be important for mechanistic considerations in the dehydrogenation of amine boranes (see section 2.2.1). The dehydrogenation of DMAB proceeds via the diammoniate of diborane to the cyclic dimer.(Zahmakiran & Ozkar, 2009) The group of Özkar found DMAB to reduce the dimeric rhodium(II)-hexanoate to rhodium(0) forming Rh-clusters and dimethyl ammonium hexanoate (Figure 8).

The formed nanoclusters (~2 nm) were tested for catalytic activity in the dehydrogenation of dimethyl amine borane and found to be highly active. The *in-situ* generated IL dimethyl ammonium hexanoate was found to act as protecting agent for the nanoparticles.

2.3. ILs as active/supporting solvents

Another interesting option is the employment of an ionic liquid (IL) as active solvent supporting the dehydrogenation of AB.(Bluhm et al., 2006; Himmelberger et al., 2009)

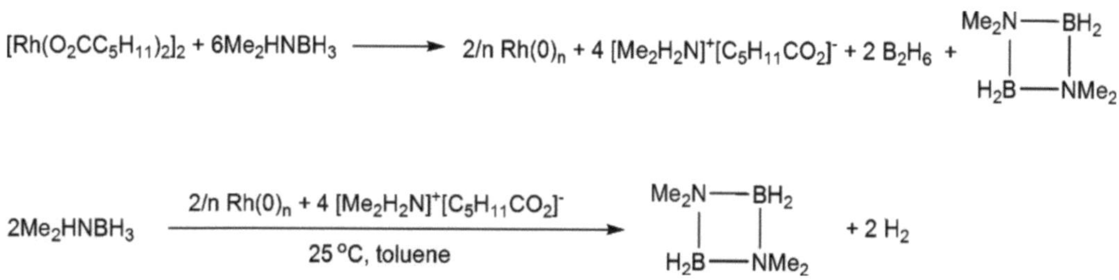

Figure 8. Rhodium nanoparticle formation and catalytic decomposition of dimethyl amine borane. Reprinted with permission from (Zahmakiran & Ozkar, 2009). Copyright (2009) American Chemical Society.

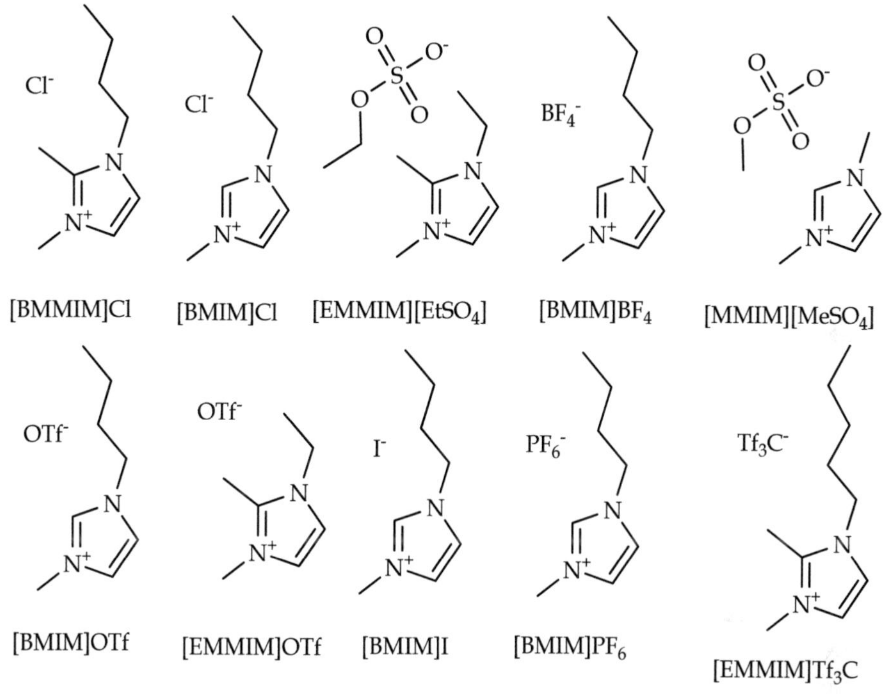

Figure 9. Structures of employed ILs by Sneddon and co-workers.(Himmelberger et al., 2009)

Figure 10 displays the hydrogen release results from Toepler pump measurements of the decomposition reactions of 50 wt% of AB in different ILs at 85 °C. Whereas the decomposition in pmmimTf3C is comparable to the decomposition of neat AB (compare Figure 11(left)), every other tested IL improves the hydrogen yield significantly, though initial rates of hydrogen release do not relate strictly to the amount of overall released hydrogen.

Deeper investigations were conducted on the decomposition of AB in 1-butyl-3-methyl-imidazolium chloride ([BMIM]Cl). The application of this IL results in enhanced kinetics (Figure 11) and improved hydrogen yield. As shown in figure 11 (left) the hydrogen yield at a temperature as low as 85 °C can be more than doubled by addition of [BMIM]Cl implicating that a 50 wt% mixture of AB and [BMIM]Cl has a better weight-efficiency than pure AB.

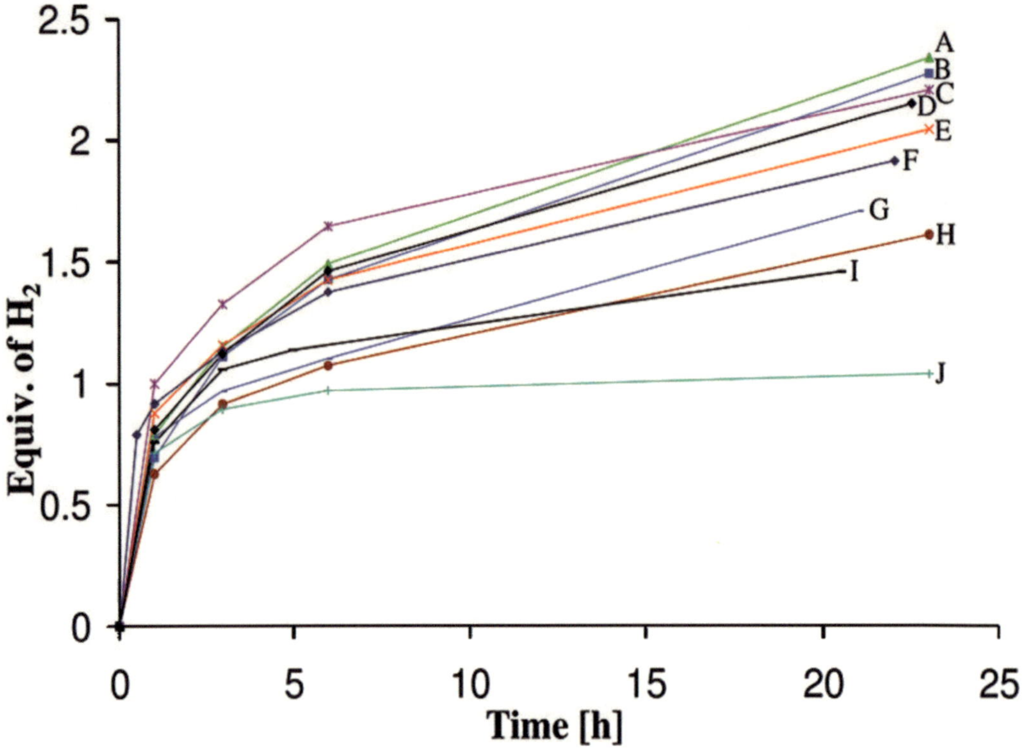

Figure 10. H_2-release measurements (Toepler pump) of the reaction of 50 wt% AB (250 mg) at 85 °C in 250 mg of (A) [BMMIM]Cl, (B) [BMIM]Cl, (C) [EMMIM][EtSO$_4$], (D) [BMIM]BF$_4$, (E) [MMIM][MeSO$_4$], (F) [BMIM]OTf, (G) [EMMIM]OTf, (H) [BMIM]I, (I) [BMIM]PF$_6$, and (J) [PMMIM]Tf$_3$C. Reprinted with permission from (Himmelberger et al., 2009). Copyright (2009) American Chemical Society.

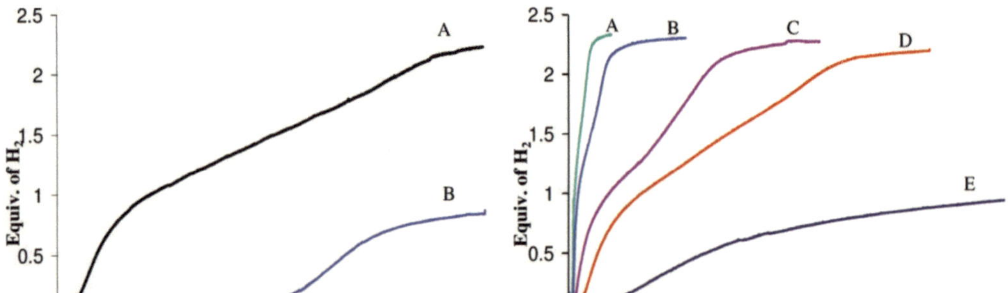

Figure 11. (Left) H_2-release measurements at 85 °C of: (A) 50-wt% AB in [BMIM]Cl and (B) solid-state AB (Right) H_2-release measurements of 50-wt % AB in [BMIM]Cl at (A) 110 °C, (B) 105 °C, (C) 95 °C, (D) 85 °C, and (E) 75 °C. Reprinted with permission from (Himmelberger et al., 2009). Copyright (2009) American Chemical Society.

Only trace amounts of volatile borazine are identified in the gas stream indicating that its formation is suppressed or borazine is soluted readily by the IL. As can be seen, the addition of [BMIM]Cl prevents induction times and promotes dehydrogenation. This effect persists, if the amount of IL is reduced to 20 wt%. In direct comparison, the promoted dehydrogenation at 75 °C releases an equal amount of hydrogen as neat AB at 85 °C but without the induction time. An increase in temperature leads to a slight increase in quantity of hydrogen and an intense acceleration: The dehydrogenation at 85 °C results in a release of about 2.1 equivalents of hydrogen in 250 minutes and at 110 °C approximately 2.3 equivalents are released in less than 30 minutes.

2.4. ILs as solvents for catalytic AB dehydrogenation

Employing a metal complex as catalyst is an option to improve reaction kinetics and optimize low temperature application for the demands of hydrogen-storage materials for transportation applications.(Alcaraz & Sabo-Etienne, 2010; Jaska et al., 2003) This idea was combined with the idea of using ILs as supporting solvents. Sneddon and co-workers reported that ILs, loaded with different precious metal (Rh, Ru, Pd) or Ni-based pre-catalysts, decrease the onset temperature (45–85°C) of AB-decomposition significantly improving hydrogen yield at the same time (Figure 12).(Wright et al., 2011)

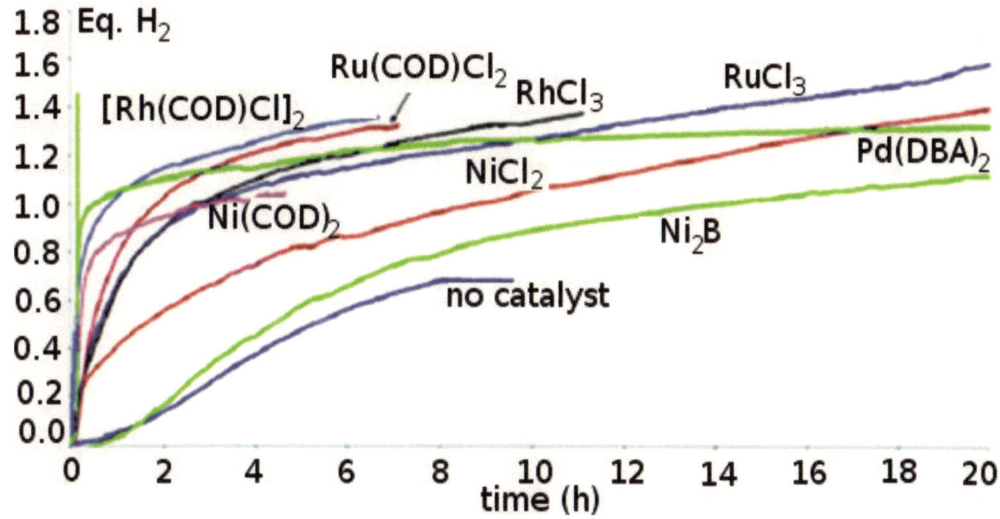

Figure 12. Hydrogen evolution from AB-decomposition at 65 °C catalyzed by different metal catalyst-precursors. Reproduced from (Wright et al., 2011) with permission of The Royal Society of Chemistry.

Though there were no deeper investigations concerning the active species, many of the precursors are known to form nanoparticles when being treated with a reducing agent.(Migowski & Dupont, 2007) This leads to the assumption that metal(0)-nanoparticles are involved in at least some of the catalyses. It is clearly visible that many of the employed precursors evoke a severe acceleration of hydrogen release as well as improving yield significantly as well. The shown results were obtained in experiments with 5 mol% of the respective catalyst.

In recent work Baker and co-workers employed another substrate blend for catalytic decomposition in IL.(Mal et al., 2011) The utilized mixture consisted of AB and sec-butyl amine borane (SBAB). The decomposition of this blend in the IL 1-ethyl-3-methylimidazolium ethyl sulfate ([EMIM][EtSO$_4$]) leaves only a solution of products behind. The formation of insoluble polyamino borane is effectively hindered. Instead the formation of the cyclic (formal trimeric) borazine derivative is observed (Figure 13).

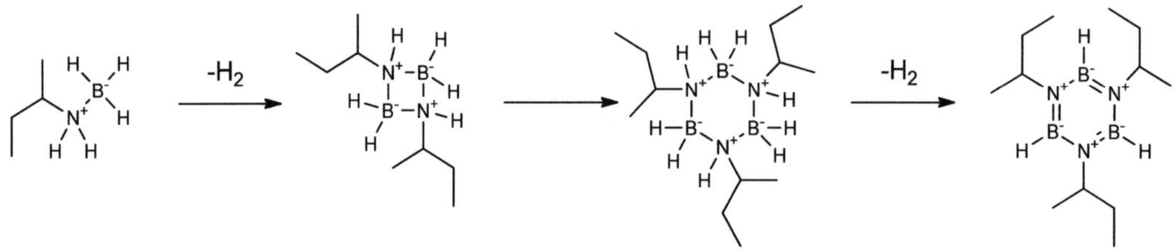

Figure 13. Observed decomposition products of SBAB decomposition.(Mal et al., 2011)

The alkyl group decreases the gravimetric efficiency to 5 wt%. This value was achieved by a metal catalyzed dehydrogenation reaction at 80 °C employing 1% RuCl$_2$(PMe$_4$)$_3$ as catalyst-precursor. Concerning the purity of the released hydrogen no investigations have been reported yet.

2.5. Functionalized ILs for catalytic decomposition of formic acid

A very interesting application of ILs is the utilization as solvent and base at the same time. In the decomposition of formic acid this can be realized by the employment of the amine-functionalized imidazolium-IL **1** (Figure 14). (Li et al., 2010; Scholten et al., 2010)

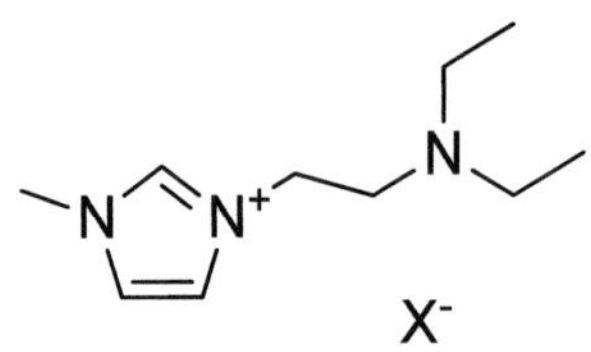

Figure 14. Structure of amino-functionalized IL **1**.(Li et al., 2010; Scholten et al., 2010)

While the Dupont group reported only about the application of the chloride IL **1**, the Shi-group used various anions, i.e. BF_4^-, OTf^-, NTf_2^- and $HCOO^-$. Both groups used the catalyst (-precursor) [(p-Cymene)RuCl$_2$]$_2$, investigations of the Dupont group suggested the formation of hydrogen bridged and hydrogen/formate bridged Ru(p-Cymene)-Dimers. The group of Shi, Deng and co-workers reported varying TOFs up to 36 h^{-1} without additional base and up to 627 h^{-1} with additional base. The group of Dupont and co-workers reported TOFs of up to 1684 h^{-1} without additional base.

In investigations of the Dupont group the amount of IL in the system seemed to play an important role, since after a required amount additional IL decreases the efficiency of the system remarkably. Addition of more formic acid proved the durability of this easy system in 6 recycling experiments. The high activity of the first run could not be sustained in subsequent runs. Nevertheless the slightly reduced activity remained constant for the subsequent runs except the 5th (Figure 15).

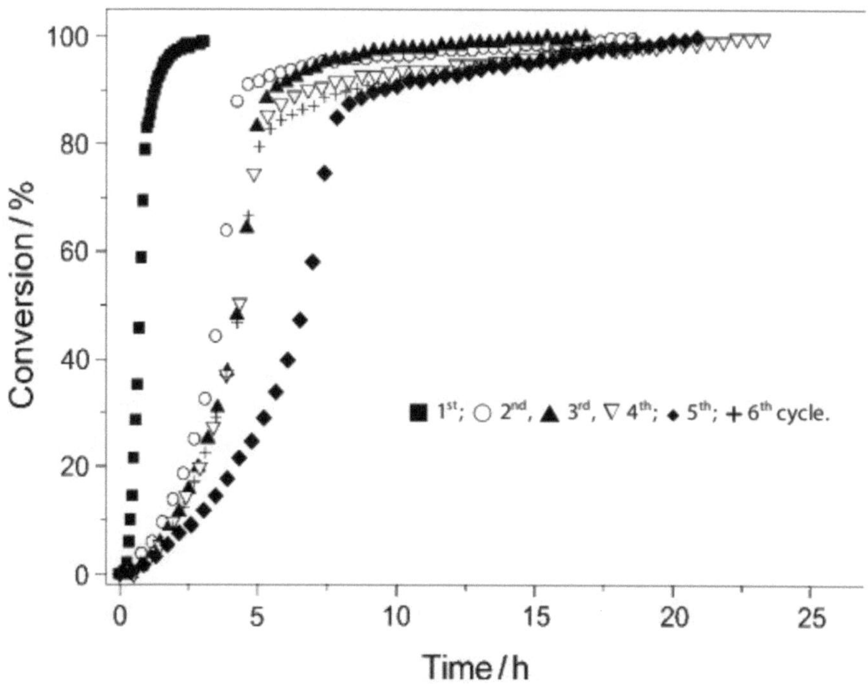

Figure 15. Reaction profiles for formic acid decomposition catalyzed by [(p-cymene)RuCl2] in IL **1** (Cl^-) at 80 °C for 6 cycles. Reprinted with permission from (Scholten et al., 2010). Copyright (2010) John Wiley and Sons.

The same catalyst (-precursor) was investigated in another context: The solubilization and decomposition of biomass especially cellulose-based ones. Different ILs can be employed to solvate the carbohydrate feedstock and the [(p-Cymene)RuCl2]2 complex (-precursor) is suitable for dehydrogenation.(Taccardi et al., 2010) Detailed investigations by Wasserscheid et al. showed that this reaction proceeds via the thermal decomposition of glucose to formic acid and subsequent catalytic decomposition of formic acid by the Ru complex. Some ILs, in this case for example 1-ethyl-3-methylimidazolium methyl-methylphosphonate, are able to solute wood. The soluted biomass can be thermally decomposed at 180 °C and the resulting formic acid can be split into hydrogen and CO_2 by the Ru-complex (Figure 16).

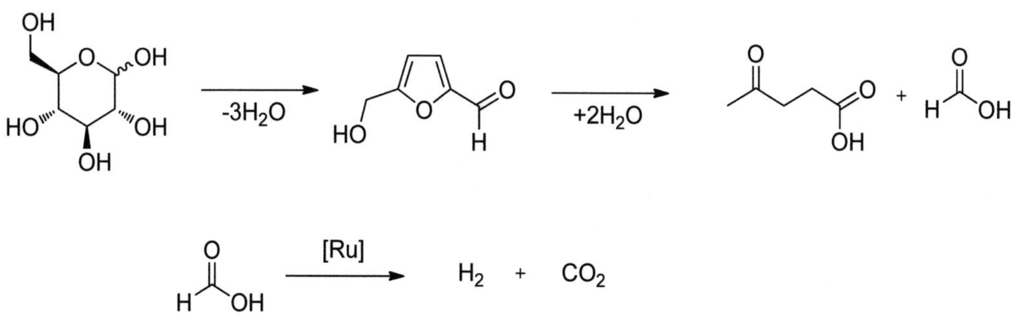

Figure 16. Thermal decomposition pathway of glucose and subsequent catalytic splitting of formic acid in IL.(Taccardi et al., 2010)

The thermal stability of the employed IL is remarkable and important in this application.

3. Conclusion and outlook

In the field of molecular hydrogen storage materials, ILs add interesting aspects in different meanings. Besides the application of ILs as (molecular) hydrogen storage materials, they can act in several other ways: They can be utilized as inert solvents, as solvents which support dehydrogenation, i.e. active solvents, as promoting agent/base or as a combination of these aspects. Especially the unique solubilization properties of ILs show considerable promise to find a liquid hydrogen storage system that remains liquid after dehydrogenation. One important aspect one has to keep in mind, when considering the application of ILs, is the addition of weight into the system, which, in general, lowers the gravimetric efficiency. This 'weight penalty' is the most severe drawback in the discussed applications of ILs.

Author details

Sebastian Sahler and Martin H.G. Prechtl*
University of Cologne, Cologne, Germany

4. References

Ahluwalia, R. K., Peng, J. K. & Hua, T. Q. (2011). Hydrogen release from ammonia borane dissolved in an ionic liquid *International Journal of Hydrogen Energy*, Vol. 36, No. 24, (Dec), pp. 15689-15697, ISSN: 0360-3199.

Al-Kukhun, A., Hwang, H. T. & Varma, A. (2011). A Comparison of Ammonia Borane Dehydrogenation Methods for Proton-Exchange-Membrane Fuel Cell Vehicles: Hydrogen Yield and Ammonia Formation and Its Removal *Industrial & Engineering Chemistry Research*, Vol. 50, No. 15, (Aug 3), pp. 8824-8835, ISSN: 0888-5885.

Alcaraz, G. & Sabo-Etienne, S. (2010). Coordination and Dehydrogenation of Amine-Boranes at Metal Centers *Angewandte Chemie-International Edition*, Vol. 49, No. 40, pp. 7170-7179, ISSN: 1433-7851.

Anthony, J. L., Maginn, E. J. & Brennecke, J. F. (2002). Solubilities and thermodynamic properties of gases in the ionic liquid 1-n-butyl-3-methylimidazolium hexafluorophosphate *Journal of Physical Chemistry B*, Vol. 106, No. 29, (Jul 25), pp. 7315-7320, ISSN: 1520-6106.

Basu, S., Zheng, Y. & Gore, J. P. (2011). An experimental study of neat and ionic liquid-aided ammonia borane thermolysis *Journal of Power Sources*, Vol. 196, No. 2, (Jan 15), pp. 734-740, ISSN: 0378-7753.

Bluhm, M. E., Bradley, M. G., Butterick, R., Kusari, U. & Sneddon, L. G. (2006). Amineborane-based chemical hydrogen storage: Enhanced ammonia borane dehydrogenation in ionic liquids *Journal of the American Chemical Society*, Vol. 128, No. 24, (Jun 21), pp. 7748-7749, ISSN: 0002-7863.

* Corresponding Author

Blum, J., Sasson, Y. & Iflah, S. (1972). Hydrogen transfer from formyl compounds to α,β-unsaturated ketones catalyzed by Ru, Rh and Ir complexes *Tetrahedron Letters*, Vol. 13, No. 11, pp. 1015-1018, ISSN: 0040-4039.

Boddien, A., Mellmann, D., Gartner, F., Jackstell, R., Junge, H., Dyson, P. J., Laurenczy, G., Ludwig, R. & Beller, M. (2011). Efficient Dehydrogenation of Formic Acid Using an Iron Catalyst *Science*, Vol. 333, No. 6050, (Sep 23), pp. 1733-1736, ISSN: 0036-8075.

Bowden, M. E., Brown, I. W. M., Gainsford, G. J. & Wong, H. (2008). Structure and thermal decomposition of methylamine borane *Inorganica Chimica Acta*, Vol. 361, No. 7, (May 20), pp. 2147-2153, ISSN: 0020-1693.

Bulushev, D. A. & Ross, J. R. H. (2011). Vapour phase hydrogenation of olefins by formic acid over a Pd/C catalyst *Catalysis Today*, Vol. 163, No. 1, (Apr 12), pp. 42-46, ISSN: 0920-5861.

Chiappe, C., Pieraccini, D., Zhao, D. B., Fei, Z. F. & Dyson, P. J. (2006). Remarkable anion and cation effects on Stille reactions in functionalised ionic liquids *Advanced Synthesis & Catalysis*, Vol. 348, No. 1-2, (Jan), pp. 68-74, ISSN: 1615-4150.

Chua, Y. S., Chen, P., Wu, G. T. & Xiong, Z. T. (2011). Development of amidoboranes for hydrogen storage *Chemical Communications*, Vol. 47, No. 18, pp. 5116-5129, ISSN: 1359-7345.

Cui, Y. G., Biondi, I., Chaubey, M., Yang, X., Fei, Z. F., Scopelliti, R., Hartinger, C. G., Li, Y. D., Chiappe, C. & Dyson, P. J. (2010). Nitrile-functionalized pyrrolidinium ionic liquids as solvents for cross-coupling reactions involving in situ generated nanoparticle catalyst reservoirs *Physical Chemistry Chemical Physics*, Vol. 12, No. 8, pp. 1834-1841, ISSN: 1463-9076.

Doroodian, A., Dengler, J. E., Genest, A., Rosch, N. & Rieger, B. (2010). Methylguanidinium Borohydride: An Ionic-Liquid-Based Hydrogen-Storage Material *Angewandte Chemie-International Edition*, Vol. 49, No. 10, pp. 1871-1873, ISSN: 1433-7851.

Dupont, J. (2004). On the solid, liquid and solution structural organization of imidazolium ionic liquids *Journal of the Brazilian Chemical Society*, Vol. 15, No. 3, (May-Jun), pp. 341-350, ISSN: 0103-5053.

Dupont, J. (2011). From Molten Salts to Ionic Liquids: A "Nano" Journey *Accounts of Chemical Research*, Vol. 44, No. 11, (Nov), pp. 1223-1231, ISSN: 0001-4842.

Dupont, J., de Souza, R. F. & Suarez, P. A. Z. (2002). Ionic liquid (molten salt) phase organometallic catalysis *Chemical Reviews*, Vol. 102, No. 10, (Oct), pp. 3667-3691, ISSN: 0009-2665.

Dupont, J. & Scholten, J. D. (2010). On the structural and surface properties of transition-metal nanoparticles in ionic liquids *Chemical Society Reviews*, Vol. 39, No. 5, pp. 1780-1804, ISSN: 0306-0012.

Dupont, J., Scholten, J. D. & Prechtl, M. H. G. (2011). Green Processes. In: *Handbook of Green Chemistry*, pp., Wiley Interscience, Weinheim.

Dupont, J. & Suarez, P. A. Z. (2006). Physico-chemical processes in imidazolium ionic liquids *Physical Chemistry Chemical Physics*, Vol. 8, No. 21, pp. 2441-2452, ISSN: 1463-9076.

Eberle, U., Felderhoff, M. & Schuth, F. (2009). Chemical and Physical Solutions for Hydrogen Storage *Angewandte Chemie-International Edition*, Vol. 48, No. 36, pp. 6608-6630, ISSN: 1433-7851.

Enthaler, S., von Langermann, J. & Schmidt, T. (2010). Carbon dioxide and formic acid-the couple for environmental-friendly hydrogen storage? *Energy & Environmental Science*, Vol. 3, No. 9, (Sep), pp. 1207-1217, ISSN: 1754-5692.

Fei, Z. F., Zhao, D. B., Pieraccini, D., Ang, W. H., Geldbach, T. J., Scopelliti, R., Chiappe, C. & Dyson, P. J. (2007). Development of nitrile-functionalized ionic liquids for C-C coupling reactions: Implication of carbene and nanoparticle catalysts *Organometallics*, Vol. 26, No. 7, (Mar 26), pp. 1588-1598, ISSN: 0276-7333.

Felderhoff, M., Weidenthaler, C., von Helmolt, R. & Eberle, U. (2007). Hydrogen storage: the remaining scientific and technological challenges *Physical Chemistry Chemical Physics*, Vol. 9, No. 21, (Jun 7), pp. 2643-2653, ISSN: 1463-9076.

Fellay, C., Dyson, P. J. & Laurenczy, G. (2008). A viable hydrogen-storage system based on selective formic acid decomposition with a ruthenium catalyst *Angewandte Chemie-International Edition*, Vol. 47, No. 21, pp. 3966-3968, ISSN: 1433-7851.

Groshens, T. J. & Hollins, R. A. (2009). New chemical hydrogen storage materials exploiting the self-sustaining thermal decomposition of guanidinium borohydride *Chemical Communications*, Vol., No. 21, pp. 3089-3091, ISSN: 1359-7345.

Hallett, J. P. & Welton, T. (2011). Room-Temperature Ionic Liquids: Solvents for Synthesis and Catalysis. 2 *Chemical Reviews*, Vol. 111, No. 5, (May), pp. 3508-3576, ISSN: 0009-2665.

Hamilton, C. W., Baker, R. T., Staubitz, A. & Manners, I. (2009). B-N compounds for chemical hydrogen storage *Chemical Society Reviews*, Vol. 38, No. 1, pp. 279-293, ISSN: 0306-0012.

Himmelberger, D. W., Alden, L. R., Bluhm, M. E. & Sneddon, L. G. (2009). Ammonia Borane Hydrogen Release in Ionic Liquids *Inorganic Chemistry*, Vol. 48, No. 20, (Oct 19), pp. 9883-9889, ISSN: 0020-1669.

Hugle, T., Kuhnel, M. F. & Lentz, D. (2009). Hydrazine Borane: A Promising Hydrogen Storage Material *Journal of the American Chemical Society*, Vol. 131, No. 21, (Jun), pp. 7444-7446, ISSN: 0002-7863.

Jacquemin, J., Gomes, M. F. C., Husson, P. & Majer, V. (2006). Solubility of carbon dioxide, ethane, methane, oxygen, nitrogen, hydrogen, argon, and carbon monoxide in 1-butyl-3-methylimidazolium tetrafluoroborate between temperatures 283 K and 343 K and at pressures close to atmospheric *Journal of Chemical Thermodynamics*, Vol. 38, No. 4, (Apr), pp. 490-502, ISSN: 0021-9614.

Jaska, C. A., Temple, K., Lough, A. J. & Manners, I. (2003). Transition metal-catalyzed formation of boron-nitrogen bonds: Catalytic dehydrocoupling of amine-borane adducts to form aminoboranes and borazines *Journal of the American Chemical Society*, Vol. 125, No. 31, (Aug), pp. 9424-9434, ISSN: 0002-7863.

Johnson, T. C., Morris, D. J. & Wills, M. (2010). Hydrogen generation from formic acid and alcohols using homogeneous catalysts *Chemical Society Reviews*, Vol. 39, No. 1, pp. 81-88, ISSN: 0306-0012.

Kawasaki, I., Tsunoda, K., Tsuji, T., Yamaguchi, T., Shibuta, H., Uchida, N., Yamashita, M. & Ohta, S. (2005). A recyclable catalyst for asymmetric transfer hydrogenation with a formic acid-triethylamine mixture in ionic liquid *Chemical Communications*, Vol., No. 16, pp. 2134-2136, ISSN: 1359-7345.

Krossing, I., Slattery, J. M., Daguenet, C., Dyson, P. J., Oleinikova, A. & Weingartner, H. (2006). Why are ionic liquids liquid? A simple explanation based on lattice and solvation energies *Journal of the American Chemical Society*, Vol. 128, No. 41, (Oct 18), pp. 13427-13434, ISSN: 0002-7863.

Li, X. L., Ma, X. Y., Shi, F. & Deng, Y. Q. (2010). Hydrogen Generation from Formic Acid Decomposition with a Ruthenium Catalyst Promoted by Functionalized Ionic Liquids *Chemsuschem*, Vol. 3, No. 1, pp. 71-74, ISSN: 1864-5631.

Loges, B., Boddien, A., Junge, H. & Beller, M. (2008). Controlled generation of hydrogen from formic acid amine adducts at room temperature and application in H(2)/O(2) fuel cells *Angewandte Chemie-International Edition*, Vol. 47, No. 21, pp. 3962-3965, ISSN: 1433-7851.

Mal, S. S., Stephens, F. H. & Baker, R. T. (2011). Transition metal catalysed dehydrogenation of amine-borane fuel blends *Chemical Communications*, Vol. 47, No. 10, pp. 2922-2924, ISSN: 1359-7345.

Marder, T. B. (2007). Will we soon be fueling our automobiles with ammonia-borane? *Angewandte Chemie-International Edition*, Vol. 46, No. 43, pp. 8116-8118, ISSN: 1433-7851.

Mateus, N. M. M., Branco, L. C., Lourenco, N. M. T. & Afonso, C. A. M. (2003). Synthesis and properties of tetra-alkyl-dimethylguanidinium salts as a potential new generation of ionic liquids *Green Chemistry*, Vol. 5, No. 3, pp. 347-352, ISSN: 1463-9262.

Migowski, P. & Dupont, J. (2007). Catalytic applications of metal nanoparticles in imidazolium ionic liquids *Chemistry-a European Journal*, Vol. 13, No. 1, pp. 32-39, ISSN: 0947-6539.

Neiner, D., Karkamkar, A., Bowden, M., Choi, Y. J., Luedtke, A., Holladay, J., Fisher, A., Szymczak, N. & Autrey, T. (2011). Kinetic and thermodynamic investigation of hydrogen release from ethane 1,2-di-amineborane *Energy & Environmental Science*, Vol. 4, No. 10, (Oct), pp. 4187-4193, ISSN: 1754-5692.

Prechtl, M. H. G., Scholten, J. D. & Dupont, J. (2010). Carbon-Carbon Cross Coupling Reactions in Ionic Liquids Catalysed by Palladium Metal Nanoparticles *Molecules*, Vol. 15, No. 5, (May), pp. 3441-3461, ISSN: 1420-3049.

Prechtl, M. H. G., Scholten, J. D. & Dupont, J. (2011). Palladium Nanoscale Catalysts in Ionic Liquids: Coupling and Hydrogenation Reactions, Ionic Liquids: Applications and Perspectives. In: *Ionic Liquids: Applications and Perspectives*. A. Kokorin, pp. 393-414, InTech, 978-953-307-248-7, Vienna.

Schlapbach, L. & Zuttel, A. (2001). Hydrogen-storage materials for mobile applications *Nature*, Vol. 414, No. 6861, (Nov), pp. 353-358, ISSN: 0028-0836.

Scholten, J. D., Leal, B. C. & Dupont, J. (2012). Transition Metal Nanoparticle Catalysis in Ionic Liquids *Acs Catalysis*, Vol. 2, No. 1, (Jan), pp. 184-200, ISSN: 2155-5435.

Scholten, J. D., Prechtl, M. H. G. & Dupont, J. (2010). Decomposition of Formic Acid Catalyzed by a Phosphine-Free Ruthenium Complex in a Task-Specific Ionic Liquid *Chemcatchem*, Vol. 2, No. 10, (Oct), pp. 1265-1270, ISSN: 1867-3880.

Staubitz, A., Robertson, A. P. M. & Manners, I. (2010). Ammonia-Borane and Related Compounds as Dihydrogen Sources *Chemical Reviews*, Vol. 110, No. 7, (Jul), pp. 4079-4124, ISSN: 0009-2665.

Stephens, F. H., Pons, V. & Baker, R. T. (2007). Ammonia - borane: the hydrogen source par excellence? *Dalton Transactions*, Vol., No. 25, pp. 2613-2626, ISSN: 1477-9226.

Stracke, M. P., Ebeling, G., Cataluna, R. & Dupont, J. (2007). Hydrogen-storage materials based on imidazolium ionic liquids *Energy & Fuels*, Vol. 21, No. 3, (May-Jun), pp. 1695-1698, ISSN: 0887-0624.

Sutton, A. D., Burrell, A. K., Dixon, D. A., Garner, E. B., Gordon, J. C., Nakagawa, T., Ott, K. C., Robinson, P. & Vasiliu, M. (2011). Regeneration of Ammonia Borane Spent Fuel by Direct Reaction with Hydrazine and Liquid Ammonia *Science*, Vol. 331, No. 6023, (Mar), pp. 1426-1429, ISSN: 0036-8075.

Taccardi, N., Assenbaum, D., Berger, M. E. M., Bosmann, A., Enzenberger, F., Wolfel, R., Neuendorf, S., Goeke, V., Schodel, N., Maass, H. J., Kistenmacher, H. & Wasserscheid, P. (2010). Catalytic production of hydrogen from glucose and other carbohydrates under exceptionally mild reaction conditions *Green Chemistry*, Vol. 12, No. 7, pp. 1150-1156, ISSN: 1463-9262.

Tedsree, K., Li, T., Jones, S., Chan, C. W. A., Yu, K. M. K., Bagot, P. A. J., Marquis, E. A., Smith, G. D. W. & Tsang, S. C. E. (2011). Hydrogen production from formic acid decomposition at room temperature using a Ag-Pd core-shell nanocatalyst *Nature Nanotechnology*, Vol. 6, No. 5, (May), pp. 302-307, ISSN: 1748-3387.

Welton, T. (1999). Room-temperature ionic liquids. Solvents for synthesis and catalysis *Chemical Reviews*, Vol. 99, No. 8, (Aug), pp. 2071-2083, ISSN: 0009-2665.

Wilkes, J. S. (2002). A short history of ionic liquids - from molten salts to neoteric solvents *Green Chemistry*, Vol. 4, No. 2, (Apr), pp. 73-80, ISSN: 1463-9262.

Wright, W. R. H., Berkeley, E. R., Alden, L. R., Baker, R. T. & Sneddon, L. G. (2011). Transition metal catalysed ammonia-borane dehydrogenation in ionic liquids *Chemical Communications*, Vol. 47, No. 11, (2011), pp. 3177-3179, ISSN: 1359-7345.

Xu, L., Chen, W. & Xiao, J. (2000). Heck Reaction in Ionic Liquids and the in Situ Identification of N-Heterocyclic Carbene Complexes of Palladium *Organometallics*, Vol. 19, No. 6, (2000/03/01), pp. 1123-1127, ISSN: 0276-7333.

Yang, X., Fei, Z. F., Geldbach, T. J., Phillips, A. D., Hartinger, C. G., Li, Y. D. & Dyson, P. J. (2008). Suzuki coupling reactions in ether-functionalized ionic liquids: The importance of weakly interacting cations *Organometallics*, Vol. 27, No. 15, (Aug 11), pp. 3971-3977, ISSN: 0276-7333.

Yasaka, Y., Wakai, C., Matubayasi, N. & Nakahara, M. (2010). Controlling the Equilibrium of Formic Acid with Hydrogen and Carbon Dioxide Using Ionic Liquid *Journal of Physical Chemistry A*, Vol. 114, No. 10, (Mar 18), pp. 3510-3515, ISSN: 1089-5639.

Zahmakiran, M. & Ozkar, S. (2009). Dimethylammonium Hexanoate Stabilized Rhodium(0) Nanoclusters Identified as True Heterogeneous Catalysts with the Highest Observed Activity in the Dehydrogenation of Dimethylamine-Borane *Inorganic Chemistry*, Vol. 48, No. 18, (Sep 21), pp. 8955-8964, ISSN: 0020-1669.

Zhao, D. B., Fei, Z. F., Geldbach, T. J., Scopelliti, R. & Dyson, P. J. (2004). Nitrile-functionalized pyridinium ionic liquids: Synthesis, characterization, and their application in carbon - Carbon coupling reactions *Journal of the American Chemical Society*, Vol. 126, No. 48, (Dec 8), pp. 15876-15882, ISSN: 0002-7863.

Zhao, Y., Deng, L., Tang, S.-Y., Lai, D.-M., Liao, B., Fu, Y. & Guo, Q.-X. (2011). Selective Decomposition of Formic Acid over Immobilized Catalysts *Energy & Fuels*, Vol. 25, No. 8, (Aug), pp. 3693-3697, ISSN: 0887-0624.

Section 2

Physical Hydrogen Storage Materials

Postsynthesis Treatment Influence on Hydrogen Sorption Properties of Carbon Nanotubes

Michail Obolensky, Andrew Basteev, Vladimir Beletsky, Andrew Kravchenko, Yuri Petrusenko, Valeriy Borysenko, Sergey Lavrynenko, Oleg Kravchenko, Irina Suvorova, Vladimir Golovanevskiy and Leonid Bazyma

Additional information is available at the end of the chapter

http://dx.doi.org/10.5772/50151

1. Introduction

It has been shown in the previous investigations (Pradhan et al., 2002; Obolensky et al., 2011a, 2011b) that the postsynthesis treatment (e.g. chemical, heat treatment, milling, irradiation etc.) of single-walled carbon nanotubes (SWCNT) can essentially change their sorption properties. Methods of SWCNT synthesis and features of the postsynthesis treatment of the SWCNT show in numerous experiments a wide range of gas (e.g. hydrogen etc.) mass contents in SWCNT ranging from extremely low (less than 0.1 wt%) to significantly high (more than 6 wt%) values. Unfortunately, none of the claims of hydrogen storage exceeding the DOE limit have been confirmed. For recent review of effect of synthesis method and post-treatment of CNTs, see Ref. (Chang & Hui-Ming, 2005).

So, it is well established that hydrogen storage capacity of CNTs depends on many parameters including their pretreatment, types and structural modifications etc. One of the ways seems to be highly influential and it is the generation of appreciable structural defects in the tube walls. However, the increase in defects in graphitized layers of CNTs should be limited otherwise their structure would be destroyed which might lead to lowering of interaction or potential energy between the hydrogen molecules and carbon atoms. For instance, non-optimized electron irradiations severely destroy the graphitic network of CNTs (Banhart et al., 2002) which is not desirable for application purposes.

It has been reported that postsynthesis treatment can easily induce defects in the wall of SWCNT (Pradhan et al., 2002; Banhart et al., 2002; Hashimoto et al., 2004). The possibility of molecular hydrogen adsorption by the defect sites can also be considered. The experimental investigations of above mentioned factors as well as the result of irradiation on to SWCNT sorption capability are reported in this study.

2. Experimental

2.1. Materials

The arc-derived as-prepared (AP) SWNT material used in the present studies was obtained from Carbon Solutions, Inc.. All data reported here are collected using materials from the same batch of SWNT material.

The impurities in AP-SWNTs may be divided into two categories: carbonaceous (amorphous carbon and graphitic nanoparticles) and metallic (typically transition-metal catalysts). Thermogravimetric analysis (TGA) can be used to analyze the amount of metallic components quantitatively whereas Raman spectroscopy can be used to estimate qualitatively the carbonaceous impurities in AP and purified-SWNTs.

The AP SWNT (Carbon Solutions, Inc.) material has 21 to 31 percent of impurities which is consistent with the previously reported (Itkis et al., 2003) 34.5 ±1.8 percent impurities range. For purification of SWNTs the following technique has been used (Pradhan et al., 2002; Obolensky et al., 2011b). The SWNT (Carbon Solutions, approximately 1g amount) was oxidized in open air at 350^{0}C temperature for 30 minutes and then heated in 2.6 M HNO_3 for 20 hours. After above procedure the treated CNTs were flushed in methanol and then in water until the neutral reaction was achieved, dried in air and then were annealed in deep vacuum (10^{-8} bar) for 20 hours at 800°C temperature. Identification of the gases allocated in an act of warming of a sample at 850°C was carried by mass-spectrometry (MX-7203). Working pressure in the chamber of the mass-spectrometer was 10^{-4} Па. Approbation of nanotubes purify technology was carried out and 700 milligrams of the purified nanotubes were obtained (metal 6-9 at. %).

After chemical treatment, CNTs were milled at liquid nitrogen temperature in a stainless steel ball mill. Milling duration was 60 minutes with interruption after each 5 minutes for CNT samples' microscopic control. After the milling procedure finishing the "ball mill" was heated up to the room temperature and CNT samples were sieved through a 63 μm sieve (Obolensky et al., 2011b).

2.2. Experimental facility

The method for SWNT electron bombardment irradiation was elaborated. In contrast to (Obolensky et al., 2011a) where the exposure of γ – quantum 10^5 Rad was used (this dose is equal to regime of electron bombardment ~$3*10^{12}$ e^-/cm^2) the fluence in the work reported here was 10^{13}-$2*10^{15} e^-/cm^2$. The electron bombardment procedure was carried out at room temperature on the ELIAS linear accelerator (National Science Center "Kharkov Institute of Physics and Technology"). The electron energy was 2.3 MeV and beam intensity was 0.2 μA/cm^2. The main parameters of ELIAS accelerator are specified in Table 1.

Study of sorption/desorption process was carried out using the standard volumetric method (Obolensky et al., 2011a, 2011b). Hydrogen desorption out from SWNT was also studied on the mass spectrometer MX7203. The schematics of the experimental facility used with both measurement methods is shown in Figure 1.

Parameters	Values
Energy of accelerated electrons	0.5- 3.0 MeV
Beam current (without scanning)	0.5-150 μA
Max beam current (with scanning):	up to 500 μA
Electron beam disperse	10^{-4} radian
Diameter of electron beam without focusing	1,0 cm
Diameter of electron beam with focusing, for 90% capacity	<1 mm
Vacuum in electron beam line	$1x10^{-7}$ mbar
Power consumption	20 κW

Table 1. Parameters of ELIAS accelerator

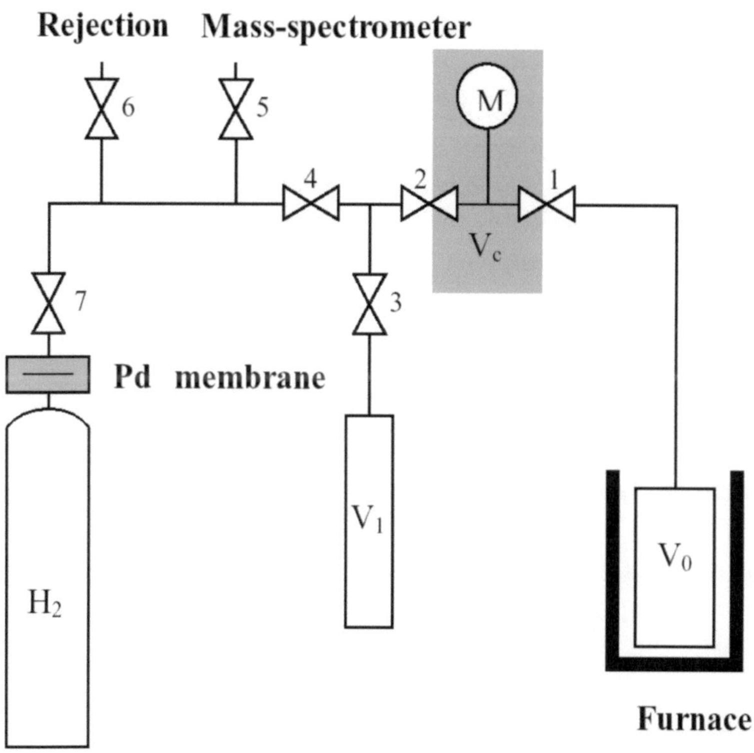

Figure 1. Experimental facility schematics (Obolensky et al., 2011b). Vc – total accurate including manometer and supply pipelines; V_1 – vessel with pure gas; V_0 – ampoule with carbon nanotubes sample; H_2 – non-purified hydrogen source

Raman spectra studies at ambient conditions with 514 nm laser line were carried out for some samples at the Physics Institute of the Penn State University.

3. Results

3.1. Adsorption studies of non-irradiated samples

Study of sorption/desorption process was carried out using the standard volumetric method on a custom designed and manufactured manufactured vacuum stainless steel facility which included the known good unit volume V_E and measuring vessel with volume V_M. Part of measuring vessel V_{78} was cooled down to liquid nitrogen vaporisation temperature of ~ 78 K. Before measurements the facility was calibrated. The good unit with volume V_E was inflated by gaseous helium or hydrogen with pressure P_I ~ 10 bar at room temperature which has been measured with ± 0.5 K precision. The pressure was controlled by electronic manometer GE Druck 104 with 1 mbar resolution. After the total facility volume V = VE + VM was filled up, the pressure dropped to the value of PF = VEPI(VE + VM - VA). This circumstance allowed exact determination of the volumes relation, taking into account the ampoule volume V_A without sample. Above relation for hydrogen and helium was found equal with approximately ~ 0.01 percent accuracy and therefore the hydrogen sorption by vacuum system elements could be excluded from consideration. After this procedure, the container with ampoule was cooled to temperature 78 K and relation P_W/P_F was determined, where P_W is the pressure in the system after cooling.

The measurement of physically absorbed hydrogen was conducted in accordance with the following procedure. Firstly, after cooling down the container to temperature 78 K the system was evacuated to pressure ~ 0.1mbar for gaseous hydrogen elimination. The next step was pressure increase registration in the system during container heating, caused by hydrogen desorption process.

The samples of SWNT with mass of 80 – 300 mg (estimated bulk density ~ 1000 kg/m³) were used for the experiments on the volumetric facility. The pressure drop character dynamics during container with non-irradiated samples cooling are shown in Figure 2.

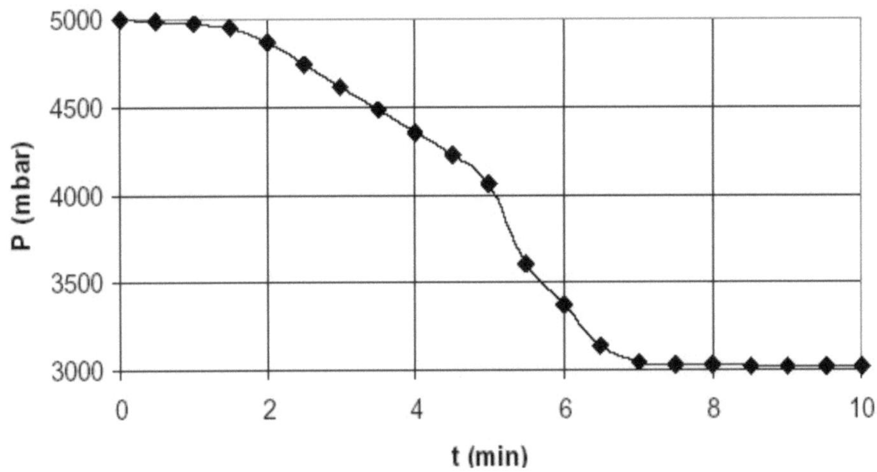

Figure 2. The dependence of pressure drop in the container with non-irradiated CNT samples upon time during container with samples cooling (P_F = 5035 mbar) (Obolensky et al., 2011b)

In Figure 2, one can see three sections with different sorption dynamics. On the base of comparison of this pressure drop dynamics with pressure drop dynamics for calibration stroke the following can be noted. The pressure drop relative to the empty container was observed already at room temperature during the letting-to-hydrogen and this pressure was noted at a ~ 20 mbar level. The additional pressure decrease was observed during further container cooling to 78 K temperature and at different cooling cycles this value was 10 – 15 mbar.

The pressure in the system during whole system heating was less than the pressure before the beginning of cooling by 10 – 12 mbar. Apparently this fact could be explained by any amount of hydrogen staying in binding state at room temperature after desorption. We didn't register this phenomenon on repeat of the heating-cooling cycles.

Figure 3 illustrates the correlation of pressure and temperature in the system for two evacuating regimes: fast (3 min) and gradual (15 min). Apparently already at 78 K temperature the increase of the duration of evacuation causes significant desorption and consequently elimination of part of hydrogen out from the system even before the beginning of heating. The duration of container heating for both regimes was approximately 25 – 30 min.

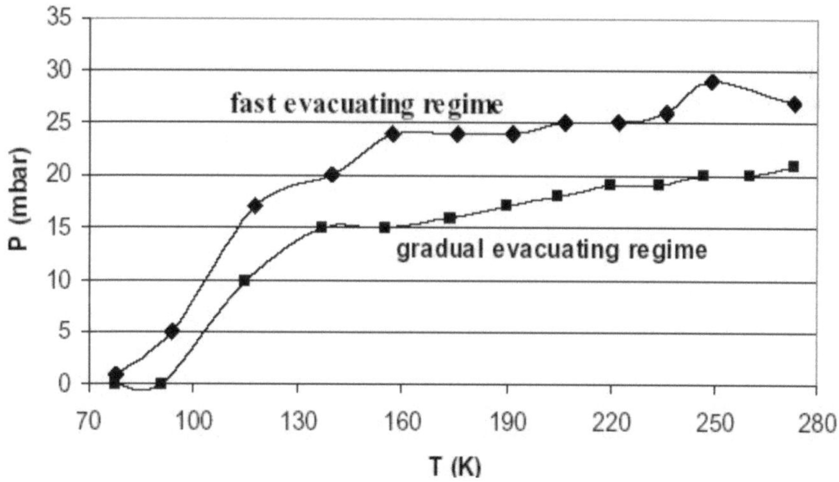

Figure 3. The dependence of pressure in the container with non-irradiated CNT samples upon temperature during container heating for different evacuation rates (Obolensky et al., 2011b)

The temperature dependences of hydrogen density in volume situated at heating regime (V_{78}) and in the rest facility volume at ambient temperature are shown in Figure 4. It could be seen that the main hydrogen mass is exuded at temperature T< 160 K that is caused by relatively low value of physical absorption activation energy. Pressure difference occurring at container heating without evacuating and initial pressure P_F as noted earlier is caused by other mechanism with higher values of desorption energies and it can be related to chemosorption regime with higher characteristic temperatures.

Our estimations show that at different sorption/desorption cycles the amount of hydrogen located in non-irradiated SWNT was equal to 0.12 ±0.2 mass percent.

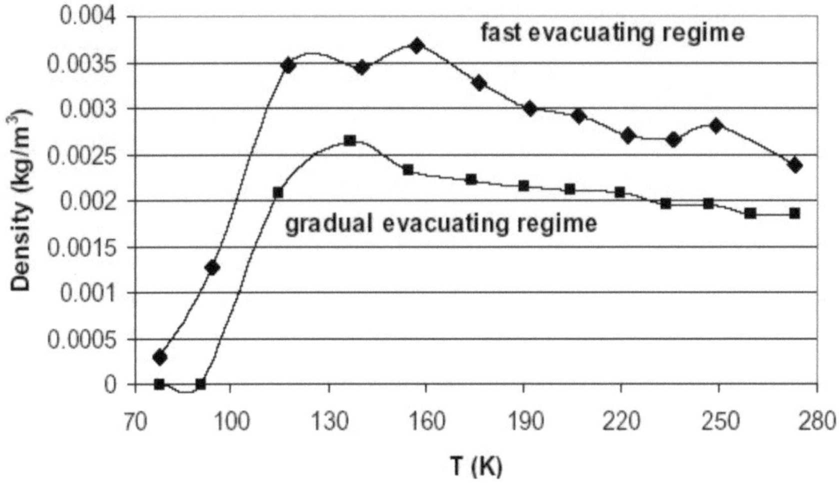

Figure 4. The dependence of hydrogen density in container with non-irradiated CNT samples upon temperature during container heating for different evacuation rates (Obolensky et al., 2011b)

3.2. Adsorption studies of irradiated samples

Hydrogen storage was studied with 2.5-90 mg samples. Saturation was carried out at liquid nitrogen temperatures (~78 K), and 300 K and at pressures between 3 and 10 bar. The measurement of physically absorbed hydrogen was conducted in accordance with the following procedure.

First, after cooling the container down to 78 K temperature the system was evacuated to ~ 0.1 mbar of pressure for gaseous hydrogen elimination. Next, pressure in the system during container heating up to room temperature caused by hydrogen desorption process was registered. Hydrogen desorption out from SWNT was also studied in the 0-900 C temperature range by mass-spectrometry method on the MX 7203 mass spectrometer.

Our estimations (Obolensky et al., 2011b) show that at different sorption/desorption cycles the amount of hydrogen located in non-irradiated SWNT was equal to 0.12 ±0.2 mass percent. Hydrogen desorption from treated and exposed to physical sorption/desorption procedure material at pressures ~3000 mbar as it is described above has been studied with the use of mass-spectrometry method within the 0 – 900 °C temperature interval on the MX 7203 mass spectrometer. The dependencies of the amount of hydrogen extracted from non-irradiated (a) and irradiated up to the fluence of 10^{14} e-/cm^2 (b) SWNT samples upon temperature are shown in arbitrary units in Figure 5.

As can be seen from Figure 5 the amount of hydrogen desorbed from irradiated material is approximately 2.5 times higher compared with that desorbed from the non-irradiated material. We have to draw attention to the fact that all procedures of the sorption/desorption processes were carried out at the 78 – 300 K temperature interval but at the same time significant amount of hydrogen is desorbed at temperatures higher than 300 K. There are additional peaks on the desorption curve which apparently correspond with different sorption sites appearing as the result of irradiation and can be characterized by various activation energies. It should be noted that over a long period of time of samples staying in air between irradiation procedures and hydrogen saturation, complete saturation generated by irradiation sites filling by the molecules of other gases has not been detected. The data in Figure 5 can be presented in another form (Figure 6) i.e. as additional amount of hydrogen desorbed from irradiated sample (Δ - the difference between the irradiated and not irradiated samples).

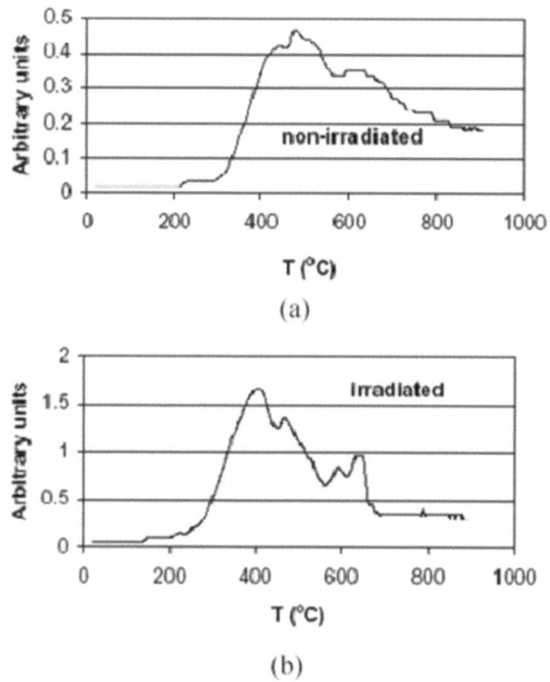

Figure 5. The dependencies of amount of hydrogen (arbitrary units) exuded from non-irradiated (a) and irradiated up to fluence 10^{14} e^{-}/cm^2 (b) samples upon temperature (Obolensky et al., 2011b)

In order to control the structural transformations of CNTs, it is important to clearly understand the defect generation mechanism and to realize the extent of defects and their influence on sorption properties of CNTs. Acid or alkali treatment does not routinely improve the hydrogen storage capacity of CNT samples. A suitable ball-milling treatment and activation process can open the caps of CNTs and produce more structural defects, and therefore may be beneficial in improving their hydrogen storage properties (Chang & Hui-Ming, 2005). In this connection we investigated the effect of electron irradiation on the hydrogen adsorption property of SWNTs without mechanical milling.

Five AP samples (cf., Table 2) have been annealed in vacuum during 10 hours at 800°C temperature. The first sample (#1) was retained as the control sample. Sample #2 has remained for control and three samples (#3-5) were irradiated with various fluence ($5*10^{14}$, 10^{15} and $2*10^{15}e^{-}/cm^2$). After that, four samples (#2-5) were saturated at 10 bar pressure for 3 hours at 78 K temperature..

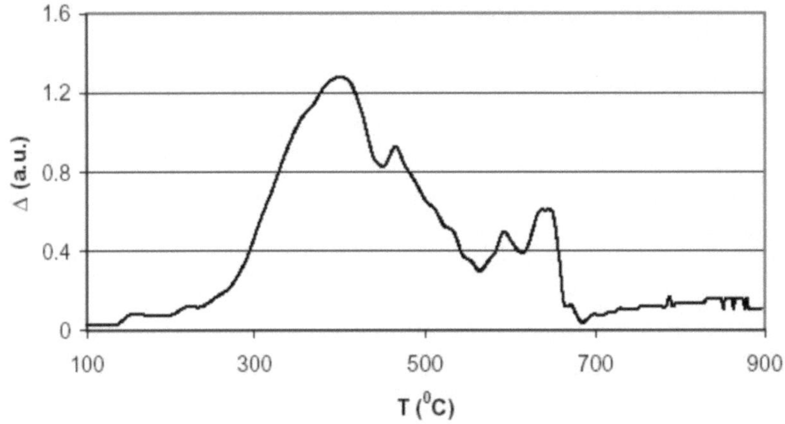

Figure 6. The dependence of additional amount of hydrogen (arbitrary units) desorbed from irradiated up to fluence 10^{14} e^-/cm^2 sample upon temperature (Obolensky et al., 2011b)

Sample	Vacuum anneal *T* (°C)/time (h)	Electron irradiation fluence (e^-/cm^2)/time (sec)	Hydrogen sorption *T* (K)/P (bar)	ΔH_2 (wt%)
1	800/10	-	-	-
2	800/10	-	~78/10	0.13
3	800/10	$5*10^{14}$/327	~78/10	0.12
4	800/10	10^{15}/706	~78/10	0.16
5	800/10	$2*10^{15}$/1340	~78/10	0.22

Note: For sample #2 Δ H_2 - the difference between irradiated and non-irradiated samples. For samples #3-5, ΔH_2 - thedifference between saturated and non-saturated samples.

Table 2. Sample history for adsorption studies

The dependencies of the amount of hydrogen extracted from the samples upon temperature are shown in arbitrary units in Figure 7. As can be seen from this Figure, the desorption curves were displaced in comparison with those for the samples exposed to all stages of treatment (i.e. oxidation, chemical treatment, annealing and milling shown in Figure 5..

For not irradiated samples the curves have remained similar, with some displacement of the maximum peak in the area of lower temperatures for the non-chemically treated and non-milled sample (2, Table 2). At the same time, the character of the curves for the irradiated samples (3-5, Table 2) essentially differs from the Ref. (Obolensky et al., 2011b). Characteristic peaks in the vicinity of 400 K and 650 K temperatures for the Ref. (Obolensky et al., 2011b) practically do not become visible for samples 3-5 (Table 2). With increase in fluence, the amount of hydrogen desorbed from irradiated samples is reduced for the given range of temperatures (Figure 8 and Figure 9).

As the temperature of samples at irradiation did not exceed 40°C (Figure 10), most likely the temperature factor could not affect decrease in the amount of hydrogen desorbed from the irradiated samples in comparison with the non-irradiated samples. This feature needs additional study.

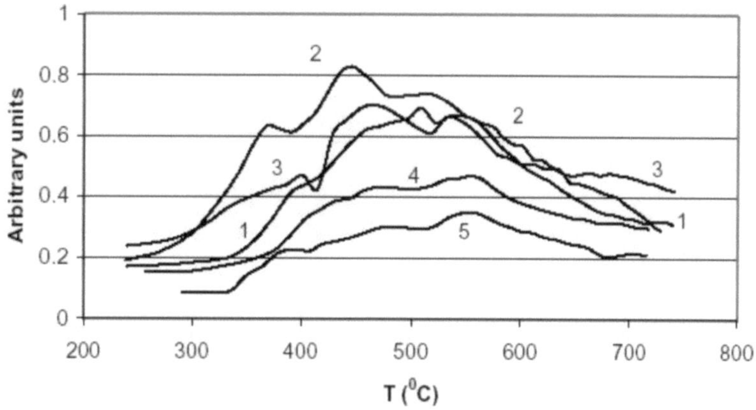

Figure 7. The dependencies of the amount of hydrogen (arbitrary units) exuded from samples (Table 1) upon temperature

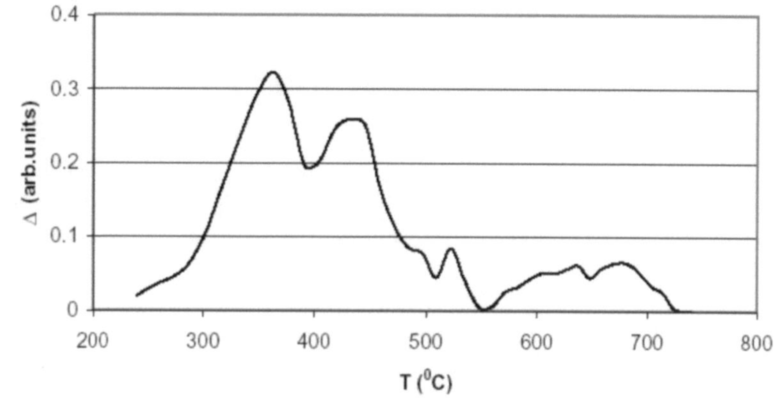

Figure 8. The dependence of the additional amount of hydrogen (the difference between saturated and non-saturated samples; arbitrary units) exuded from non-irradiated samples (Table 2) upon temperature

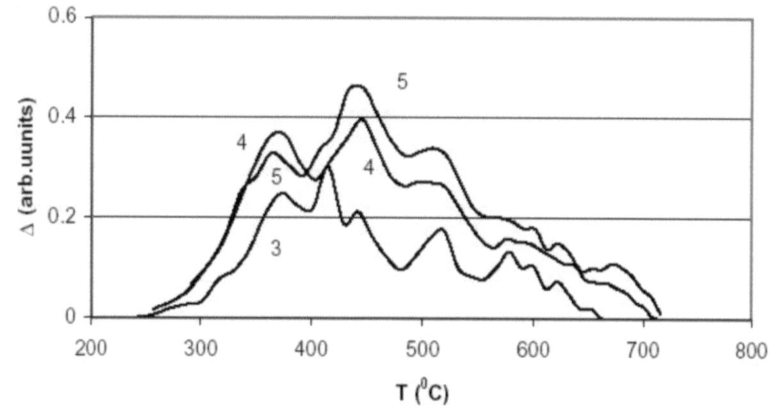

Figure 9. The dependencies of the additional amount of hydrogen (the difference between saturated non-irradiated sample and saturated irradiated samples; arbitrary units) for irradiated samples (Table 2) upon temperature

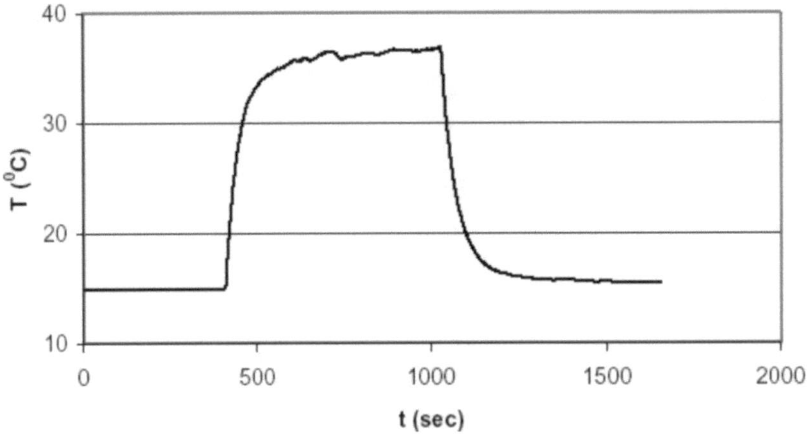

Figure 10. The dependence of samples temperature on the irradiation time (fluence $10^{15}e^{-}/cm^{2}$)

3.3. Raman scattering studies

Raman scattering is a sensitive probing of the structure and bonding in carbon materials, particularly carbon nanotubes. Raman scattering spectra for seven sets of samples (cf., Table 2) taken at room temperature in the range 100 to 3200 cm^{-1} are given in Figures 11-15..

The dominant spectral features include the low-frequency radial breathing modes (RBM) in the range approximately 150 to 200 cm^{-1} and the higher frequency modes in the range 1300 to 2700 cm^{-1} (Figure 11).

The SWNT tangential displacement modes observed near approximately 1600 cm^{-1} (G band) are related to the high frequency vibrational modes of a flat graphene sheet. The band (at 1300 cm^{-1}), commonly called the D band, has been observed in many sp^2-bonded carbon materials and is associated with disorder in the hexagonal carbon network.

Sample	Sample history	Vacuum anneal *T* (°C)/time (h)	Metal (at. %)
1	AP SWNTs [a)], irradiation [b)]	-	21-31
2	AP SWNTs [a)] irradiation [b)], sorption [c)]	-	21-31
3	AP SWNTs [a)], sorption [c)]	1000/20	21-31
4	Treated [d)], milling, sorption [c)]	1000/20	6-9
5	Treated [d)], milling, irradiation [b)]	-	6-9
6	Treated [d)], milling, irradiation [b)], sorption [c)]	-	6-9
7	Treated [d)], milling, irradiation [e)], sorption [c)]	1000/20	6-9

[a)] AP-SWNT "Carbon Solution"
[b)] Electron irradiation – 10^{13} e^{-}/cm^{2} (time – 375 sec)
[c)] Sorption – 78 K and 10 bar
[d)] Heated in 2.6 M HNO_3 during 20 hours
[e)] Electron irradiation – 2*10^{13} e^{-}/cm^{2} (time – 625 sec)

Table 3. Sample history for adsorption studies

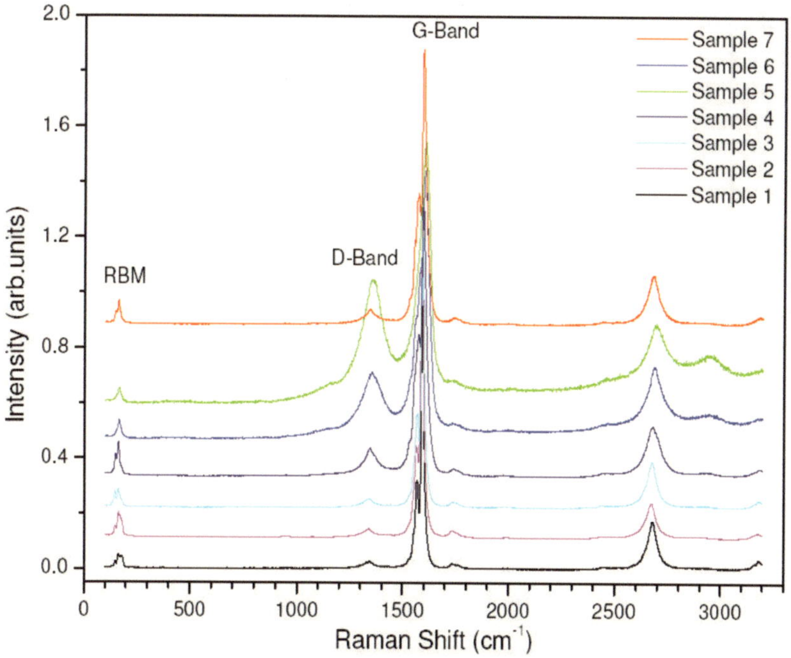

Figure 11. Room-temperature Raman spectra of bundles of arc-derived carbon nanotubes at various stages of post synthesis (Table 3). The Raman spectra were taken using 514-nm excitation

From the analysis of Raman spectra, it follows that the RBM frequency, ω_R, is inversely proportional to the tube diameter, d, while its value is up-shifted owing to the intertube interaction within a SWNT bundle (Pradhan et al., 2002). One of the empirical relations between d and ω_R, applicable for bundled SWNT is (Pradhan et al., 2002) is d (nm) = [224 cm^{-1} nm]/ [ω_R (cm^{-1}) - 12 cm^{-1}]. According to this equation, the main RBM peaks (Figure 12) at 163-166 cm^{-1} correspond to SWNT with a diameter of ~1.46 nm, whereas the shoulder at ~150 cm^{-1} is related to SWNT with a diameter of ~1.62 nm.

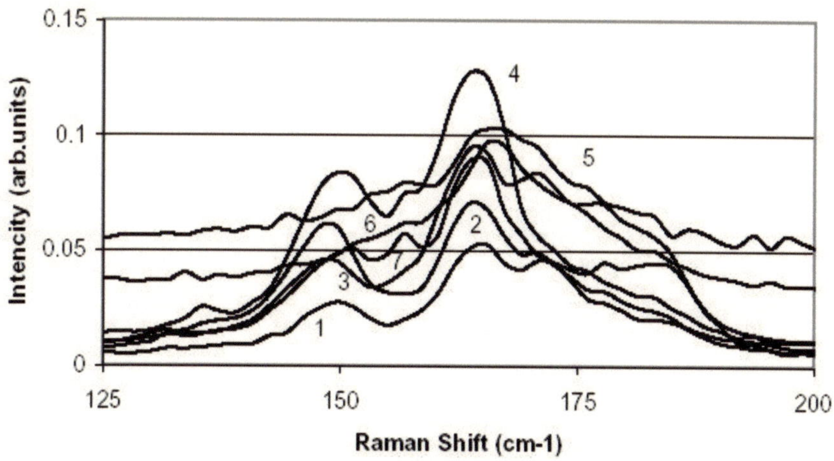

Figure 12. Raman spectra (RBM) at 514nm laser excitation energy (Table 3)

The intense Raman G-band at a higher energy corresponds to the C–C stretching vibrations in tangential and axial directions of the SWNT that splits to G^- (tangential) and G^+ (axial) bands located at 1566-1571 cm^{-1} and 1588-1602 cm^{-1}, respectively. The shape of the G^--band is sensitive to the electronic properties (strongly related to chirality) of SWNT.

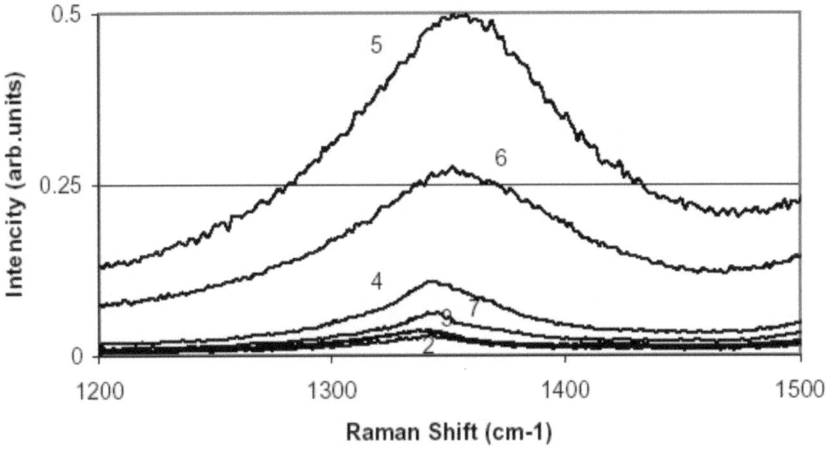

Figure 13. Raman spectra (D band) at 514nm laser excitation energy (Table 3)

D band (1341-1352 sm^{-1}, Figure 13) testifies to presence in samples of amorphous carbon and defects in structure of SWCNT. The ratio of peak intensity D band to peak intensity G band allows to judge as a level of crystallization and quantity of defects. It is possible to see that for samples which were not exposed to chemical treating and cryogenic milling (#1-3), intensity of peak D is minimal (0.03-0.036; I(D)/I(G)= 0.333-0.793, fig.14) and quantity of defects in structure of SWCNT was small.

Intensity of peak D band culminates for sample #5 (0.498; I(D)/I(G)= 2.129, Figure 14) which was exposed to chemical treating and cryogenic milling, but was not exposed to high-temperature heating in vacuum, to an irradiation and to hydrogen saturation.

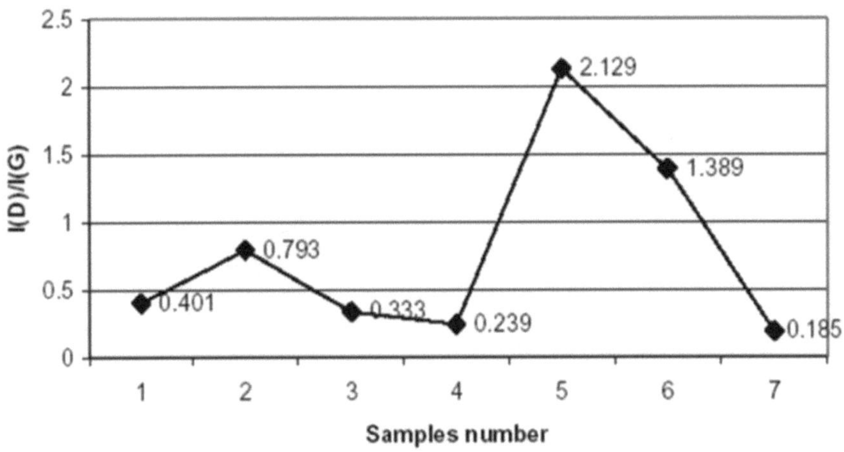

Figure 14. Distribution of the intensity ratio I(D)/I(G) for samples (Table 3)

Intensity of peak D for sample #6 which differed from sample #5 only on stages of hydrogen sorption is a little bit lower (0.275; I(D)/I(G)= 1.389, fig.14), but considerably exceeds intensity of peak D for samples #1-3. Apparently, during hydrogen sorption there was "treatment" of defects due to introduction of hydrogen in vacant sites of SWCNT.

Addition to the above-stated procedures of high-temperature heating in vacuum (samples #7 and #4) leads to essential decrease in intensity of peak D (0.062 and 0.107; I(D)/I(G)= 0.185 and 0.239, Figure 14). Most likely, at once two factors here dominate: treatment due to heating and treatment due to formation of C-H bond.

Range G' (2672-2686 cm^{-1}, Figure 15) which characterizes presence of "positive" or "negative" defects in SWCNT is submitted in spectra and has the obvious tendency: hydrogen sorption lowers intensity of this peak and leads to displacement of peak "to the left" in short-wave area of a spectrum.

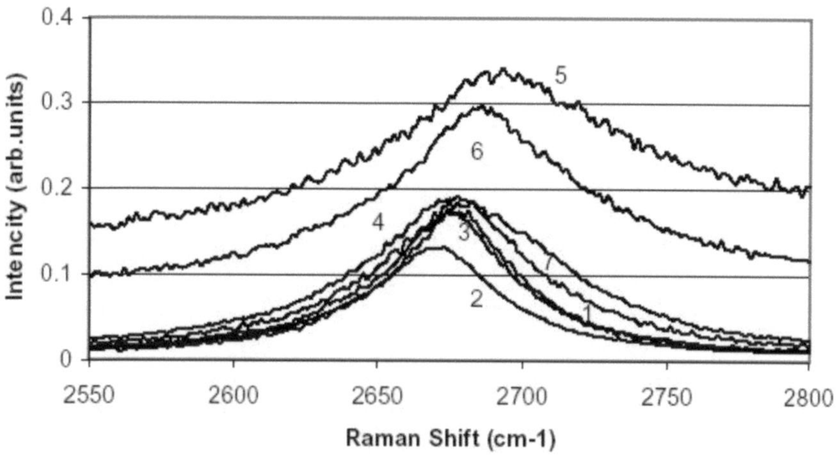

Figure 15. Raman spectra (G' band) at 514nm laser excitation energy (Table 3)

4. Conclusion

The amount of hydrogen desorbed from non-irradiated samples was 0.12 ±0.2 mass percent. The significant increase (more than two times) of hydrogen mass absorbed in irradiated SWCNT samples (10^{14} e^-/cm^2) relative to non-irradiated ones has been proved. For samples irradiated with fluences more than 10^{14} e-/cm^2 the decrease of hydrogen desorption was observed.

Author details

Andrew Basteev and Leonid Bazyma
National Aerospace University "Kharkov Aviation Institute", Ukraine

Michail Obolensky, Andrew Kravchenko and Vladimir Beletsky
V.N. Karazin Kharkiv National University, Ukraine

Yuri Petrusenko, Valeriy Borysenko, Sergey Lavrynenko
National Science Center - Kharkov Institute of Physics and Technology, Ukraine

Oleg Kravchenko and Irina Suvorova
A.N. Podgorny Institute for Mechanical Engineering Problems, Ukraine

Vladimir Golovanevskiy
Western Australian School of Mines, Curtin University, Australia

Acknowledgement

This research was supported by Science and Technological Center in Ukraine, project #4957. The Raman measurements were performed at the Physics Institute of the Penn State University by Dr. Humberto R. Gutierrez and Dr. Xiaoming Liu.

5. References

Banhart, F.; Li J. X.; & Krasheninnikov, A.V. (2005) Carbon nanotubes under electron irradiation: stability of the tubes and their action as pipes for atom transport. *Phys. Rev.* B 71: 241408-4. ISSN 1098-0121.

Carbon Solutions, Inc.; web site: http://www.carbonsolution.com

Chang Liu, & Hui-Ming Cheng. (2005) Carbon nanotubes for clean energy applications. *J. Phys. D: Appl. Phys.* 38: R231–R252, ISSN 0022-3727

Itkis, M.E.; Perea, D.E.; Niyogi, S.; Rickard, S.M.; Hamon, M.A.; Hu, H.; Zhao, B. & Haddon R. C. (2003) Purity Evaluation of As-Prepared Single-Walled Carbon Nanotube Soot by Use of Solution-Phase Near-IR Spectroscopy. *Nano Lett.*, 3(3): 309-314, c.

Hashimoto, A.; Suenaga, K.; Gloter, A.; Urita, K.; & Iijima, S. (2004) Direct evidence for atomic defects in graphenelayers. *Nature* 430: 870-873, ISSN 0028-0836

Obolensky, M.A.; Basteev, A.V.; & Bazyma, L.A. (2011) Hydrogen storage in irradiated low dimensional structures. Fullerenes, *Nanotubes, and Carbon Nanostructures*, 19(1): 133-136, ISSN 1536-383X

Obolensky, M.; Kravchenko, A.; Beletsky, V.; Petrusenko, Yu.; Borysenko, V.; Lavrynenko, S.; Andrew Basteev, A. & Leonid Bazyma, L. (2011) Thermal, Chemical and Radiation Treatment Influence on Hydrogen Adsorption Capability in Single Wall Carbon Nanotubes. Fundamentals of Low-Dimensional Carbon Nanomaterials. Edited by John J. Boeck. *Mater. Res. Soc. Symp. Proc.* 1284: 125-130, ISBN 978-1-605-11261-9.

Pradhan, B.K.; Harutyunyan, A.R.; Stojkovic, D.; Grossman, J.C.; Zhang, P.; Cole, M.W.; Crespi, V.; Goto, H.; Fujiwara, J. & Eklund, P.C., (2002) Large cryogenic storage of hydrogen in carbon nanotubes at low pressures. *J. Mater. Res.* 17(9): 2209-2016, ISSN 0884-2914.

Chapter 8

Electrospun Nanofibrous Materials and Their Hydrogen Storage

Seong Mu Jo

Additional information is available at the end of the chapter

http://dx.doi.org/10.5772/50521

1. Introduction

The hydrogen is a clean fuel source, which produces water vapor as the only exhaust gas when it is burnt with oxygen. The chemical energy density of hydrogen (142 MJ/kg) is at least three times larger than that of other chemical fuels. When the hydrogen is electrochemically burnt using a fuel cell system, the efficiency can reach 50~60%, twice as much as the thermal process because the efficiency of the direct process of electron transfer from oxygen to hydrogen in a fuel cell system is not limited by the Carnot efficiency in [1]. However, the hydrogen volume is 3000 times higher than that of gasoline at room temperature and atmosphere because it is a molecular gas. Therefore, on-board hydrogen energy storage need compact, light, safe and affordable containment. The condensation of a monolayer of hydrogen on a solid leads to a maximum of 1.3×10^{-5} mol/m^2 of adsorbed hydrogen. For automotive applications, the US DOE required a hydrogen storage capacity of greater than 6.5 wt% and ambient temperatures for hydrogen release and moderate storage pressures for industrial applications. Hydrogen storage in carbon materials is a very attractive field since high gravimetric storage capacities may be possible owing to the low specific weight and high specific surface area of carbon. The reversibly adsorbed quantity of hydrogen on nanostructure graphitic carbon amounts to 1.5 mass% per 1000 m^2/g of specific surface area at 77 K (liquid nitrogen temperature) in [1]. On active carbon with the specific surface area of 1315 m^2/g, 2 mass% of hydrogen was reversibly adsorbed at a temperature of 77 K in [2]. Carbon materials with different nanostructures are available for hydrogen storage, e.g. carbon nanofibers (CNF), graphite nanofiber (GNF), carbon nanohorns, multiwalled carbon nanotubes (MWNT), and single-walled carbon nanotubes (SWNT).

Since the excellent 6 to 8 wt% hydrogen storage using carbon nano-materials at room temperature and with atmospheric pressure was first reported in [3], several studies on hydrogen storage using SWNT, MWNT, GNF, active carbon, and active carbon fibers etc., have been conducted in [4,5]. The hydrogen adsorption of 4.2 wt% (0.5-H/C) at 100 bars and at room temperature was observed using SWNT synthesized through the arc electric discharge method in [6]. A hydrogen adsorption of approximately 3 wt% at 3~100 MPa and room temperature was reported using a well-aligned SWNT bundle in [4]. The hydrogen storage of SWNTs measured using volumetric method, however, showed scattered capacities within the range of 0.03~4 wt% at room temperature because of introduction of some error during the measurement in [7,8]. As listed in Table 1, SWNTs in more accurate volumetric measurements showed low capacity within the range of 0.14~0.43 wt% and results for MWNTs and graphite powder were less than 0.04 wt% in [9]. Despite the large volume of data from studies conducted on SWNTs, MWNTs, GNFs, etc., as potential hydrogen storage materials, these data are scattered and are thus inconclusive.

Carbon nanomaterials	Sources	Evaluation Temperature/Pressure	H_2 Storage capacity
SWNT	Carbon nanotechnology Inc.	298 K / 80 bars	0.43 wt%
SWNT	MTR, Ltd (20~40% purity)	298 K / 80 bars	0.14 wt%
MWNT	ground core, Strem Chemicals, Inc	298 K / 80 bars	< 0.04 wt%
SWNT	arc-discharge	300 K /145 bars	0.2~0.4 wt%
MWNT	acetylene pyrolysis	300 K /145 bars	0.2~0.6 wt%
MWNT	arc-discharge	300 K /145 bars	2.6 wt%
aligned MWNT bundle	ferrocene pyrolysis	300 K /145 bars	1.0~3.3wt%
aligned MWNT bundle	ferrocene/acetylene pyrolysis	300 K /145 bars	3.5~3.7wt%
GNF	acetylene pyrolysis	-	2.4 wt%
GNF	hexane/ferrocene pyrolysis	298 K /100 bars	1.29~3.98 wt%
Commercial ACF	A-20(Osaka Gas Chemicals Co. Ltd) /FT300-20(Kuraray Chemical Co., Ltd)_	298 K / 80 bars	0.35~0.41 wt%
vitreous carbon	80-200μm, 99.5% purity (Goodfellow Cambridge, Ltd.)	298 K / 80 bars	< 0.04 wt%
Graphite powder	200μm, 99.997% purity (Goodfellow Cambridge Ltd.)	298 K / 80 bars	< 0.04 wt%

Table 1. The hydrogen storage capacities of several carbon nano-materials evaluated by using the PCT and gravimetric method.

Typical carbon materials, such as active carbon, active carbon fiber, and graphite powder, were also investigated as potential materials for hydrogen storage. Purified SWNT (285 m^2/g) and saran carbon (1600 m^2/g) with a high BET surface area were also reported to have a hydrogen adsorption of approximately 0.04 and 0.28 H/C, respectively, at 0.32 MPa and 80 K in [5]. Large hydrogen adsorption was also observed by a micro porous zeolite and active carbons at 77 K under atmospheric pressure in [10]. In the case of highly porous carbon (AX-21 carbon), very high hydrogen adsorption of 5.3 wt% (0.64 H/C) was observed at 77 K and 1 MPa in [11]. However, active carbon materials with very high surface areas showed very low capacities at room temperature. This may be due to the very low levels of the effective pore size for hydrogen storage in spite of their high surface areas. The hydrogen storage capacity of materials surface greatly depended on the adsorption potential energy between the materials and hydrogen molecules. But too high adsorption potential energy may give to irreversible storage with chemisorptions of hydrogen molecule. The potential fields from opposite walls may overlap so that the attractive force acting on hydrogen molecules is greater than that on an open flat surface. Therefore, in micro porous carbon materials the pores with a width not exceeding a few hydrogen molecules may be more effective pores for hydrogen storage because of the dynamic diameter of hydrogen molecule with 0.41 nm, in [1].

TiO_2 nanotubes could reproducibly store up to about 2 wt% H_2 at room temperature and 60 bars in [12]. However, only 75% of the H_2 is physisorbed and can be reversibly released upon pressure reduction. Approximately 13% is weakly chemisorbed and can be released at 70 °C as H_2, and 12% is bonded to oxide ions and released only at temperatures above 120 °C as H_2O. The sorption of hydrogen between the layers of the multilayered wall of nanotubular TiO_2 was also investigated in the temperature range of -195 to 200 °C and at pressures of 0 to 6 bar and it got a 1~2.5 wt% hydrogen sorption at 1 bar and temperatures in the range 80 to 125 °C, in [13]. The hydrogen storage capacity of 0.83 wt% using ZnO nanowires with the mean diameter of 20 nm was found under the pressure of about 3.03 MPa and at room temperature, and about 71% of the stored hydrogen can be released under ambient pressure at room temperature, in [14]. And also hydrogen storages using MoS_2 nanotube and TiS_2 nanotube were investigated in [15,16]. So we need the study on the effective pore size of materials with appropriate adsorption potential energy for hydrogen storage at room temperature rather than large surface area of materials.

Carbon fibers have been used in a variety of fields as high-performance and functional materials. Woven or nonwoven carbon fibers are used as absorbed materials because of better adsorption capacity than conventional activated granular and powder carbon materials. They are being applied to gas separation and liquid adsorption. Saran carbons showed higher the hydrogen storage capacity than that of other active carbon materials, in [5]. Poly(vinylidene fluoride) (PVdF) also can be used in obtaining meso porous carbon similar to saran polymers, in [17]. The carbonization of PVdF can also produce a polyacetylene or carbyne structure and its pore size may be smaller than that of saran carbon due to the small size of the fluorine atom. Nano-sized fibers may be more helpful in obtaining carbon materials with a well-defined pore structure compared to the carbonization of micro-sized polymer fibers. Because thinner fibers are expected to be more desirable for those separation and adsorption applications, there has been growing interest in electrospinning for producing ultrafine fibers.

The recent electrospinning process for polymer or metal oxide sole gel solutions is a powerful method for producing ultrafine fibers within the range of a few to a few hundred nanometers in diameter, core-shell nanostructure nanofibers, etc., which cannot be easily obtained using traditional methods. There is a growing interest in the electrospinning process of polymer or metal oxide sol-gel solution because of their several potential applications such as ultrasensitive gas sensors, polymer electrolytes for lithium ion polymer battery, dye-sensitized solar cell etc., in [18~22]. Thus, such electrospun polymeric nanofibers can be used as effective precursors for carbon nanofibers. In addition, a transition metal would promote carbonization and graphitization of polymeric precursor, which was verified for Kapton films by various research groups, in [23~25]. Recent reports showed the effects of a transition metal on the carbonization behavior of the electrospun polyimide nanofiber, PAN nanofibers and PVdF nanofibers, resulting in graphite nanofiber (GNF), in [26,27]. And also electrospinning of metal oxide sol-gel solution provide metal oxide nanofibers with various morphologies after calcinations. They are also expected to have some hydrogen storage because of higher adsorption potential energy between the metal oxide materials and hydrogen molecules although they had much lower surface area than those of electrospun polymer based carbon nanofibers.

In this chapter, the preparation of carbon nanofibers, graphite nanofiber, and metal oxide nanofibers such as titanium oxide and lithium titanate nanofiber through heat-treatment of electrospun precursor nanofibers, their structural properties such as surface area and pore size, and morphologies were investigated. And their hydrogen storage capacities discussed with their pore size and surface area.

2. Results and discussion

2.1. Morphology and crystalline structure of CNFs, GNFs, and lithium titanate nanofiber

Polyacrylonitrile (PAN) typically has been used for preparation of carbon fiber with high performance. Carbonizations of PVdF or saran polymers also give to meso porous or micro porous carbon materials. Ultrafine structure of electrospun PVdF nanofibers is expected to be suitable for the formation of pores which are effective for hydrogen adsorption, compared to PVdF films or microfibers. So, micro porous carbon nanofibers as hydrogen storage materials were prepared through the carbonization of as-electrospun PAN and PVdF nanofibers. As-electrospun PAN or PVdF-based nanofibers were prepared from the typical electrospinning of the polymer solution containing several contents of iron (III) acetylacetonate (IAA) on the weight of the polymer as a catalyst for graphitization. Carbon nanofibers (CNF) and graphite nanofiber (GNF) were prepared from carbonization after stabilization of them, in [28~31].

Firstly polyacrylonitrile (PAN) solutions for electrospinning were prepared by dissolving PAN (Mw 150,000, polyscience) in N,N'-dimethylacetamide. The PAN solutions contained 0 wt%, 2 wt%, 5wt%, and 7.5 wt% of iron (III) acetylacetonate (IAA) on the weight of the polymer as a catalyst for graphitization. As-electrospun PAN-based nanofibers were stabilized by heating them at a rate of 1 °C/min up to 260 °C, and by holding them for 2 hrs under air atmosphere. Carbonization was performed at a given temperature within the range of 900 to 1500 °C under nitrogen atmosphere. The samples were kept for 1 hr sequentially at 400 and 600 °C and then heated up to final temperature at a rate of 3 °C/min. As-electrospun PVdF nanofibers were also obtained from electrospinning of 11 wt% PVdF solution (Kynar 761) in acetone/N,N'-dimethylacetamide (=7/3, wt. ratio) mixture. They were slightly dehydrofluorinated (DHF) in the methanol/ N, N'-dimethylacetamide (=9/1 wt. ratio) solution containing 10 ml of 1,8-diazabicyclo[5.4.0] undec-7-ene (DBU) at 50 °C for 5 hours. They were also highly dehydrofluorinated in the methanol/DBU (=1/2 wt. ratio). A PVdF solutions for GNFs were prepared by dissolving 11 wt% of PVdF in 100 ml of acetone/N,N'-dimethylacetamide (=7/3, wt. ratio) mixture containing 25 ml of 1,8-diazabicyclo[5.4.0] undec-7-ene for partial dehydrofluorination and also contained 5.5 wt% of IAA based on the weight of the polymer as a catalyst for graphitization. As-electrospun PVdF nanofibers for GNFs were chemically dehydrofluorinated with a 4 M aqueous NaOH solution containing 0.25 mmole of tetrabutylammonium bromide at 70 °C for 1 h. Carbonization was performed to induce micro pore structures without a further activation process at a given temperature within the range 800~1800 °C under a nitrogen atmosphere. The samples were heated at a rate of 3 °C/min and were maintained for 1 h at the final temperature.

Figure 1 shows the SEM images of electrospun PAN- and PVdF-based CNFs. In the case of as-electrospun PAN-based nanofiber with a diameter of about 90 nm the fiber diameter hardly changed during the carbonization at 1300 °C, while that with a diameter of about 240 nm remarkably shrank to 110 nm and showed roughened surfaces. Solvent evaporation during the electrospinning process greatly has an effect on internal structure of the resulting PAN nanofibers. Thin fiber is denser and has a higher orientation than thick fiber because it is formed at a much higher draw ratio and with a much faster solvent evaporation during the electrospinning process. Oxidation and carbonization of the above as-electrospun PAN nanofibers were carried out under a tensionless condition or with a slight tension. Therefore, the dimensions of dense, highly oriented ultrathin fibers are thought to hardly change during carbonization and the PAN-based CNFs showed very smooth surfaces. In the case of as-electrospun PVdF nanofibers, it is also necessary to make the nanofibers infusible to maintain their fibrous shape through dehydrofluorination (DHF) treatment before the carbonization. The effect of DHF treatment in the carbonization of PVdF polymer has been reported in previous studies, in [17]. The structure of the PVdF-based CNFs greatly depends on the DHF condition, in [30]. The carbon nanofibers carbonized after a high DHF treatment had very smooth surfaces and dense, nonporous structures in Figure 1(c), while slightly DHF treatment gave to the micro porous carbon nanofibers with granular-shaped surfaces, and with an internal structure consisted of 20 to 30-nm carbon granules after carbonization at above 800 °C, as shown in Figure 1(d). The pore structure became dense with the increase of carbonization temperature, indicating the formation of micro pores at higher temperature. The slight DHF treatment of PVdF nanofibers induced inhomogeneous structures consisting of the DHF-treated, amophorous region and the non-reacted crystalline region. The non-reacted crystalline region melted at a high temperature while the dehydrofluorinated region maintained its fibrous shape during the carbonization. The onset temperature and the amount of volume reduction during carbonization differed between the non-reacted and reacted regions. Thus, relatively large pores were produced between these regions.

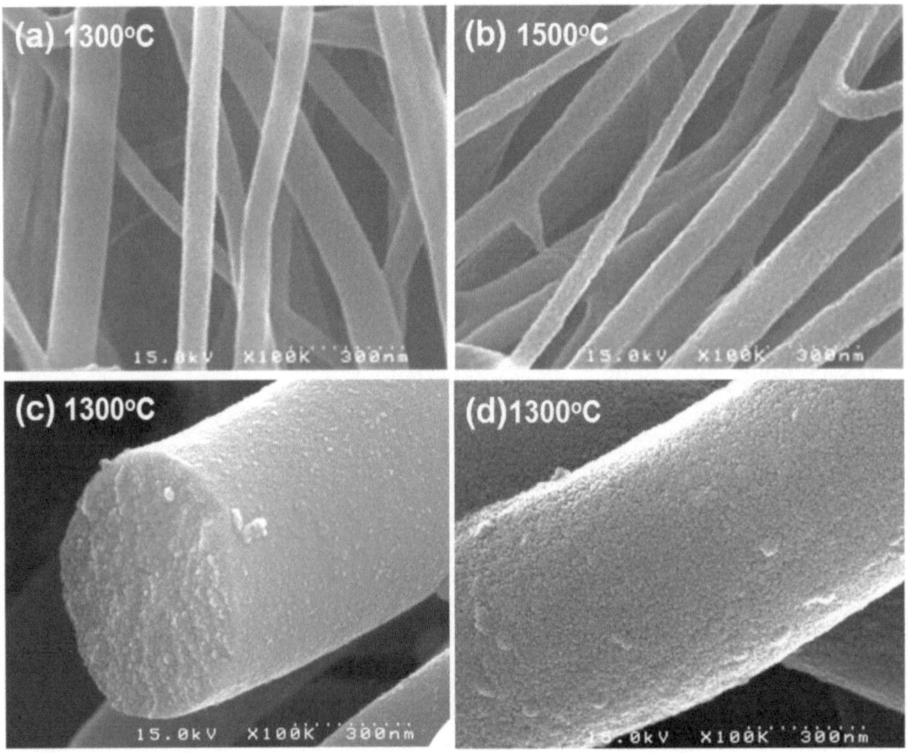

Figure 1. SEM images of carbon nanofibers prepared through carbonization of electrospun PAN nanofibers (a), (b), and PVdF nanofibers after (c) high DHF, (d) low DHF- treatment. in [28~30] (*These data were reproduced under permissions of The Polymer Society of Korea and Cambridge University Press*)

As shown in Figure 2, the PAN-based CNFs had disordered, amorphous carbon structures with d_{002}>0.37 nm, and had broad peaks structures in the XRD regardless of the carbonization temperature. The PVdF-based CNFs also showed disordered carbon structures in the XRD and Raman spectra regardless of the carbonization temperature. The CNFs prepared after low DHF treatment were expected to have higher surface areas than those prepared after high DHF treatment. Their high surface areas are thought to be due to their micro porous granular surfaces. However, this does not indicate high hydrogen storage capacity because micro pores with a width not exceeding 1 nm are thought to be much more effective for hydrogen storage when compared to the kinetic diameter of hydrogen molecule sizes of about 0.41 nm.

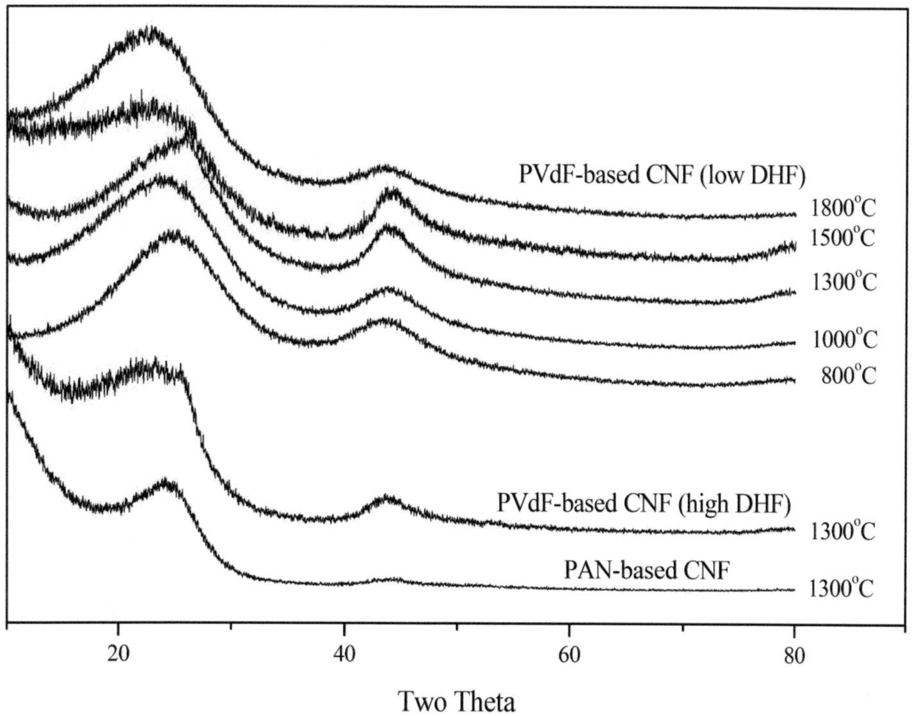

Figure 2. XRD patterns of (a) the PAN- and (b) the PVdF-based CNFs. in [28] (*These data were reproduced under permissions of The Polymer Society of Korea*)

Catalytic graphitization using volatile hydrocarbon fractions during the carbonization may be helpful in increasing the carbon yield and in forming effective ultra micro pores for hydrogen storage. For these purpose, as-electrospun PAN and PVdF nanofibers containing IAA were carbonized to induce catalytic graphitization within the range 800~1800 °C under a nitrogen atmosphere. PAN based graphite nanofibers (GNF) with a diameter of 150-300 nm were prepared by carbonization at 900~1500 °C after stabilization of as-electrospun PAN nanofibers containing 2, 5, and 7.5 wt% of IAA at air atmosphere, respectively. Figure 3 and 4 shows SEM and TEM images of PAN-based GNF. White spots were observed on the surface of the GNF fibers (1100 °C), indicating the development of graphite crystal structures centered on the Fe catalyst. This catalytic graphitization was accelerated at above 1300 °C. The TEM image around the white spot on the surface of the GNF showed a well-ordered graphite structure similar to natural graphite.

In the case of PVdF-based GNF, a notable grainy structure was observed on the surface and cross-section of the GNFs. The partial dehydrofluorination of PVdF nanofibers induced inhomogeneous structures consisting of a dehydrofluorinated amorphous region and an non-reacted crystalline region. Therefore, carbonization of them produced porous GNFs with a high surface area due to their porous granular surface. Figure 5 shows SEM and TEM images of PVdF-based GNFs. Clusters of Fe catalyst and the development of graphite structures centered on the Fe catalyst are clearly observed in TEM images of PVdF-based GNFs. The size of the Fe catalyst is from a few tens to a few hundreds of nanometers.

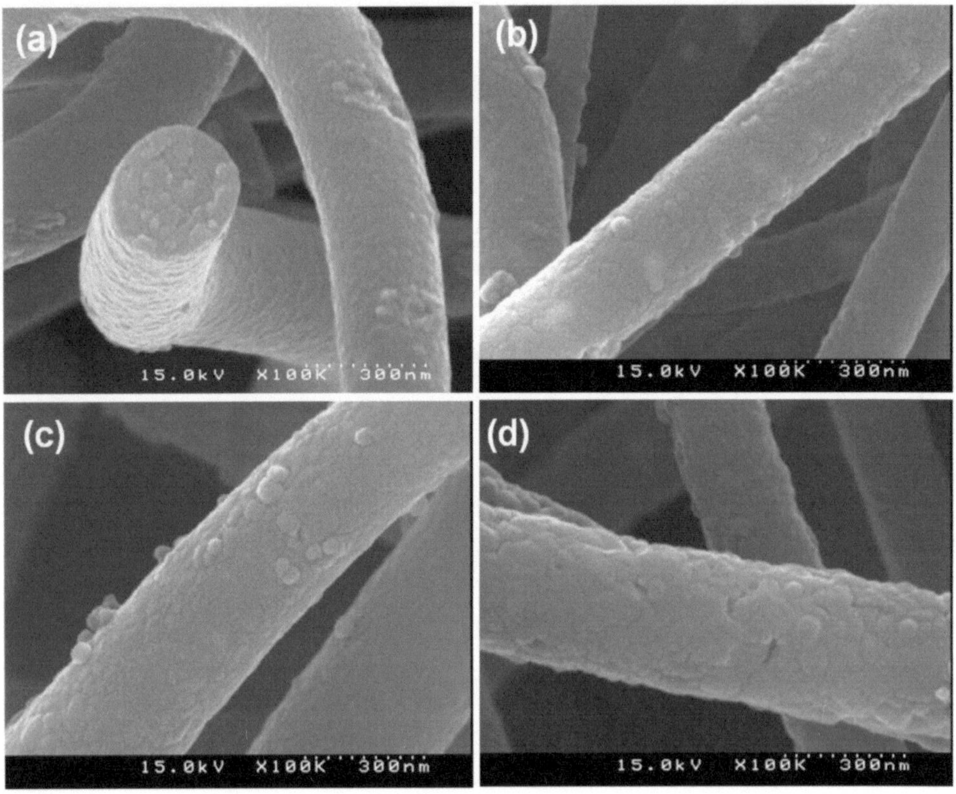

Figure 3. SEM images of the GNFs prepared from electrospun PAN nanofibers containing IAA 2 wt% ; carbonization; (a) 900°C (b) 1100°C (c) 1300°C (d) 1500°C. in [28] (*These data were reproduced under permissions of The Polymer Society of Korea*)

Figure 6 and 7 shows XRD patterns of PAN and PVdF-based GNFs. The catalytic graphitization of electrospun PAN nanofibers intensively started to proceed from 900 °C, while PVdF nanofibers intensively started to proceed from 800 °C. The sharp peaks in the PAN-based GNF at 1100 °C were observed at around 26° (002) and 42-46° (100), respectively. New peaks also appeared at 35° and 50°, corresponding to Fe_3O_4. In the case of PVdF-based GNFs, the GNFs at 800 °C show sharp peaks at approximately 26° and 44°, corresponding to the diffraction of the (002) plane and (100)/ (101) of the graphite structure, respectively. The presence of the (112) peak at 83° is also indicative of a graphite structure. The intensities of these peaks increased and sharpened with the carbonization temperature. New peaks also appeared at 36° and 48°, corresponding to Fe_3O_4. It was assumed that IAA was converted into Fe_3O_4 *via* *a*-FeO(OH) during carbonization, and that the reduction of Fe_3O_4 at above 800~900 °C resulted in the production of the α-Fe catalyst to be able to induce the graphitization reaction of the PAN or PVdF based nanofibers. In the case of the PAN-based GNF prepared at above 1300 °C, most of the Fe_3O_4 transformed to the α-Fe. As shown in Figure 6, the PAN-based GNFs using higher contents of IAA and higher carbonization temperature obviously showed α-Fe peak at around 42-44° (110) and 65° (200). The PVdF-based GNF prepared at 1500 °C obviously showed α-Fe peaks at approximately 42~44° (110)

and 65° (200). Since the GNFs, however, did not entirely have this graphite structure, the d_{002} of PAN-based GNFs and PVdF-based GNFs were almost 0.34 nm and in the range 0.333~0.343 nm, respectively. The net structure of these GNFs consists of a graphite-like structure, which forms a turbostratic-oriented graphite layer. Generally, this type of structure has been obtained from carbonization of rigid polymers such as Kapton imides, in [32]. However, the electrospun thermoplastic nanofibers also transformed to form a well ordered graphite structure similar to natural graphite through catalytic graphitization during carbonization. Single hexagonal crystal graphite shows a Raman active peak at 1582 cm^{-1} (G mode) and a band around 1357 cm^{-1} can be attributed to the D mode of disorder induced scattering, which is due to imperfection or lack of hexagonal symmetry in the carbon structure. A wide Gaussian band (M mode) is considered to represent an amorphous carbon contribution. $L_a=4.2(I_G/I_D)$ in Raman spectra reflects the crystallite planar size of the graphite structure. As listed in Table 2, the Raman spectra of the PAN-based GNFs show that the relative intensity of the G band (1580 cm^{-1}) over the D band (1360 cm^{-1}) increased with the increase of the carbonization temperature. L_a (nm) greatly increased to 4.1 nm (900 °C), 4.75 nm (1100 °C), and 6.54 nm (1300 °C). As shown in Figure 8, the I_G/I_D of the PVdF-based GNFs also rapidly increased with increasing carbonization temperature. L_a (nm) greatly increased from 4.32 nm (800 °C) to 72.5 nm (1800 °C) with increasing carbonization temperature, in [31].

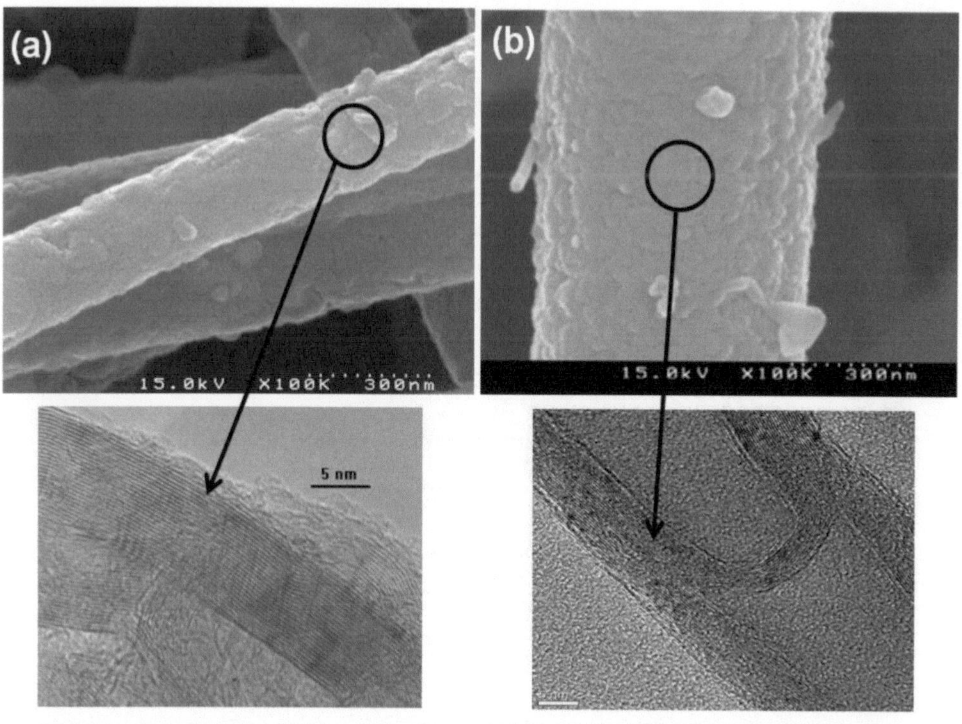

Figure 4. SEM and TEM images of the GNFs prepared through carbonization of electrospun PAN nanofibers containing IAA (a) 5 wt%, (b) 7.5 wt%. in [28] (*These data were reproduced under permissions of The Polymer Society of Korea*)

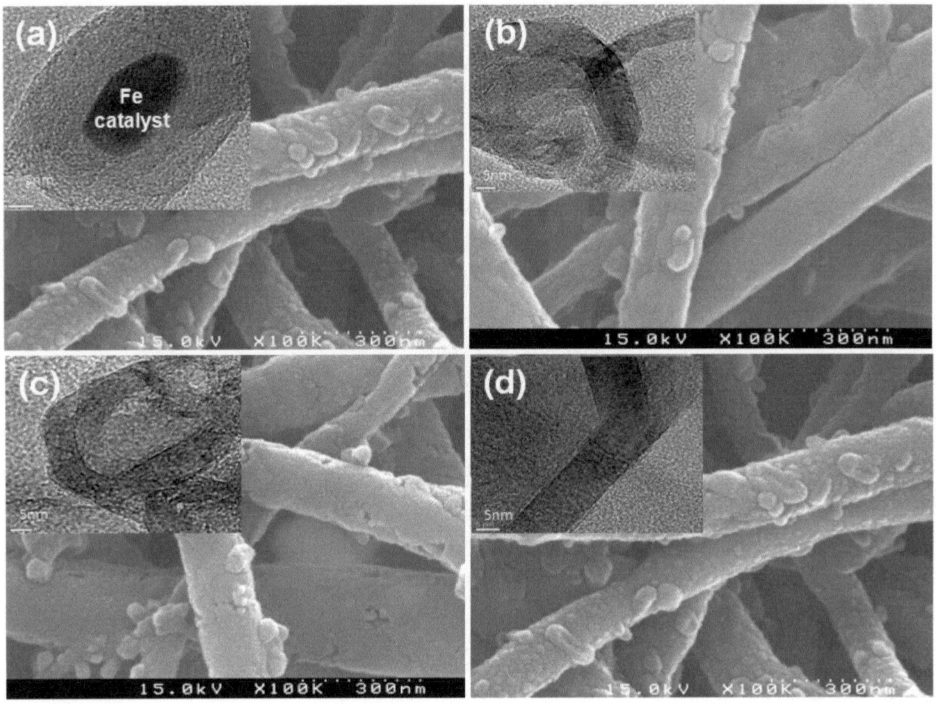

Figure 5. SEM and TEM images of the GNFs prepared through carbonization of electrospun PVdF nanofibers containing IAA 5.5 wt%. (a) 800ºC; (b) 1000ºC; (c) 1500ºC; and (d) 1800ºC. in [31] (*These data were reproduced under permissions of Elsevier)*

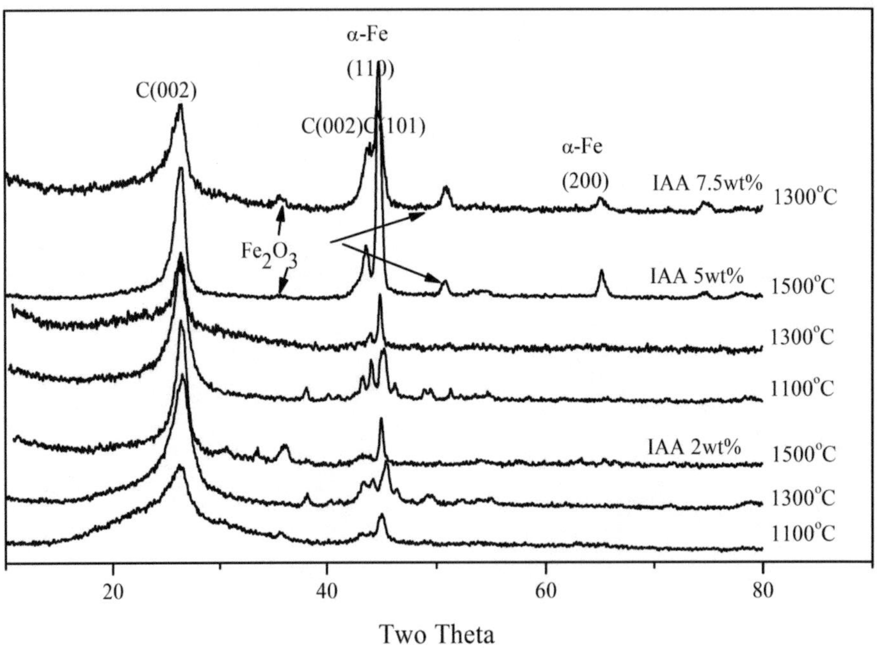

Figure 6. XRD patterns of the PAN-based GNFs. in [28] (*These data were reproduced under permissions of The Polymer Society of Korea*)

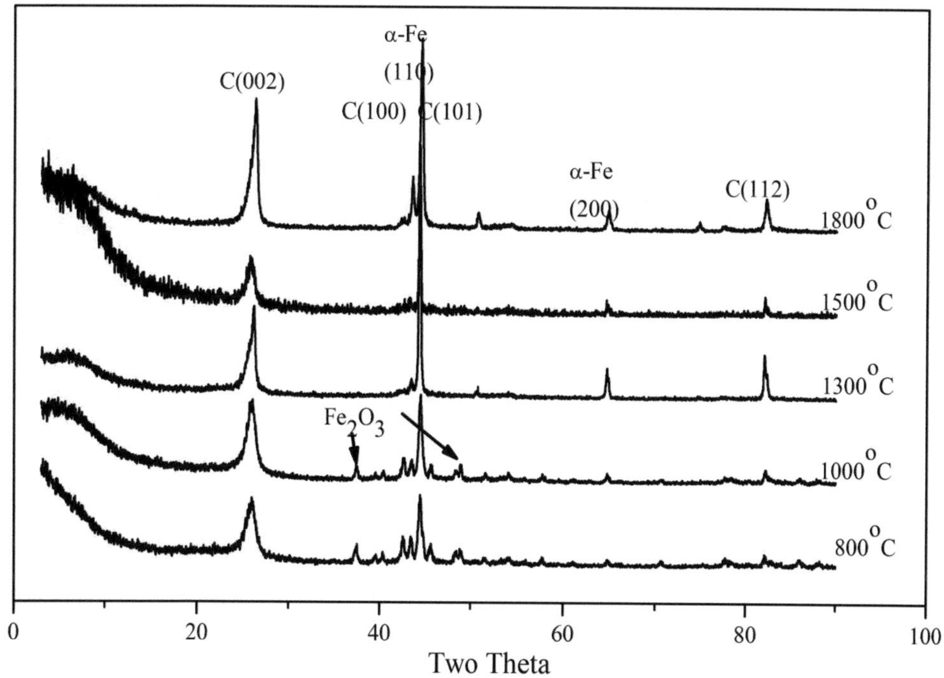

Figure 7. XRD patterns of the PVdF-based GNFs. in [31] (*These data were reproduced under permissions of Elsevier)*

Samples	Carbonization Temperature(°C)	XRD		Raman
		2 theta(°)	d_{002}*(nm)*	L_a*(nm)*[a]
PAN-based CNF	1300	25.98	> 0.370	-
PAN-base GNF	900	25.90	0.344	4.10
	1100	26.10	0.341	4.75
	1300	26.12	0.341	6.54
PVdF-based GNF	800	26.11	0.341	4.32
	1000	25.96	0.343	4.83
	1300	26.12	0.341	7.50
	1500	26.28	0.339	10.9
	1800	26.27	0.333	72.5

a : L_a *(Raman)* = 4.4 (I_G/I_D)

Table 2. The (002) spacing values and in-plane sizes of small graphite crystals, L_a, of electrospun PAN- and PVdF-based graphitic carbon nanofibers. in [31] (*These data were reproduced under permissions of Elsevier)*

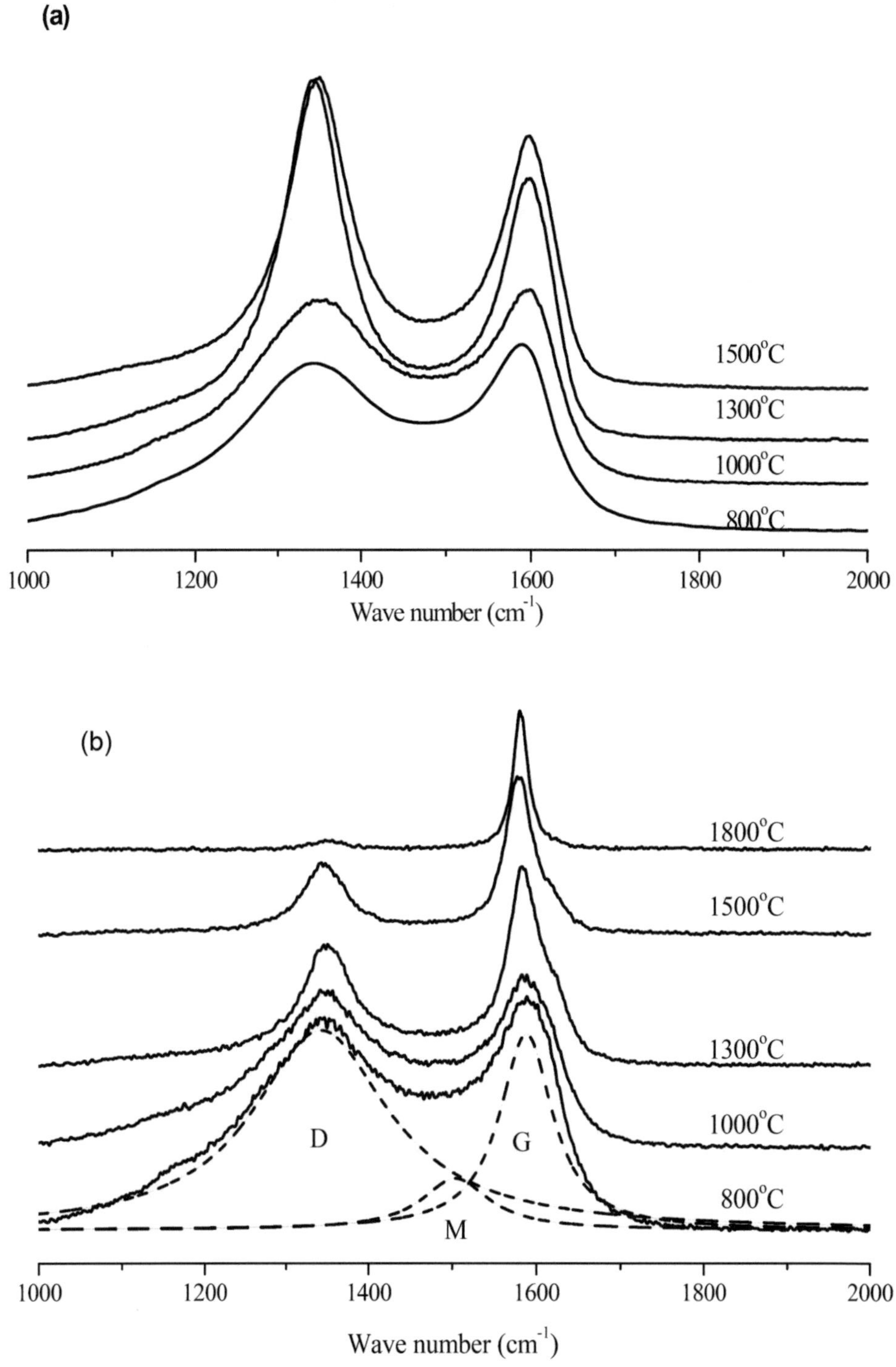

Figure 8. Raman spectra of the PVdF-based (a) CNFs and (b) GNFs at several carbonization temperatures. in [31] (*These data were reproduced under permissions of Elsevier*)

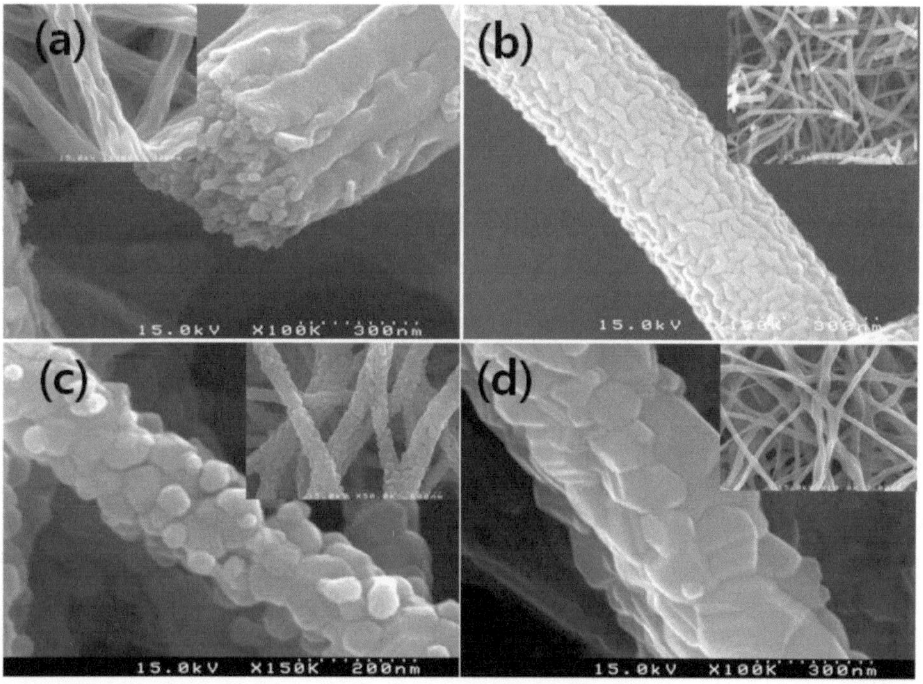

Figure 9. SEM images of (a) electrospun TiO_2 nanofiber after calcinations at 450°C, and electrospun $LiTi_2O_4$ nanofibers after calcinations at (b) 450°C (c) 600°C, and (d) 700°C.

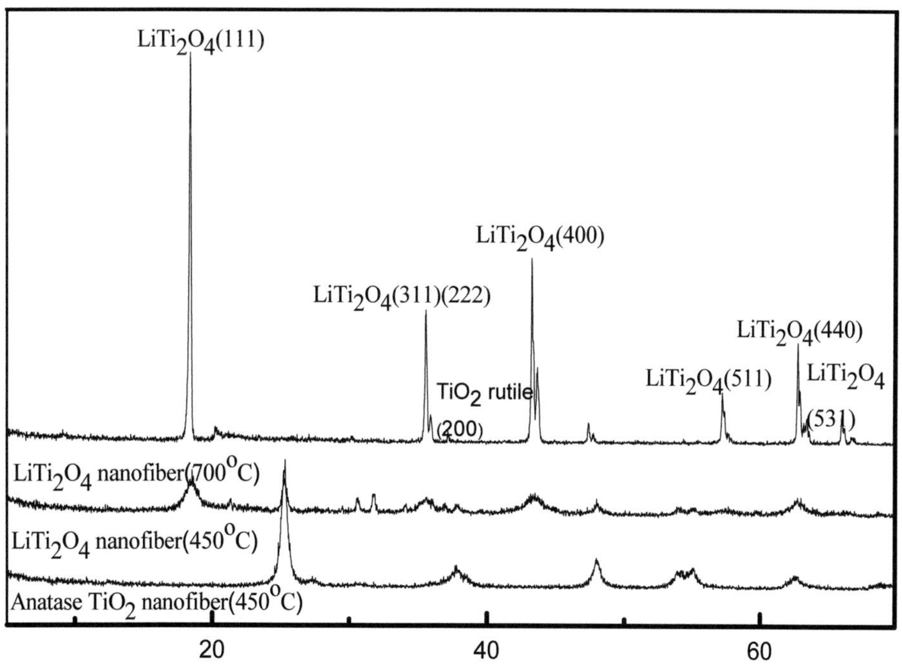

Figure 10. XRD patterns of electrospun TiO_2 nanofiber (a) and $LiTi_2O_4$ nanofibers after calcinations at 450°C and 700°C.

TiO_2 nanofiber was also prepared from typical electrospinning of a mixture solution of titanium tetraisoproxide and polyvinyl acetate (PVAc, Mw 500,000) in N, N'-dimethylacetamide (DMF). As-electrospun TiO_2/PVAc nanofiber was calcined at 450°C to completely remove the PVAc component by thermal decomposition and to give to TiO_2 nanofiber. As shown in Figure 9, the resulting TiO_2 nanofiber showed a smooth surface and internal structure composed of 20- to 50-nm TiO_2 granules. Figure 10 indicated this TiO_2 nanofiber was composed of typical anatase crystalline. Lithium titanate nanofiber was also prepared by electrospinning using a mixture of $LiNO_3$ and titanium tetraisoproxide (1:2 mole ratio) instead of titanium tetraisoproxide similar to preparation of TiO_2 nanofiber. As-electrospun lithium titanate/PVAc nanofibers were calcined at 450°C, 600°C, and 700°C to remove the PVAc component by thermal decomposition and to give to lithium titanate nanofibers. As shown in Figure 9, the lithium titanate nanofiber calcined at 450 °C showed lots of wrinkled structure in the surface and it looked like the surface composed of 20 to 60-nm nanorods. Crystalline size of lithium titanate was increased with increase of calcinations temperature. The lithium titanate nanofibers calcined at 600 °C and 700 °C showed the fibrous morphology composed of lithium titanate granules. These lithium titanate nanofibers calcined at above 450°C were typical crystalline structure of $LiTi_2O_4$ as show in Figure 10, in [33, 34].

2.2. Specific surface area and pore structure

The electrospun PAN-based CNFs showed the typical adsorption curves very similar to that of nonporous carbon in the nitrogen gas adsorption-desorption isotherms, while the PAN-based GNFs showed the typical curve of micro porous carbon in addition to a hysteresis loop that indicates existence of the meso pore, as shown in Figure 11. The PAN-based CNFs and GNFs had low surface areas within the range of 22~31 m^2/g and 60~253 m^2/g, respectively, as listed in Table 3. The surface areas of PAN-based GNFs were much higher than the CNFs, but they decreased with increase of carbonization temperature and increased with increase of IAA content. Although this could not be fully explained at present, it may be due to the surface roughness and inhomogeneous structure of the GNFs, which resulted from the induction of the metal catalyst in the GNFs. But they still had much lower surface area compared to common active carbon. Commercial active carbons and active carbon fibers generally have very high surface areas of above 1000 m^2/g, and SWNT also has a surface area of a few hundred m^2/g. They had low storage capacities, however, within the range of 0.35-0.41 wt%, at room temperature, in [9]. So, high hydrogen adsorption by PAN-based CNFs and GNFs with very low surface area may not be expected. However, if they have effective pores for hydrogen storage when compared to hydrogen molecule sizes of about 0.41 nm, they may show high hydrogen adsorption. The electrospun PAN-based GNFs showed the adsorption curves very similar to that of meso porous carbon in the nitrogen gas adsorption-desorption isotherms in spite of their low surface area and also they had micro pores unlike those in the CNFs in the nitrogen gas adsorption-desorption isotherms. In the case of the PAN-based CNFs the change of the pore volumes with increase of carbonization temperature did not show because of their very low surface area. The micro pore volumes and meso pore volumes of the PAN-based GNFs, however, decreased with increase of carbonization temperature.

The electropun PVdF-based CNFs prepared after a slight DHF-treatment showed typical curves of micro porous carbon in the nitrogen gas adsorption-desorption isotherms. They showed high surface areas of 414~1300 m^2/g. BET surface area rapidly decreased with increase of carbonization temperature, as shown in Table 3. Micro pore volume at 1500 °C greatly decreased while meso pore volume continuously increased with an increase of carbonization temperature. However, the PVdF-based CNF at 1800 °C showed a very high surface area of 1300 m^2/g and a high volume (1.767 cm^3/g) of only ultra- or super micro pores. The PVdF-based CNFs prepared after high DHF-treatment showed very low surface area and adsorption curves similar to those of nonporous carbon in nitrogen gas adsorption-desorption isotherms. They did not have the micro pores. In the case of PVdF-based GNF nitrogen adsorption–desorption isotherms were the type II showing a hysteresis loop that indicates the existence of meso pores, as shown in Figure 11. They showed high surface areas of 377~473 m^2/g but still have low surface areas compared to typical active carbon. The BET surface area and micro pore volume decreased with increasing carbonization temperature, as listed in Table 3. Decreases in the surface area and micro pore volume are thought to be due to densification of the porous structure with increasing carbonization temperature. Nitrogen adsorption–desorption isotherms for the $LiTi_2O_4$ nanofiber calcined at 450 °C were also the type II showing a hysteresis loop that indicates the existence of meso pores, as shown in Figure 10. However, the TiO_2 nanofiber and the $LiTi_2O_4$ nanofibers calcined at 450 °C have very low surface are of about 50 m^2/g, as listed in Table 4. The surface area of the $LiTi_2O_4$ nanofibers calcined at 700 °C decreased with increase of crystalline size.

The pores in carbon materials are classified by their size into macro pores (> 50 nm), meso pores (> 2-50 nm), and micro pores (< 2 nm) according to IUPAC. Micro pores are further divided into super micro pores with a size of 0.7~2 nm and ultra micro pores of less than 0.7 nm, in [35]. Unfortunately, it is difficult to exactly analyze the ultra micro pore size distribution and volume in porous carbon through the nitrogen gas adsorption-desorption isotherms measurement when compared to the kinetic diameter of hydrogen molecules. In the case of direct observations of pores on the surface of carbon materials by scanning tunneling microscopy (STM/AFM), the problem is how to differentiate the pores from other surface defects, such as depressions and trenches. The net ultra micro pore volume of carbon material cannot obtain from the pore analysis of small area on the surface by STM. The oxidation and carbonization of PAN precursor fiber for making carbon fiber usually accompany with the release of NH_3, HCN, N_2 gases, etc., resulting in the formation of pores within the carbon fiber structure, in [35]. So the preparation of carbon fiber with high tensile strength requires the removal of pore structure by heat treatment at high temperature. The carbonization of the electrospun PVdF nanofibers is also usually accompanied by the release of HF, H_2, F_2 and other gases, resulting in the formation of pores within the carbon fiber structure. In addition, micro and meso pores are generated through the carbonization of PVdF nanofibers after partial dehydrofluorination. The calcinations of as-electrospun lithium titanate/PVAc nanofibers or TiO_2/PVAc nanofibers may also produce the pore structure by evaporation of thermally decomposition product of PVAc. Therefore, we assume the generation of ultra micro pores and super micro pore during the carbonization and calcinations of the electrospun polymeric nanofibers and metal oxide nanofiber precursors. These pore structures became dense with the increase of carbonization temperature. Therefore, increase of carbonization temperature might bring out the increase of ultra- and super micro pore volume instead of loss of large pores.

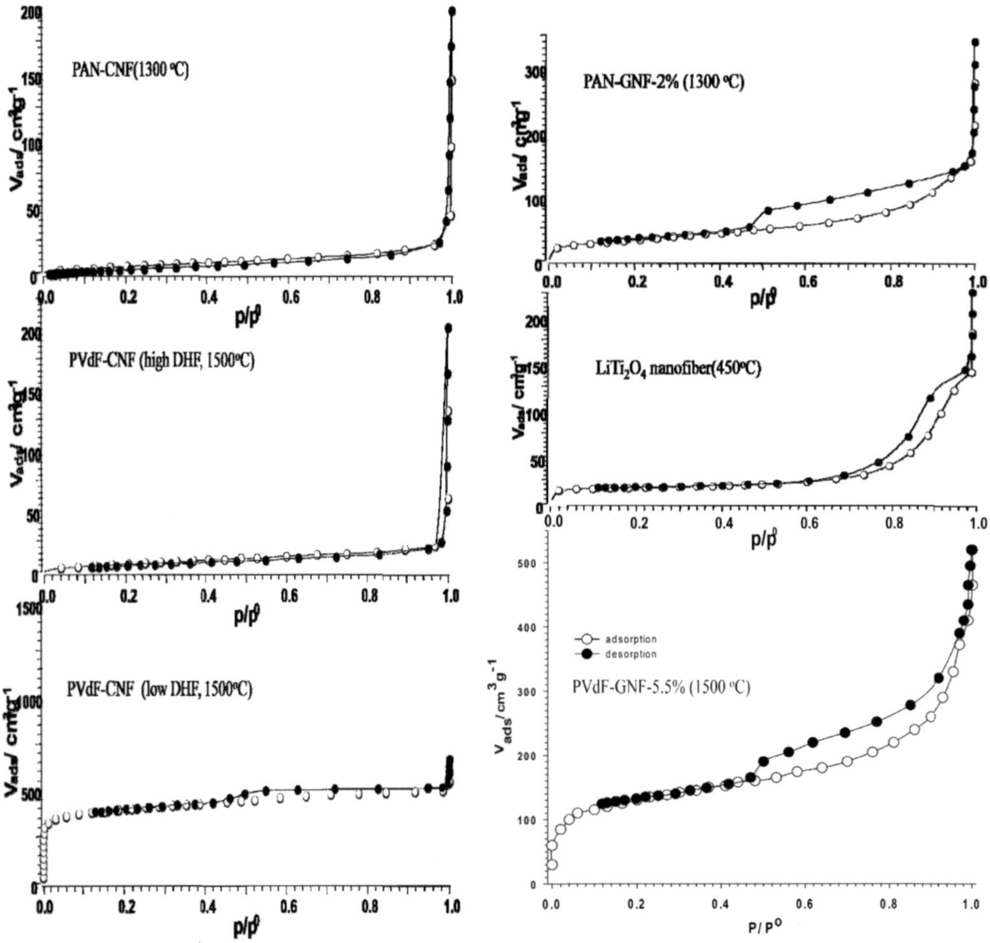

Figure 11. Typical nitrogen adsorption–desorption isotherms for the CNF, GNF and $LiTi_2O_4$ nanofiber prepared from the electrospun nanofibers. in [28] (*These data were reproduced under permissions of The Polymer Society of Korea*)

Samples	Carbonization temperature(°C)		Total pore volume(cm^3/g)	IAA (wt%)	Surface area(m^2/g)	Pore Size Distribution(cm^3/g)			
						< 1 nm[a]	1~2 nm[a]	2~4 nm[b]	4~10 nm[b]
PVdF-CNF	low DHF	800	-	0	967	0.146	0.052	0.075	0.070
		1000	-	0	921	0.166	0.056	0.084	0.101
		1300	-	0	865	0.249	0.065	0.092	0.094
		1500	-	0	414	0.057	0.072	0.103	0.109
		1800	0.63	0	1300	1.767[c]	-	-	-
	high DHF	1000	-	0	33	-	-	0.017	0.014
		1300	-	0	16	-	-	0.006	0.004
		1500	-	0	26	-	-	0.012	0.010
PVdF-GNF		800	0.74	5.5	473	0.162	0.042	0.132	0.294
		1000	0.91	5.5	445	0.158	0.040	0.133	0.315
		1500	1.01	5.5	431	0.143	0.048	0.132	0.186
		1800	2.02	5.5	377	0.115	0.044	0.125	0.180
PAN-CNF		1000	-	0	32	-	-	0.011	0.012
		1300	-	0	22	-	-	0.012	0.012
		1500	-	0	22	-	-	0.008	0.007
PAN-GNF		900	-	2	198	0.048	0.027	0.048	0.039
		1100	-	2	198	0.032	0.024	0.050	0.059
		1300	-	2	60	-	-	0.021	0.046
		1500	-	2	65	0.008	0.007	0.023	0.039
		900	-	5	243	0.044	0.027	0.047	0.060
		1100	-	5	247	0.065	0.030	0.050	0.084
		1300	-	5	116	0.025	0.017	0.040	0.066
		1500	-	5	109	0.030	0.012	0.034	0.056
		1300	-	7.5	163	0.047	0.019	0.041	0.062

[a] determined by applying the Horvath Kawazoe pore sizes for micro porous samples.
[b] determined by applying the B.J.H. pore sizes for meso porous samples.
[c] ultra micro pores of below 0.8 nm

Table 3. Surfacearea and pore analysis of electrospun PVdF- and PAN-based CNFs and GNFs. in [28~31] (These data were reproduced under permissions of The Polymer Society of Korea, Cambridge University Press, and Elsevier)

Samples	Calcinations temperature (°C)	Surface area (m^2/g)	Pore Size Distribution(cm^3/g) (1~2 nm[a])
TiO_2 nanofiber	450	49.4	0.0157
$LiTi_2O_4$ nanofiber	450	50.2	0.0187
$LiTi_2O_4$ nanofiber	700	26.3	0.0090

Table 4. Surface area and pore analysis of electrospun TiO_2 and $LiTi_2O_4$ nanofibers.

2.3. Hydrogen storage capacities

The hydrogen storage capacity of electrospun nanofibrous materials in this chapter was evaluated through the gravimetric method using magnetic suspension balance (MSB, Rubotherm), as show in Figure 12. First, the blank test chamber containing samples was evacuated, to remove the impurities and water, at 150 °C/10^{-6} torr for 6 hrs. The weights of the sample basket and samples were then measured at 10^{-6} torr/25 °C (±0.5 °C) and at a He gas atmosphere of 10 bars, respectively. It was assumed that He gas was not adsorbed by the nanofiber samples in this condition. The weight difference between the vacuum and the 10 bars He gas, which indicates buoyancy due to the He gas, was used to determine the volume of the nanofiber samples, as follows:

$$Vs = W_1/d_{He} \tag{1}$$

where Vs is the volume of the samples, W_1 is the weight difference of the samples between in the vacuum and at the 10-bar He gas atmosphere, and d_{He} is the density of He at a specific pressure and temperature. The weights of the samples were measured under different H_2 pressures (10~100 bars) at 25 °C (±0.5 °C). The weights of the absorbed hydrogen were determined after the correction of the buoyancy due to the hydrogen gas atmosphere, using the sample volume (Vs), as follows:

$$\text{The weight of the adsorbed } H_2 = W_2 + Vs\, d_{H2} \tag{2}$$

where W_2 is the weight difference of the samples between in the vacuum and at the specific H_2 pressure, Vs is the volume of the samples, and d_{H2} is the density of H_2 at a specific pressure and temperature. The densities of He and H_2 gas for buoyancy correction were calculated from a real gas equation using the Thermodynamic and Transport Properties of Pure Fluid Program (NIST-supported).

In the case of monolayer condensation of hydrogen on carbon absorbents, theoretical quantity of absorption is 1.3×10^{-5} mol/m^2, in[1,13], and the quantity of reversible hydrogen absorption is known to proportional to specific surface area of absorbents. Commercial active carbons and active carbon fibers generally have very high surface areas of above 1000 m^2/g, and SWNT also has a surface area of a few hundred m^2/g. They showed low storage capacities, however, within the range of 0.35~0.41 wt%, at room temperature, in [7]. Low storage capacity of carbon materials at ambient temperature is due to too low absorption potential between carbon and hydrogen. If the tendency that hydrogen is going to escape from carbon absorbent is smaller than absorption potential, hydrogen will be absorbed as condensed phase by whole micro pore. It is predicted there are the optimum pore size and pore geometry for hydrogen absorption. Therefore, when the kinetic diameter of hydrogen molecule (0.41 nm) is considered, ultra micro pores (< 0.7 nm) are expected by doing important contribution for hydrogen storage by means of nanocapillary mechanism and superposing of potential on the pore wall substantially. Hydrogen adsorption of the CNFs with very low surface area may not be expected because of their low surface area. If they have ultra micro pores, however, which may be effective for hydrogen storage, they will show hydrogen adsorption.

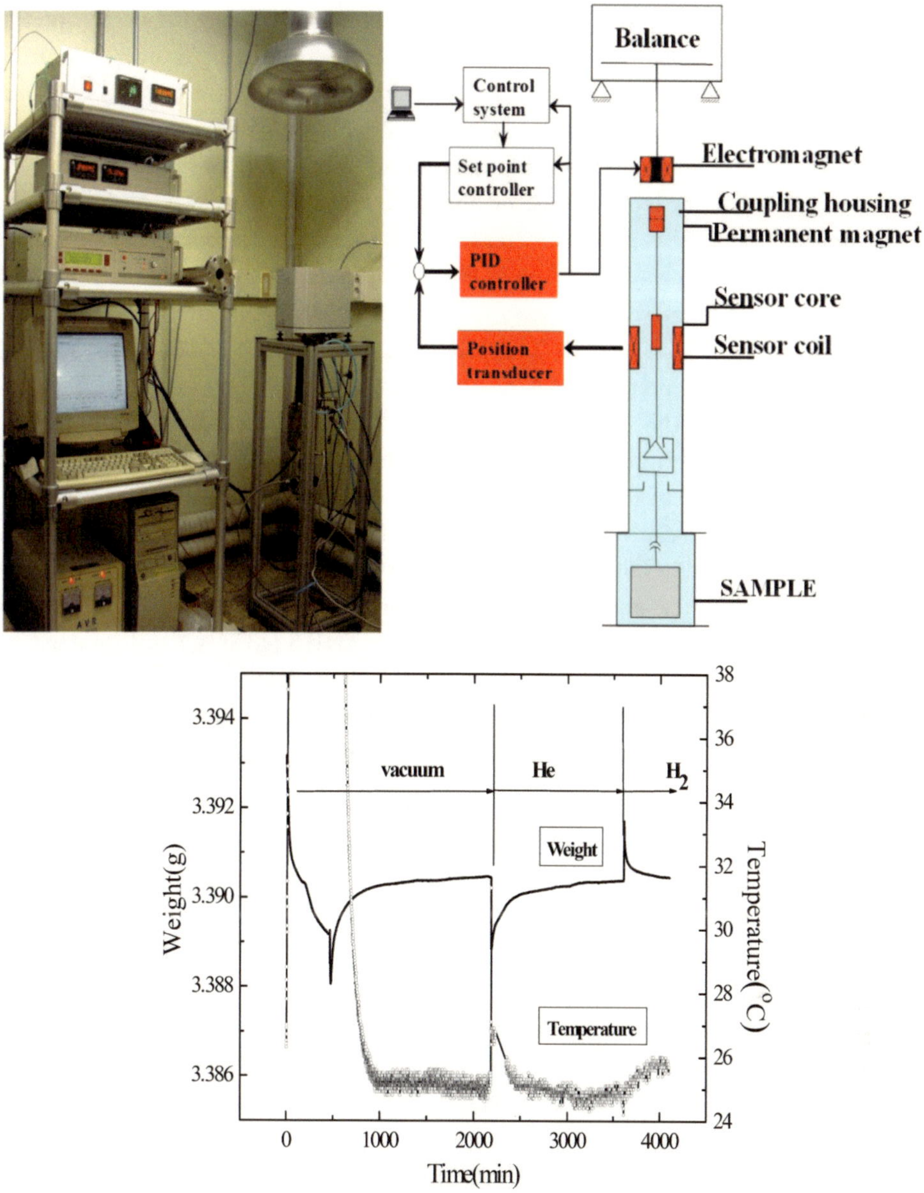

Figure 12. Procedure for evaluation of hydrogen storage by Gravimetric method using Magnetic Suspension Balance.

Figure 13 shows the hydrogen storage results of the electrospun PAN-based CNFs and GNFs under several hydrogen pressures and at room temperature. The hydrogen storage was measured after 2 hrs under a specific hydrogen pressure. Their hydrogen adsorption continuously increased even after 2 hrs, and also increased with increase of hydrogen pressure. The dotted line in Figure 13(a) indicates the increased hydrogen storage capacities after 16 hrs. The hydrogen storage capacities of the PAN-based CNFs obtained from carbonization at 1000°C, 1100 °C and 1300°C, and 1500°C were 0.16 wt%, 0.37 wt%, 0.50 wt%, and 0.26 wt% respectively, although they were nonporous carbon with very low surface areas in the nitrogen gas adsorption-desorption isotherms. These may indicate the presence of ultra micro pores that cannot differentiate by using nitrogen gas adsorption-desorption isotherms. The reduction of hydrogen storage in the CNF obtained from carbonization at 1500 °C was thought to be due to the disappearance of pore structure including ultra micro pore at high carbonization temperature of above 1300 °C. In the case of the PAN-based GNFs the hydrogen storage capacities increased with increase of carbonization temperature and the content of IAA catalyst, as shown in Figure 13(b), and were higher than those of the CNFs. The hydrogen storages of PAN-based GNF-5 showed highest capacity of 1.01 wt% at 1300 °C and lowest capacity of 0.14 wt% at 1500 °C_similar to those of the PAN based CNF samples. Increase of the content of IAA resulted in increase of the hydrogen storage. The hydrogen storage of the PAN-based GNF-7.5 at 1300 °C, however, showed very low storage of 0.32 wt% though it had higher surface area and higher micro-, meso pore volume than those of GNF-2 and GNF-5 at 1300 °C. So, the hydrogen storage of the PAN-based CNF and GNF did not show the correlation with surface area, and micro-, meso pore volume in Table 3. Fe metal catalyst in the GNFs may contribute to the hydrogen adsorption. Figure 14 showed the cycle property about the hydrogen adsorption of the PAN-based GNF-5 (1300°C), which showed highest storage capacity. The GNF-5 (1300 °C) still retained initial hydrogen capacity storages, indicating physisorption of hydrogen. However, about 0.078 wt% of hydrogen did not desorbed under atmosphere and vacuum of 10^{-6} torr at room temperature. This is thought to be chemisorptions by Fe metal catalyst.

Figure 15 shows the hydrogen storage capacities of the electrospun PVdF-based CNFs. The PVdF-based CNFs (high DHF) also showed some hydrogen absorption although they had no micro pore volumes. The hydrogen absorptions of about 0.3 wt% (100 H_2 bar) were observed in the PVdF-based CNFs (high DHF) prepared at 1,300~1500 °C although they have very low specific surface area of 16~33 m^2/g. But the PVdF-based CNFs (low DHF) prepared at 800~1300 °C showed the hydrogen storage capacities of only 0.05~0.2 wt% in spite of high specific surface area of 865~967 m^2/g. The hydrogen adsorption of the carbon nanofibers with high surface areas decreased with the increase of hydrogen pressure. This may be due to the buoyancy effect of hydrogen gas adsorbed on the samples. Hydrogen storage capacities of the electrospun PVdF-based CNF (low DHF) increased with increase of carbonization temperature and showed the maximum value of about 0.39 wt% at 1500 °C in spite of its lowest surface area. And also the PVdF-based CNF (low DHF) carbonized at 1800 °C showed hydrogen storage capacity of 0.39 wt% even though it had highest surface area of 1300 m^2/g and highest ultra micro pore volume of 1.767 cm^3/g, similar to the activated carbon fibers having 0.35~0.41 wt% hydrogen storage, in [9]. This is thought to be due to the disappearance of the pore structure including ultra micro pores at high carbonization temperatures. Figure 16 shows hydrogen storage results for the PVdF-based GNFs under several hydrogen pressures at room temperature. Their hydrogen adsorption increased with increase of carbonization temperature while specific surface area and micro pore volume ($<$ 1 nm) were decreased, but they showed very low storage capacities of about 0.1~0.2 wt% although they have highly graphite crystalline structure.

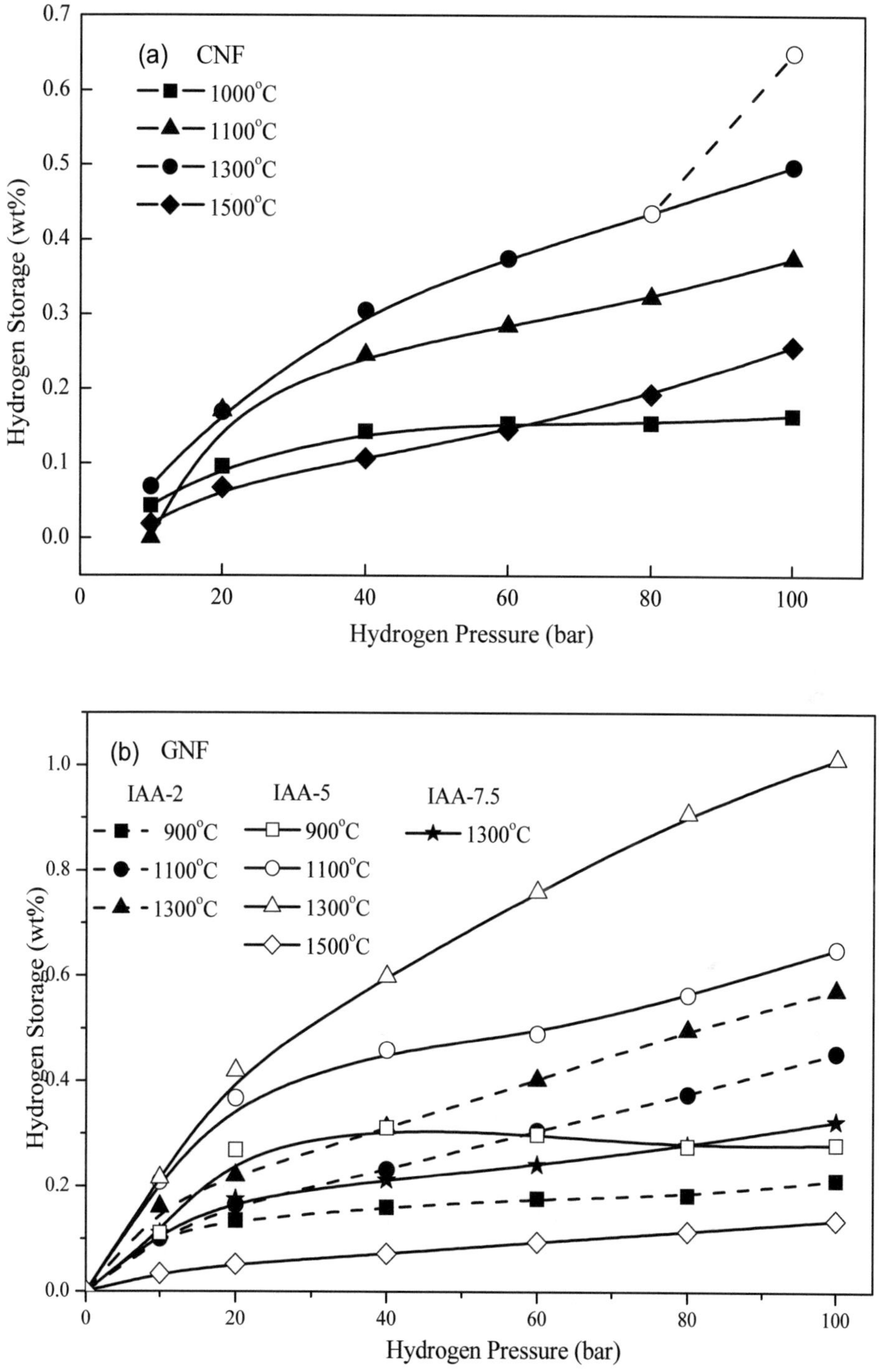

Figure 13. The hydrogen storage of the PAN-based (a) CNFs and (b) GNFs under several hydrogen pressures and at room temperature. in [28,29] (*These data were reproduced under permissions of The Polymer Society of Korea and Cambridge University Press*)

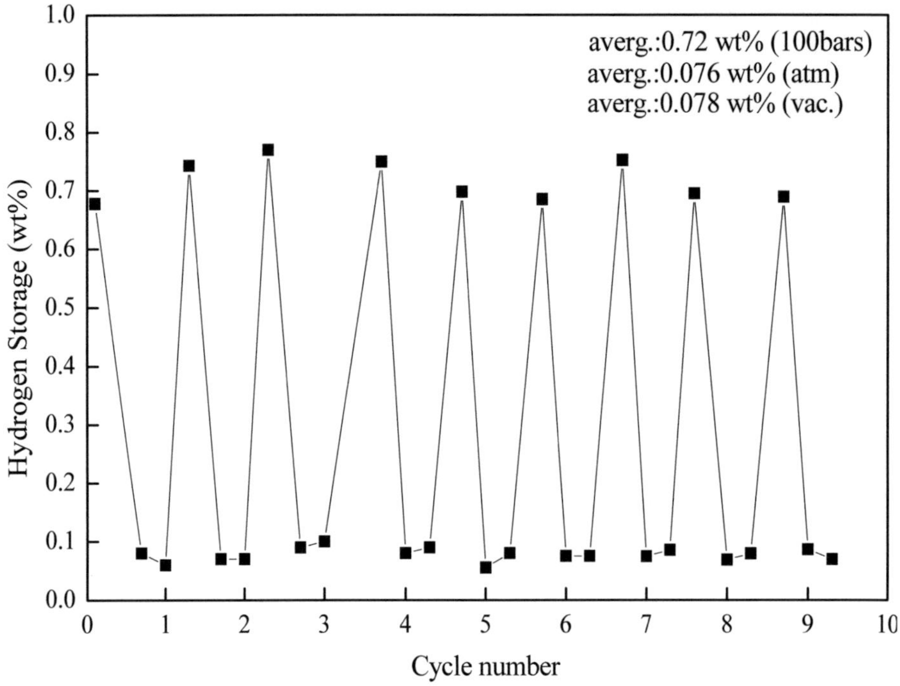

Figure 14. The hydrogen adsorption/desorption cycle property of the PAN-based GNF-5 (1300°C) under hydrogen pressures of 100 bars and at room temperature. in [28] (*These data were reproduced under permissions of The Polymer Society of Korea*)

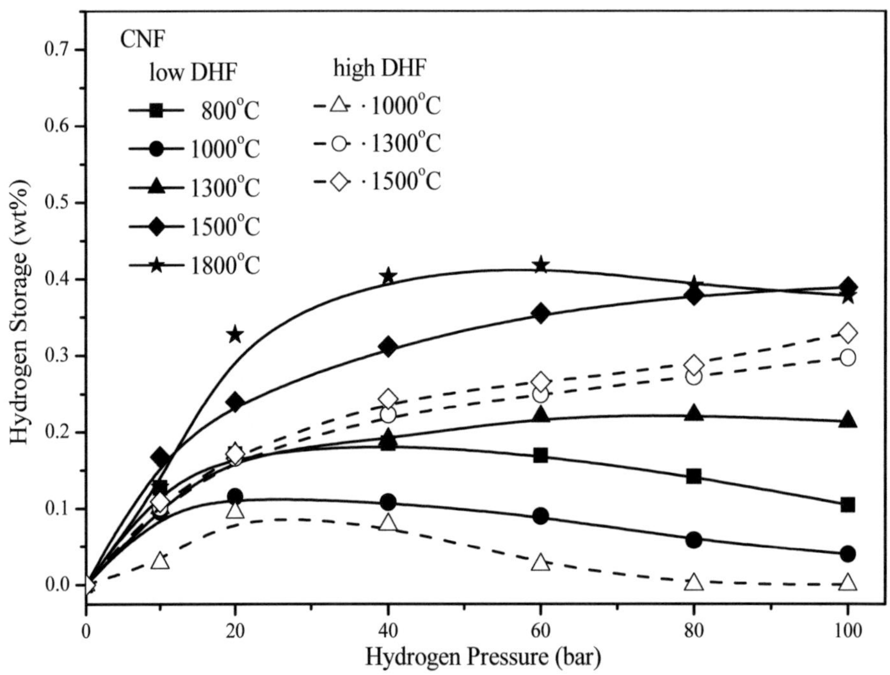

Figure 15. The hydrogen storage of the PVdF-based CNFs under several hydrogen pressures and at room temperature. in [30] (*These data were reproduced under permissions of Cambridge University Press*)

The hydrogen storage capacity of the GNFs and CNFs did not show correlation with surface area or micro- and meso pore volume, as shown in Table 3. The quantity of adsorbed hydrogen on nanostructure graphitic carbon as well active carbon materials at 77 K is proportional to the specific surface area of carbon materials, in [10]. However, because at ambient temperature the thermal motion of hydrogen molecules overcomes van der Waals-type weak physisorption of molecular hydrogen, their hydrogen storage capacities were very low. So the hydrogen adsorption on the GNFs and CNF samples may be influenced by pore structure as well as specific surface area. Therefore, we think that micro- and meso pores that are calculated using the nitrogen gas adsorption–desorption isotherms are not the effective pore for hydrogen storage. The effective pore for hydrogen storage may require small pore size not exceeding 1 nm, when compared to the kinetic diameter of hydrogen molecule of about 0.41 nm. It is assumed that these micro pores are different from the micro pores calculated using nitrogen adsorption–desorption isotherms. Thus, hydrogen adsorptions by the electrospun PAN or PVdF-based CNFs and GNFs may be due to the presence of ultra micro pores rather than micro- and meso pores, even though they have very low surface areas compared to commercially available active carbons and active carbon fibers.

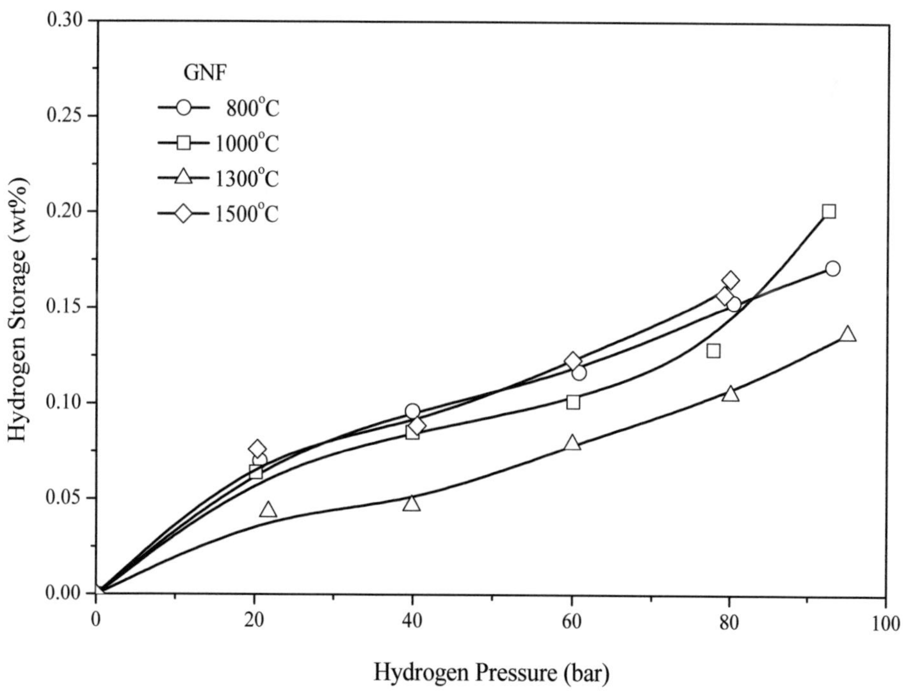

Figure 16. The hydrogen storage of the PVdF-based GNFs under several hydrogen pressures and at room temperature. in[31] (*These data were reproduced under permissions of Elsevier*)

The multilayered TiO_2 nanotubes with a surface area of 199 m^2/g (a pore size; 8 nm, and a pore volume; 0.70 cm^3/g) had known to store a 1~2.5 wt% hydrogen at 1 bar and room temperatures in the range 80 to 125 °C, in [13]. This high storage capacity may result from much higher adsorption potential energy between the multilayered TiO_2 nanotubes and hydrogen molecules than those of carbon materials in consideration of very low hydrogen storage of typical meso porous carbon materials with high surface area. Figure 17 shows hydrogen storage results for the electrospun TiO_2 nanofiber and $LiTi_2O_4$ nanofibers. The hydrogen storage was measured after 2 hrs under a specific hydrogen pressure. The TiO_2 nanofiber and $LiTi_2O_4$ nanofibers calcined at 450 °C showed high hydrogen storages of 1.11 wt% and 0.74wt% in spite of their low surface area of 49.4 m^2/g and 50.2 m^2/g, respectively. Their hydrogen absorptions were higher than those of the electrospun CNFs and GNFs. Although we presently cannot determine effective ultra micro pore volumes for hydrogen storage in the TiO_2 nanofiber and $LiTi_2O_4$ nanofibers, these are thought to be due to higher adsorption potential energy between the metal oxide materials and hydrogen molecules than those of carbon nanofibers. The $LiTi_2O_4$ nanofiber showed higher hydrogen storage than TiO_2 nanofiber with similar surface area. This is also thought to be due to higher adsorption potential energy of $LiTi_2O_4$ nanofiber than TiO_2 nanofiber. The hydrogen storage of the $LiTi_2O_4$ nanofiber calcined at 700 °C was greatly reduced to 0.41 wt%. Increase of the calcination temperature resulted in decrease of hydrogen storage with great reduction of surface area, indicating loss of effective pores with increase of $LiTi_2O_4$ crystalline size, as shown in Figure 9. Figure 18 showed the hydrogen adsorption/desorption cycle of electrospun $LiTi_2O_4$ nanofibers. Their hydrogen adsorption continuously increased even after 2 hrs under hydrogen pressure of 100 bars. The hydrogen storage of the $LiTi_2O_4$ nanofibers calcined at 450°C and 600°C were about 1.50wt% and 1.23 wt% at the equilibrium state under hydrogen pressure of 100 bars, respectively. However, about 0.06 wt% and 0.054 wt% of hydrogen were not desorbed under atmosphere and vacuum of 10^{-6} torr at room temperature, respectively. These are thought to be chemisorptions.

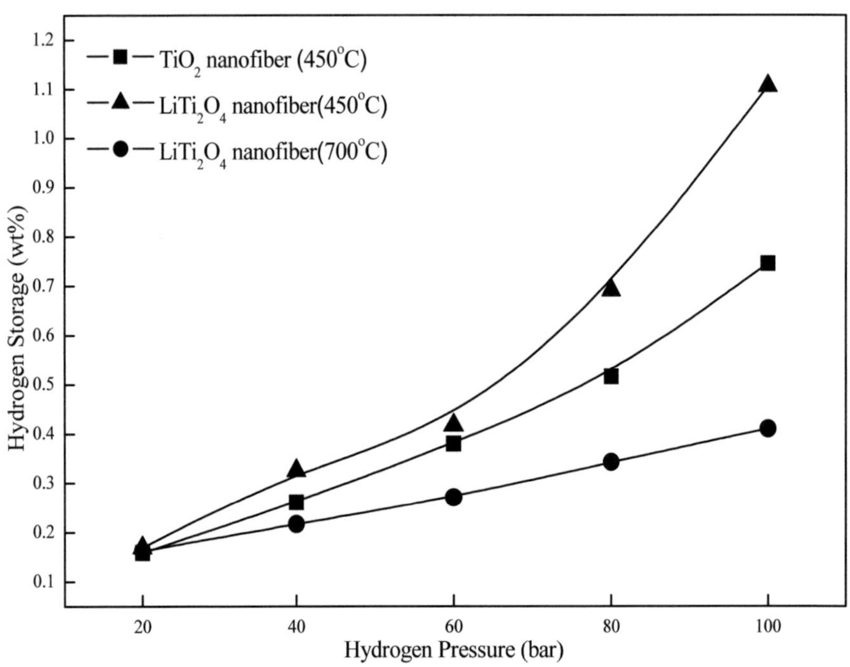

Figure 17. The hydrogen storage capacity of the electrospun TiO_2 nanofiber and $LiTi_2O_4$ nanofibers.

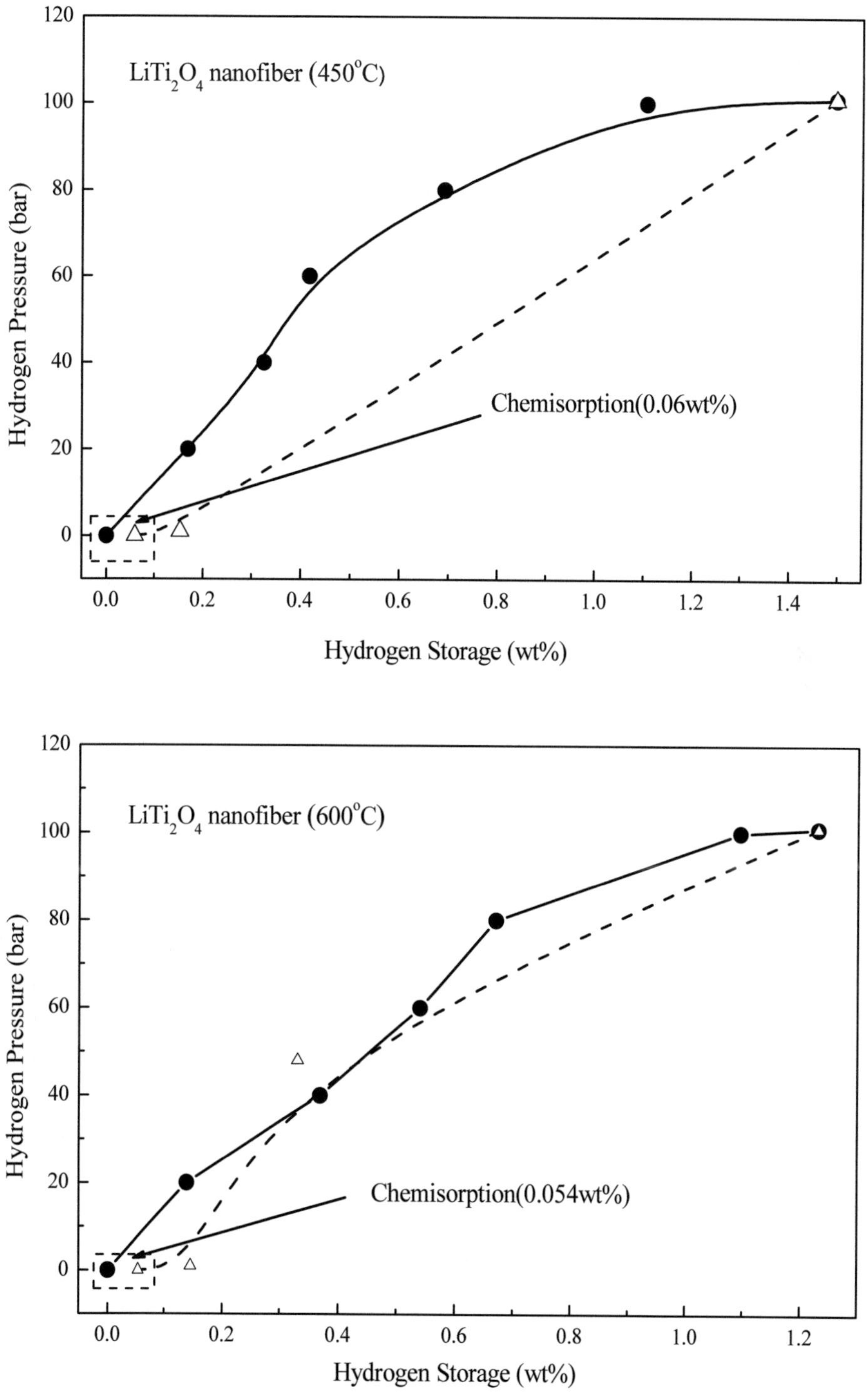

Figure 18. The hydrogen adsorption/desorption cycle of the electrospun $LiTi_2O_4$ nanofibers under hydrogen pressures of 100 bars and at room temperature.

2.4. Further work

Active carbon with the same specific surface area reversibly adsorbed 2 mass% hydrogen at a temperature of 77 K, in [1,2]. The quantity of adsorbed hydrogen on nanostructured graphitic carbon as well active carbon materials at 77 K is proportional to the specific surface area of carbon materials, in [10]. However, because at ambient temperature the thermal motion of hydrogen molecules overcomes van der Waals-type weak physisorption of molecular hydrogen, high surface area and large micro- and meso pores volumes of active carbon does not greatly contribute to the hydrogen adsorption. The effective pores for hydrogen storage are assumed to be ultra micro pores with small pore size not exceeding 1 nm, when compared to the kinetic diameter of hydrogen molecule of about 0.41 nm. The electrospun CNF and GNF were prepared by carbonization without further activation process to induce increase of ultra micro pore through heat-treatment for densification of large pores structure at high temperature. Thus, hydrogen adsorptions results of the electrospun CNFs and GNFs indicated the presence of ultra micro pores even though they have very low surface area and micro-, meso pores, that are calculated using the nitrogen gas adsorption–desorption isotherm, when compared to commercially available active carbons or active carbon fibers. For automotive and industrial applications, the solid absorbent with hydrogen storage capacity of greater than 6.5 wt% and ambient temperatures for hydrogen release are presently required. The hydrogen storage capacities of the electrospun CNFs and GNFs, however, still showed the limitations in overcoming this requirement. That is, it may not be possible to increase effective ultra micro pores for above 6.5 wt% hydrogen storage even though we can find more improved process for the CNF and GNF in future.

On the other hands, the hydrogen storage results in TiO_2 nanofiber and $LiTi_2O_4$ nanofibers gave to some encouragement for overcoming hydrogen storage target. They showed 1.2~1.5 wt% hydrogen in spite of very low surface area of about 50 m^2/g. These were higher than those of the electrospun CNFs and GNFs because the TiO_2 nanofiber and $LiTi_2O_4$ nanofibers are thought to be due to higher adsorption potential energy between the metal oxide materials and hydrogen molecules than those of carbon nanofibers. So, future works for the high hydrogen storage capacity need new solid absorbent materials with a high ultra micro pore volume, high surface area, and appropriate hydrogen adsorption potential energy for only reversible physisorption.

Therefore, further work for this purpose will be tried to prepare new nanostructures metal oxide nanofibers with high reversible physisorption at ambient temperature, which are controlled to a high surface area and high ultra micro pore volumes.

3. Conclusions

The hydrogen storage capacities of electrospun nanofibrous materials were discussed in view of their pore size, surface area, and adsorption potential energy for hydrogen molecules. Carbon nanofibers (CNF) and graphite nanofibers (GNF) were prepared through the carbonization of the electrospun PAN- and PVdF-based nanofibers. The TiO_2 nanofiber and $LiTi_2O_4$ were also prepared through typical electrospinning of precursor solutions.

The hydrogen storage capacities of the PAN-based CNFs prepared by carbonization at 1000 °C, 1100 °C and 1300°C, and 1500°C were 0.16 wt%, 0.37 wt%, 0.50 wt%, and 0.26 wt% respectively, although they were nonporous carbon with very low surface areas of about 22~32 m^2/g in the nitrogen gas adsorption-desorption isotherms. The PVdF-based CNFs (high DHF) prepared by carbonization at 1300°C and 1500°C showed very low surface area of about 16~26 m^2/g without the micro pores and showed the adsorption curves similar to those of nonporous carbon in nitrogen gas adsorption-desorption isotherms. However, they also stored the hydrogen of 0.30 wt% and 0.33 wt%, respectively.

The PAN-based GNFs had some micro-, meso pores and higher surface areas of 60~253 m^2/g than the CNFs though they were still much lower surface area compared to common active carbon. Their surface area decreased with increase of carbonization temperature and increased with increase of IAA content. Although this could not be fully explained at present, it may be due to the surface roughness and inhomogeneous structure of the GNFs, which resulted from the induction of the metal catalyst in the GNFs. The hydrogen storage capacities of them increased with increase of carbonization temperature and the content of IAA catalyst, and were higher than those of the CNFs. The hydrogen storages of PAN-based GNF-5 showed highest capacity of 1.01 wt% at 1300 °C and lowest capacity of 0.14 wt% at 1500 °C.

The PVdF-based CNFs (low DHF) showed typical curves of micro porous carbon in the nitrogen gas adsorption-desorption isotherms. They had high surface areas of 414~1300 m^2/g and stored the hydrogen of 0.04-0.39 wt%. The PVdF-based CNF (low DHF) at 1800 °C showed the hydrogen storage of only 0.38 wt% in spite of high surface area of 1300 m^2/g and a high volume (1.767 cm^3/g) of only ultra- or super micro pores. Nitrogen adsorption–desorption isotherms for the PVdF-based GNFs prepared from carbonization at 800~1800°C were the type II showing a hysteresis loop that indicates the existence of meso pores. They had high surface areas of 377~473 m^2/g but showed very low storage capacities of about 0.1~0.2 wt% although they have highly graphite crystalline structure.

The above hydrogen storage capacities of the GNFs and CNFs did not show any correlations with surface area or micro- and meso pore volume calculated using nitrogen adsorption–desorption isotherms. Because at ambient temperature the thermal motion of hydrogen molecules overcomes van der Waals-type weak physisorption of molecular hydrogen, their hydrogen storage capacities were very low. So the hydrogen adsorption on the GNFs and CNF samples may be influenced by pore structure as well as specific surface area. Therefore, micro- and meso pores that are calculated using the nitrogen gas adsorption–desorption isotherms are not thought to be the effective pore for hydrogen storage. The effective pore for hydrogen storage may require small pore size not exceeding 1 nm, when compared to the kinetic diameter of hydrogen molecule of about 0.41 nm. Thus, hydrogen adsorptions by the electrospun PAN or PVdF-based CNFs and GNFs may be due to the presence of ultra micro pores rather than micro- and meso pores, even though they have very low surface areas compared to commercially available active carbons and active carbon fibers.

The TiO_2 nanofiber and $LiTi_2O_4$ nanofibers calcined at 450 °C showed high hydrogen storages of 1.11 wt% and 0.74wt% in spite of their low surface areas of 49.4 m^2/g and 50.2 m^2/g, respectively. Their hydrogen storages were higher than the electrospun CNFs and GNFs. Although we presently cannot determine the effective ultra micro pore volumes for hydrogen storage, their high hydrogen adsorptions are thought to be due to higher adsorption potential energy than those of carbon nanofibers. The hydrogen storage of $LiTi_2O_4$ nanofiber was higher than that of TiO_2 nanofiber with similar surface area, indicating higher adsorption potential energy of $LiTi_2O_4$ nanofiber than that of TiO_2 nanofiber. So, the high hydrogen storage capacity need new solid absorbent materials with a high ultra micro pore volume, high surface area, and appropriate hydrogen adsorption potential energy for only reversible physisorption at ambient temperature.

Author details

Seong Mu Jo
Center for Materials Architecturing, Korea Institute of Science and Technology, Seoul, Republic of Korea

4. References

[1] Schlapbach L and Zuttel A, (2001), Hydrogen-Storage Materials for Mobile Applications, Nature, 414: 353-358.

[2] Nellis, W. J, Louis, A. A, and Ashcroft, N. W, (1998), Metallization of fluid hydrogen. Phil. Trans. R. Soc. Lond. A356: 119-135

[3] Dillon A. C, Jones K. M, Bekke-dahl T. A., Kiang H, Bethune D. S, and Heben M. J, (1997), Storage of Hydrogen in Single-Walled Carbon Nanotubes, Nature, 386: 377-379.

[4] Zhu H, Cao A, Li X, Xu C, Mao Z, Ruan D, Liang J, and Wu D, (2001), Hydrogen Adsorption in Bundles of Well-Aligned Carbon Nanotubes at Room Temperature, Appl. Surf. Sci., 178: 50-55.

[5] Ye Y, Ahn C. C, Witham C, Fultz B, Liu J, Rinzler A. G, Colbert D, Smith K. A, and Smalley R. E, (1999), Hydrogen Adsorption and Cohesive Energy of Single-Walled Carbon Nanotubes, Appl. Phys. Lett.,74: 2307-2309.

[6] Liu C, Fan Y. Y, Lyu M, Cong H. T, Cheng H. M, and Dresselhaus M. S, (1999), Hydrogen Storage in Single-Walled Carbon Nanotubes at Room Temperature, Science, 286: 1127-1129.

[7] Liu C, Yang Q. H, Tong Y, Cong H. T., and Chen H. M, (2002), Volumetric Hydrogen Storage in Single-Walled Carbon Nanotubes, Appl. Phys. Lett., 80: 2389-2391

[8] Quinn D. F, (2002), Supercritical Adsorption of 'Permanent' Gases under Corresponding States on Various Carbons, Carbon, 40: 2767-2773.

[9] Kajiura K, Tsutsui S, Kadona K, and Ata M, (2003), Hydrogen Storage Capacity of Commercially Available Carbon Materials at Room Temperature, Appl. Phys. Lett., 82: 1105-1107.

[10] Nijkamp M. G, Raaymakers J. E. M. J, Van Dillen A. J, and de Jong K. P, (2001), Hydrogen Storage using Physisorption-Materials Demands, Appl. Phys., A, 72: 619-623.
[11] Chahine R and Bose T. K, (1994), Low-pressure Adsorption Storage of Hydrogen, Int. J. Hydrogen Energy, 19: 161-164.
[12] Lim S. H, Luo J, Zhong Z, Ji W, and Lin J, (2005), Room-Temperature Hydrogen Uptake by TiO_2 Nanotubes, Inorg. Chem., 44: 4124-4126.
[13] Bavykin D. V, Lapkin A. A, Plucinski P. K, Friedrich J. M., and Walsh F. C, (2005), Reversible Storage of Molecular Hydrogen by Sorption into Multilayered TiO_2 Nanotubes, J. Phys. Chem. B 109: 19422-19427.
[14] Wan Q, Lin C. L, Yu X. B, and Wang T. H, (2004), Room-Temperature Hydrogen Storage Characteristics of ZnO Nanowires, Appl. Phys. Lett., 84 (1): 124-126.
[15] Chen J, Kuriyama N, Yuan H, Takeshita H. T, and Sakai T, (2001), Electrochemical Hydrogen Storage in MoS_2 Nanotubes, J. Am. Chem. Soc., 123: 11813-11814.
[16] Chen J, Li S.-L., Tao Z.- L, Shen Y.- T, and Cui C.- X, (2003), Titanium Disulfide Nanotubes as Hydrogen-Storage Materials J. Am. Chem. Soc., 125: 5284-5285.
[17] Yamashita J, Shioya M, Kikutani T, and Hashimoto T, (2001), Activated Carbon Fibers and Films Derived from Poly(vinylidene fluoride), Carbon, 39: 207-214.
[18] Choi S. W, Jo S. M, Lee W. S, and Kim Y. R, (2003), An Electrospun Poly(vinylidene fluoride) Nanofibrous Membrane and Its Battery Applications, Adv. Mater, 15(23): 2027-2032.
[19] Kim J. R, Choi S. W, Jo S. M, Lee W. S, and Kim B. C, (2005), Characterization and Properties of PVdF-HFP-Based Fibrous Polymer Electrolyte Membrane Prepared by Electrospinning, J. Electrochem. Soc., 152(2): A295-A300
[20] Choi S. W, Kim J. R. Jo S. M. Lee W. S. and Kim Y.-R. (2005), Electrochemical and Spectroscopic Properties of Electrospun PAN-Based Fibrous Polymer Electrolytes, J. Electrochem. Soc., 152(5), A989-A995.
[21] Song M. Y, Kim D. K, Ihn K. J, Jo S. M., and Kim D. Y, (2004), Electrospun TiO_2 Electrodes for Dye-Sensitized Solar Cells, Nanotechnology, 15 : 1861-1865.
[22] Song M. Y, Ahn Y. R, Jo S. M , Ahn J-. P, and Kim D. Y, (2005), TiO_2 Single-Crystalline Nanorod Electrode for Quasi-Solid State Dye-Sensitized Solar Cells, Appl. Phys. Lett., 87: 113113-1~113113-3.
[23] Kaburagi Y, Hishiyama Y, Oka H, and Inagaki M, (2001), Growth of Iron Clusters and Change of Magnetic Property with Carbonization of Aromatic Polyimide Film Containing Iron Complex, Carbon, 39: 593-603.
[24] Konno H, Shiba K, Kaburagi Y, Hishiyama Y, and Inagaki M, (2001), Carbonization and Graphitization of Kapton-type Polyimide Film having Boron-bearing Functional Groups, Carbon, 39: 1731-1740.
[25] Reshetenko T. V, Avdeeva L. B, Ismagilov Z. R, Pushkarev V. V, Cherepanova S. V, Chuvilin A. L, and Likholobov V. A, (2003), Catalytic Filamentous Carbon Structural and Textural Properties, Carbon, 41: 1605-1615.
[26] Chung G. S, Jo S. M, and Kim B. C, (2005), Properties of Carbon Nanofibers Prepared from Electrospun Polyimide, J. Appl. Polym. Sci., 97: 165-170.

[27] Park S. H, Jo S. M, Kim D. Y, Lee W. S, and Kim B. C, (2005), Effects of Iron Catalyst on the Formation of Crystalline Domain during Carbonization of Electrospun Acrylic Nanofiber, Syn. Met., 150: 265-270.

[28] Kim D.- K, Park S. H, Kim B. C, Chin B. D, Jo S. M, and Kim D. Y, (2005), Electrospun Polyacrylonitrile based Carbon Nanofibers and Their Hydrogen Storage, Macromolecular Research, 13(6): 521-528.

[29] Park S. H. Kim B. C. Jo S. M. Kim D. Y. and Lee W. S. (2005), Cabon Nanofibrous Materials Prepared from Electrospun Polyacrylonitrile Nanofibers for Hydrogen Storage, Mater. Res. Soc. Symp. Proc., 837: 71-76

[30] Chung H. J, Lee D. W, Jo S. M, Kim D. Y, and Lee W. S, (2005), Electrospun Poly(vinylidene fluoride)-based Carbon Nanofibers for Hydrogen Storage, Mater. Res. Soc. Symp. Proc., 837: 77-82

[31] Hong S. E, Kim D.- K., Jo S. M, Kim D. Y, Chin B. D, Lee D. W, (2007), Graphite Nanofibers Prepared from Catalytic Graphitization of Electrospun Poly(vinylidene fluoride) Nanofibers and Their Hydrogen Storage Capacity, Catalysis Today, 120(3-4): 413-419

[32] Hishiyama Y, Igarashi K, Kaneko I, Fujii T, Kaneda T, Koidezawa T, Shimazawa Y, and Yoshida A, (1997), Graphitization Behavior of Kapton-derived Carbon Film Related to Structure, Microtexture and Transport Properties, Carbon, 35: 657-668.

[33] Ra W, Nakayama M, Uchimoto Y, and Wakihara M, (2005), Experimental and Computational Study of the Electronic Structural Changes in $LiTi_2O_4$ Spinel Compounds upon Electrochemical Li Insertion Reactions, J. Phys. Chem. B, 109(3): 1130-1134.

[34] Manickam M, Takata M, (2003), Lithium Intercalation Cells $LiMn_2O_4/LiTi_2O_4$ without Metallic Lithium, J. Power Sources, 114: 298-302.

[35] Inagaki M, (2000), Control of structure and functions, in New Carbons : Elsevier Science Ltd., chapt. 4- 5.

Chapter 9

On the Possibility of Layered Crystals Application for Solid State Hydrogen Storages – InSe and GaSe Crystals

Yuriy Zhirko, Volodymyr Trachevsky and Zakhar Kovalyuk

Additional information is available at the end of the chapter

http://dx.doi.org/10.5772/50307

1. Introduction

Among the variety of chemical compounds that are available in nature, the class of layered crystals comprises a specific place. As can be seen in Fig. 1, it is rather wide and covers the range from simple compounds of the graphite type up to silicates, micas and organic compounds.

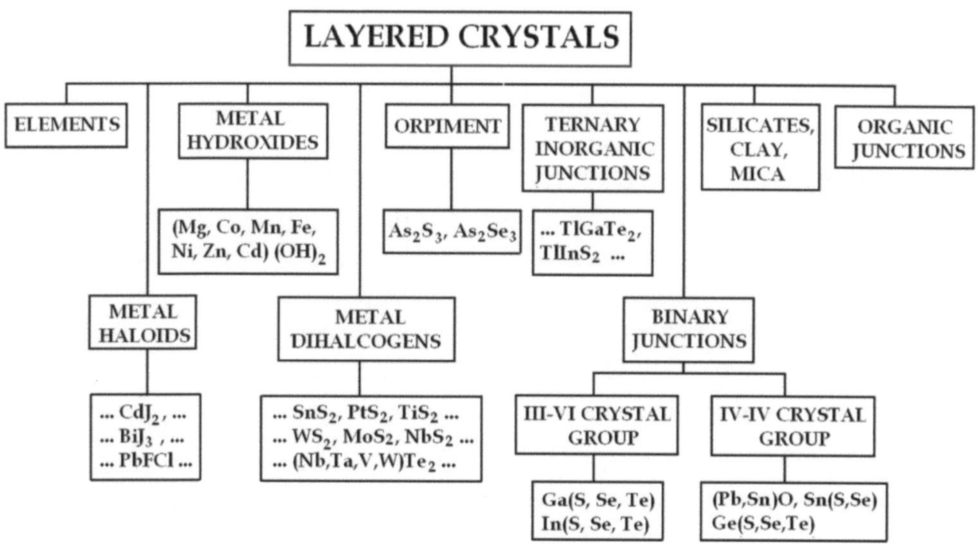

Figure 1. Class of layered crystals

Due to their natural properties, as a rule layered crystals easily dissolve in their bulk various molecular gases, beginning from hydrogen and metane up to complex organic molecules.

This circumstance makes them very attractive for application as solid-state hydrogen storages.

In this work, using the example of InSe and GaSe layered crystals and modern methods of scientific investigations, we have considered some aspects to deeper ascertain physical-and-chemical processes taking part during intercalation and deintercalation of layered crystals with hydrogen.

As is shown on a Fig.1 dicussed in this work InSe and GaSe layered crystals are related to binary compounds A3B6 where crystalline layers consist of four monolayers including Se and In(Ga) atoms and located one above another in the sequence Se-In(Ga)-In(Ga)-Se. These atoms are bound to each other by ion-covalent hybrid sp^2-bonds. Inside crystalline layers, orientation of atoms corresponds to the spatial group D_{3h}. In this case (see Fig. 2b), each In(Ga) atom inside crystalline layer yields its two valent electrons to three nearest Se atoms to fill their ***p***-shells. The remaining electron of this In(Ga) atom together with the remaining electron of the adjacent In(Ga) atom in this layer form a filled *s*-shell. This causes configuration in which orbitals of valent electrons are located inside the layer and are not practically overlapped with orbitals of valent electrons related to atoms in the neighboring layer. As a result, the adjacent layers are only linked with a weak van der Waals bond.

This clearly pronounced anisotropy of chemical bonds between layers and inside crystalline layers allows intercalation of layered crystals, i.e. introduction of foreign atoms or molecules into the interlayer space of grown crystals. For example, in InSe and GaSe crystals the volume of the van der Waals space (the so-called "van der Waals gap") comprises 40…45 % of all the crystal bulk, while the internal surface of this space is close to $(2\ldots2.5)\cdot10^3\ m^2$ per 1 cm^3.

It is known (Belenkii & Stopachinskii,1983; Polian et al., 1976; Ghosh, 1983), that single crystals grown by the Bridgman method possess four polytypes (β-, δ-, ε- and γ-) varying between each other in the sequence of crystalline layer stacking. As a rule (see Fig. 2a), InSe single crystals are of γ-polytype with rhombohedral (trigonal) crystalline structure (spatial group C^5_{3v}, in which the primitive unit cell contains one In_2Se_2 molecule comprising three layers). The non-primitive hexagonal unit cell comprises three crystalline layers and consists of three In_2Se_2 molecules, respectively. GaSe single crystals are of ε-polytype. These belong to hexagonal crystalline structure (spatial group D^1_{3h}), the unit cell of which consists of two Ga_2Se_2 molecules located within two crystalline layers.

Lattice parameters of the crystals γ-InSe and ε-GaSe are studied well. In the case of ε-GaSe, they are as follows: lattice parameters along the perpendicular to layers $C_0 = 15.95$Å and in the layer plane $a_0 = 3.755$Å (Kuhn et al., 1975). For γ-InSe, they are, respectively: $C_0 = 25.32$ Å and $a_0 = 4.001$ Å. The distances between the nearest atoms inside the layer (see Fig. 2c) are as follows: $C_{In\text{-}In} = 2.79$ Å and $C_{In\text{-}Se} = 2.65$ Å, respectively; $\angle\varphi = 119.3^0$, the layer thickness $C_1 = 5.36$Å, the distance between layers $C_i = 3.08$ Å, while the distance between adjaicent Se atoms located in different layers $C_{Se\text{-}Se} = 3.80$ Å (Goni et al., 1992; Olguin et al., 2003).

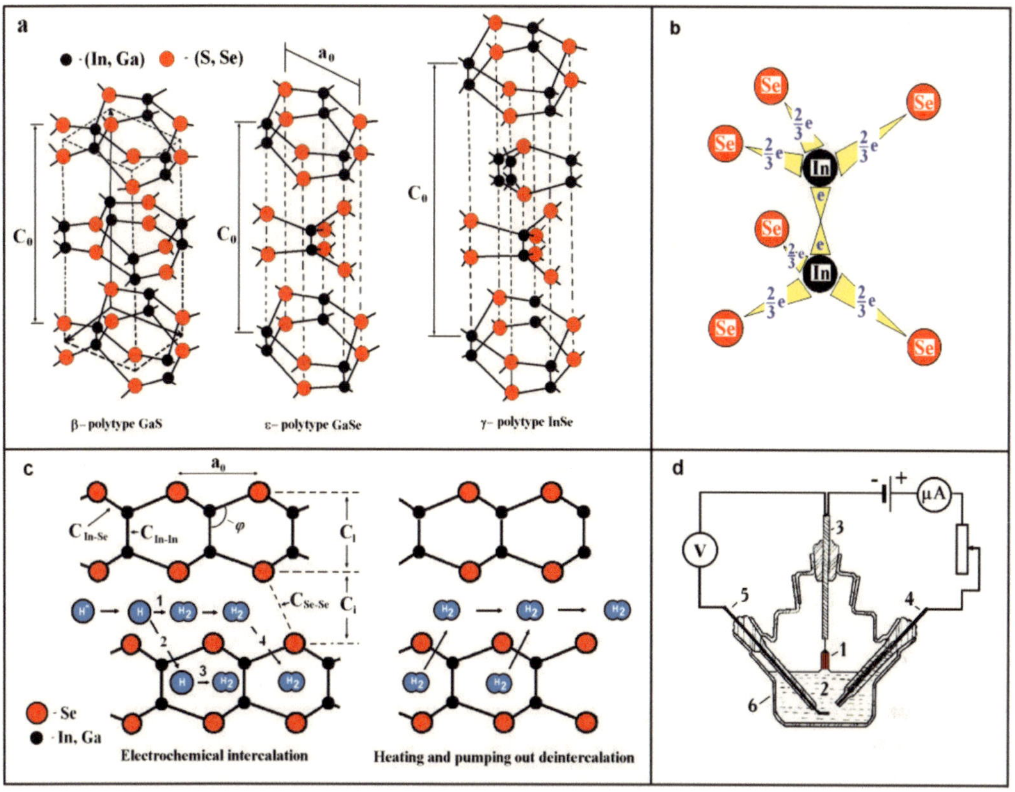

Figure 2. a – crystalline structure and layer stacking for various crystal polytypes of group; b – distribution of chemical bonds between atoms In(Ga) and Se inside the crystalline layer; c – two-dimensional scheme illustrating the intercalation and deintercalation processes for hydrogen in layered crystals InSe and GaSe; d – scheme of electrochemical cell: 1- layered crystal; 2 – electrolyte; 3 – operation platinum electrode; 4- comparative electrode (AgCl); 5 – auxiliary platinum electrode; 6 – cell body.

Undoubtedly, these unique properties of semiconductor compounds in this class, and in particular of InSe and GaSe crystals, attract special attention of researches, as heterostructures based on them possess good photosensitivity and can be used not only in solar elements (Martines-Pastor et al., 1987; Lebedev et al., 1998; Shigetomi & Ikari, 2000) and accumulators of electrical energy (Grigorchak et al., 2002), but in accord with investigations (Zhirko et al., 2012) are rather promising in creation of gamma radiation sensors.

Besides, as it was shown in a number of recent works (Kozmik et al., 1987; Zhirko et al., 2007a; Zhirko et al., 2007b), layered crystals InSe and GaSe can be applied in hydrogen energetics as operation elements in solid hydrogen storages. The hydrogen concentration in them can reach values close to $x = 5...6$, where x is the amount of embedded hydrogen atoms per one formular unit of the intercalated crystal matrix.

Described in this work is the method of hydrogen intercalation-deintercalation in InSe and GaSe crystals. Besides, being based on the performed complex electron-microscopic (SEM, EDS, HKL), radiospectroscopic (NMR, EPR), optical and electrophysical investigations, we have discussed the model describing introduction of hydrogen into layered crystals InSe and GaSe as well as considered some physical aspects related to this process.

2. The method for hydrogen intercalation and deintercalation of InSe and GaSe layered crystals

Before the process of intercalation with hydrogen, γ-InSe and ε-GaSe bulk single crystals were used to spall crystalline plates along the cleavage planes with parallel mirror surfaces that do not require any further polishing or etching. To combine investigations of processes for hydrogen intercalation with those mentioned in Introduction, the sample thickness was chosen within the range 20 to 30 μm, and linear dimensions were 5x5 mm. Powders with grain dimensions 5…20 μm were prepared from γ-InSe and ε-GaSe single crystal ingots by using an ultrasound mill. For further investigations, these powders were pressed to tablets.

Intercalation with hydrogen was performed using the electrochemical method in the three-electrode glass cell made from chemically- and thermally-stable glass. The cell (see Fig. 2d) consisted of AgCl comparative electrode, operation and auxiliary (platinum wire) electrodes. The studied samples were sweated with wires on the operation electrode and placed into electrolyte consisting of 0.1 normal solution prepared from doubly distilled water and chemically-pure concentrated HCl acid.

Intercalation of the γ-InSe and ε-GaSe samples was performed using the way described in (Kozmik et al., 1987; Zhirko et al., 2007a) from electrolytic solution by the method of "pulling field" in the galvanostatic regime. The chosen special conditions for the optimal electric voltage and current density (E = 30…50 V/cm and $j \leq 10$ μA/cm²) allowed to prepare homogeneous by their composition hydrogen intercalates H_xInSe and H_xGaSe within the ranges of concentrations $0<x\leq 5$ for plates and $0<x\leq 6$ for powders of the same crystals. The hydrogen concentration was determined via the amount of electric charge passing through the sample placed into the cell. These conditions for intercalation allowed obtaining the samples of high quality, the crystal being not destroyed.

Deintercalation of hydrogen from layered intercalates of H_xInSe and H_xGaSe crystals were made for 3 to 9 hours at T = 110 °C and continous pumping down. It enabled us to ascertain that the degree of intercalation for the samples H_xInSe increases practically in the linear manner with growth of the hydrogen concentration from 60% for $x\rightarrow 0$ up to 75-80% for $x\rightarrow 2$. Further growth of the hydrogen concentration up to $x\rightarrow 4$ shows that the deintercalation degree in H_xInSe and H_xGaSe increases up to 85%. As to powders of InSe and GaSe crystals, it was shown that the hydrogen concentration in them can reach values x = 5…6, and the deintercalation degree can do 90%. In this case, using the optical and electron-microscopic methods applied to the samples of H_xInSe and H_xGaSe intercalates, it was ascertained that repeated cycles of intercalation-deintercalation with hydrogen for concentrations $0<x<2$ do not cause any visible worsening the physical parameters of InSe and GaSe crystals.

3. SEM investigations of InSe and GaSe layered crystals

SEM, EDS and HKL investigations of ε-GaSe and γ-InSe single crystals as well as their hydrogen intercalates were performed using Zeiss EVO 50 XVP scanning electron microscope equipped with the detectors INCA ENERGY 450 and that for diffraction of back-scattered electrons HKL Channel 5 EBSD of the firm Oxford Instruments Analytical Ltd. The concentration of intercalated hydrogen in these samples was from $x = 0$ up to $x = 5$, where x is the amount of hydrogen atoms per one formular unit of the crystal.

With this aim, we prepared 12 samples of GaSe crystals, 10 of which contained hydrogen with the concentrations x = 0; 0.25; 0.5; 1.0; 1.25; 1.5; 2.5; 3.0; 4.0; 5.0; as well as non-intercalated GaSe crystal annealled in silica ampoule for 3 hours in vacuum at the temperature T = 350 °C. Besides, we prepared the sample of GaSe crystal annealled in silica ampoule filled with gas-like hydrogen under pressure for 3 hours at the same temperature. By analogy, in the case of InSe crystals, the hydrogen concentrations were as follows: x = 0; 0.1; 0.2; 0.5; 0.75; 1.0; 1.25; 1.5; 1.75; 2.0. It should be noted that, contrary to the case of InSe crystals, after annealling the GaSe crystals we observed a red color deposit on internal sufaces of ampoules, which corresponds to polycrystalline Se films.

The performed SEM studies showed that, even at $8{\cdot}10^4$-fold magnification over the whole range of chosen x concentrations, electron images of crystalline surfaces did not reveal any structural defects related with introduction of hydrogen into these crystals.

Along with it, as seen in Figs. 3a and 3b for the example of the sample No 1 (x=0), these layered GaSe crystals both non-intercalated and intercalated with hydrogen contain considerable amount of foreign spatial inclusions with dimension of 60 to 300 nm. Perfomed by us analogous investigations of InSe crystals did not reveal such foreign inclusions.

Complex EDS and XRD investigations enabled us to ascertain that these inclusions were present in the bulk of these crystals, and they were inclusions of monoclinic modification inherent to β-Se (group 2,2/m). They are located in the interlayer space of GaSe crystal. As a result of matrix pressure, lattice parameters of the foreign phase β-Se essentially deviate from their standard values. Note that after annealling the GaSe crystals in vacuum or hydrogen atmosphere under pressure their electron images (see Fig. 3c, sample No 11) did not contain any foreign inclusions.

In addition, low-angle XRD investigations showed that, with increasing the concentration of intercalated hydrogen, lattice parameter deviation in H_xGaSe matrix increases, too, and the lattice parameter differs from its standard value for GaSe (hexagonal group 9,6/mmm) as a cosequence of hydrogen pressure on the matrix. Indeed, as it was shown in (Zhirko et al., 2007b) using the direct XRD investigations of H_xGaSe crystals at T = 300 K and x = 1.0, C_0 parameter grows by 0.031±0.003 Å from C_0 = 15.94 Å up to C_0 = 15.971 Å, while the parameter a_0 does by 0.006 ±0.003 Å from a_0 = 3.753 Å up to a_0 = 3.759 Å.

In first approximation (Zhirko et al., 2007a) assume that at low concentrations atomic hydrogen enter into the van der Waals gap by the presented on the Fig. 2c schema and creates H_2 molecules that occupy an ordered positions schematically shown in Fig. 2c.

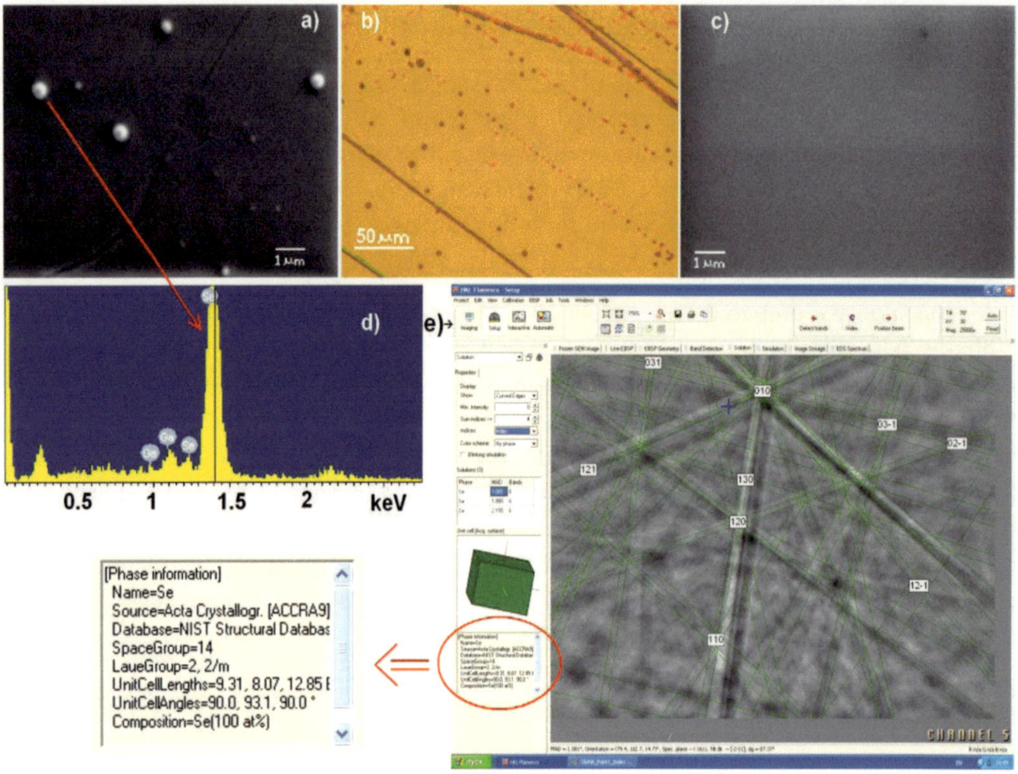

Figure 3. SEM images of GaSe crystals non-intercalated (a) and annealled (c) in silica ampoule for 3 hours in vacuum at the temperature T = 350 °C, which are obtained at 10000-fold magnification by using Zeiss EVO 50 XVP scanning electron microscope; (b) image of GaSe crystal obtained using transmission optical microscope Carl Zeiss Primo Star 5 at 1000-fold magnification; (d) EDS spectra and (e) low-angle XRD image of foreign inclusion β-Se shown in Fig. 3a.

Appearance of hydrogen molecules in the gap result in occurrence of interlayer pressure and in increasing of interlayer parameter C_0. At higher concentrations atomic hydrogen incorporate into the crystal layers.

Starting from this concept, (Zhirko et al., 2007a) estimate the concentration and pressure of H_2 molecules in the van der Waals gap of H_xInSe at x=2. As the volume of γ-InSe conventional hexagonal cell is equal V = 351 Å^3, the concentration of elementary cells is N_0 = 1/V = $2.85 \cdot 10^{21}$ cm^{-3}. At x=2, one H_2 molecule corresponds to one In_2Se_2 molecule. Hence, H_2 molecular concentration is equal $N = 3N_0$= 8.55 x 10^{21} cm^{-3}. Using Clapeyron's equation for the ideal gas pressure $P = Nk_B T$, where k_B is the Boltzmann constant and T is the absolute temperature, (Zhirko et al., 2007a) deduce that the pressure caused by H_2 molecules in InSe van der Waals gap is equal to 9.4 MPa at T = 80K and 35.4 MPa at T = 300K.

It should be mentioned that the influence of hydrostatic pressure on lattice parameters, for instance γ-InSe crystals, is studied rather well. In accord with (Olguin et al., 2003), growth of the hydrostatic pressure from 10^5 up to 8×10^9 Pa causes essential narrowing the van der Waals gap as a consequence of decreasing the distance $C_{Se\text{-}Se}$ from 3.8 Å down to 3.3 Å (13%)

(see Fig. 2c). In this case, the distance C_{In-In} decreases from 2.79 Å down to 2.69 Å (3.58%), C_{In-Se} from 2.65 Å down to 2.59 Å (2.26%). However, due to the angle $\angle\varphi$ increase from 119.3^0 up to 121.3^0 (1.5%), the thickness of the crystalline layer C_l is decreased by only 0.26%, i.e. from 5.36 Å down to 5.345 Å.

Thereof, extrapolating the data of (Olguin et al., 2003) to the case when H_2 molecules do not narrow but expand the interlayer space, in accord with (Zhirko et al., 2007a) one can estimate that for the pressure $P = 35.4$ MPa ($x = 2$, T = 300 K – the case when each molecule in the interlayer space corresponds to one molecule In_2Se_2) the distance C_{Se-Se} should increase from 3.8 Å up to 3.808 Å, while the constant C_0 for the crystal γ-InSe - by 0.024±0.01 Å, respectively.

It should be noted that, in the case when the intercalation degree is considerable, the pressure of foreign atoms or molecules in the interlayer space can cause even exfoliating the crystals, as it takes place in graphite (Anderson & Chung, 1984). The analogous phenomenon was also observed by us in fast regimes of hydrogen intercalation in layered crystals, and hard regimes of intercalation caused destruction of crystals up to powder-like states. But we used here the specially chosen regimes that allow preparation of homogeneous samples, and multiply repeated processes of intercalation-deintercalation do not worsen parameters of InSe and GaSe crystals.

4. Optical investigations of InSe and GaSe hydrogen intercalates

Optical methods of investigation as well as electron microscopy are related to non-destructive methods and bring considerable information about elementary excitations in crystals. Low-temperature measurements ($T = 4.5 \ldots 250$K) of exciton absorption spectra of H_xInSe and H_xGaSe intercalates were made using a 0.6-m monochromator MDR-23 (LOMO, Russia) with a grating of 1200 grooves/mm. The spectral width of the slit did not exceed 0.2 meV during the experiments. Investigations of absorption spectra were made using a helium cryostat (A-255) designed at the Institute of Physics, National Academy of Sciences of Ukraine. It was equipped with a UTRECS K-43 system allowing control over sample temperature within the range of 4.2 to 350 K with high accuracy (0.1 K). A tungsten-halogen lamp was used for the absorption measurements. The photomultiplier tubes FEU-79 and FEU-62 served as a radiation detectors.

Calculations of excitonic absorption spectra inherent to H_xInSe and H_xGaSe intercalates, in accord with (Zhirko, 1999; Zhirko, 2000), were made in approximation of weak interaction of Wannier excitons with effective vibrations of crystalline lattice (with the frequency Ω), energy of which $\hbar\Omega$ practically coincided with the energy of homopolar optical A'-phonon. The absorption bands for the main and excited exciton states as well as for band-to-band transitions were obtained with account of scattering by optical phonons and defects of crystalline lattice in accord with the standard convolution procedure for the theoretical value of absorption coefficient $\alpha(\hbar\omega)$ in the Elliott model (Elliott, 1957) with the Lorentz function $f(\hbar\omega) = \Gamma/\left[\pi\left(E^2 + \Gamma^2\right)\right]$ in the Toyozawa model (Toyozawa, 1958), where Γ is the halfwidth of the exciton absorption band, which is associated with the lifetime $\hbar/2\Gamma$.

4.1. InSe hydrogen intercalates

Our studies of exciton absorption have shown that the influence of hydrogen intercalation on optical properties of H_xInSe crystals is more pronounced than that in the H_xGaSe ones. Therefore, let us first consider exciton absorption spectra of H_xInSe crystals, and then generalize the obtained conclusions to the crystals H_xGaSe.

The depicted in Fig. 4a absorption spectrum of the crystal $H_{0.07}InSe$ is obtained at T = 80 K with account of light reflection from the crystal frontal face (n_0= $[n_0^{\perp 2} \cdot n_0^{\parallel}]^{1/3}$ = 3.5 (Zhirko, 1999). In the same figure, we have shown the fitted absorption bands for n = 1 (curve 1) and n = 2 (curve 2) exciton states, band-to-band transition (curve 3) and shallow donor level (with participation of two optical full-symmetric $A_1{}^{\prime 1}$-phonons) located at 20±2 meV below the conduction band bottom (curve 4). The resulting fitting curve 5 was obtained with account of contribution from n = 1 to n = 4 exciton transitions, band-to-band transition, shallow donor and acceptor levels located by 20 meV below the conduction band bottom and by 90 meV above the top of valence band.

When using hydrogen intercalation with the concentration x = 0.07 in $H_{0.07}InSe$ crystals (see Fig. 4c), one can observe the shortwave shift of n=1 exciton absorption band by 0.8 meV relatively to its position in the non-intercalated InSe crystal. Further growth of the hydrogen concentration causes a non-monotonous shift of this band as well as its maximum. For example, within the range $0<x\leq 0.5$ energy position of this peak is shifted by 4.5 meV from E_1 = 1.3275 eV up to E_1 = 1.3320 eV; while within the range $0.5<x\leq 1$ one can observe the decrease in energy down to E_1 = 1.3295 eV; further growth of the hydrogen concentration does not lead to any change in the energy position $E_1(x)$.

As can be seen in Fig. 4b (curve 6), simultaneously with the shift of E_1 caused by growth of the hydrogen concentration, one can also observe an increase in the halfwidth Γ_1 for n = 1 excitonic absorption band. Within the range of values $0<x\leq 1$, the halfwidth Γ_1 grows with x, and for $x>1$ becomes practically constant. Thus, $\Gamma_1(x)$ dependence can be approximated by the following analytic function

$$\Gamma_1(x)\ [\text{meV}] = \Gamma_1 \cdot \{1+1/\exp[1/(\Gamma_1 \cdot x)]\}. \qquad (1)$$

4.1.1. *Discussion of the model for exciton localization in hydrogen intercalates of InSe layered crystals*

Being based on the data of other authors concerning the influence of hydrostatic pressure P on the bandgap width E_g for InSe, it was shown in the work (Zhirko et al., 2007a) that the shortwave shift E_1 by the value 4.5 meV for x = 0.5 cannot be satisfactorily explained by dependences of the exciton binding energy $R_0(P)$, dielectric permittivity $\varepsilon_0(P)$ and bandgap width $E_g(P)$ on the pressure P arising with introduction of hydrogen into the interlayer space, as their contributions $R_0(P)$ and $E_g(P)$ to this shift are opposite, and for x = 0.5 (P = 2.4 MPa) their total contribution reaches only 0.01 meV. Moreover, the dependence of exciton parameters E_1 and Γ_1 on x gives some bases to believe that this shortwave shift of exciton absorption band is caused by the ε_0 growth as a cosequence of hydrogen introduction into the interlayer space. In accord with the following equation

$$R_0(x)[meV] = 13605 \cdot \mu / \varepsilon_0^2(x) \quad (2)$$

it becomes apparent in lowering the exciton binding energy R_0.

For more convenient discussion, let us describe the ε_0 growth as a result of appearance of molecular hydrogen in the interlayer space by introduction of an additional parameter $\varepsilon^*(x)$ that grows with x and characterizes growth of anisotropy in crystal dielectric permittivity within the range comprised by the exciton:

$$\varepsilon_0(x) = [\varepsilon_0^{\perp 2} \cdot \varepsilon_0^{\parallel} \cdot \varepsilon^*(x)]^{1/3} \equiv \varepsilon_0 \cdot \varepsilon^*(x)^{1/3} \quad (3)$$

where the parameter

$$\varepsilon^*(x) = \prod_{j=1}^{m} \varepsilon_j(x) \equiv \varepsilon_i(x)^m \quad (4)$$

m is the exponential index equal to the amount of interlayer spaces comprised by the exciton diameter d_{exc}; $\varepsilon_i(x)$ – dielectric permittivity of one interlayer space, which is dependent on x. In the case of non-intercalated InSe crystal, it is equal to vacuum dielectric constant $\varepsilon_i(0) = \varepsilon_v = 1$.

With account of performed estimations, the shortwave shift $E_1(x)$ for values $0<x=0.5$ (see Fig. 4c) is caused by lowering R_0 from 14.5 down to 10.0 meV as a consequence of the dielectric permittivity $\varepsilon_0(x)$ growth from 10.5 up to 12.6 (see the insert to Fig. 4b). It is rather well described by the following relation

$$\varepsilon_0(x) = 10.5 \cdot (1+1.5 \cdot x)^{1/3}, \quad (5)$$

represented by the curve 7. In this case, when $\varepsilon_0(x)$ behavior corresponds to Eq. (5), the curve 8 in Fig. 4c describes $R_0(x)$ dependence within the range of values $0<x=2$. It is seen that the curve 8 well describes experimentally observed $R_0(x)$ decrease within the range $0<x=0.5$, but further we observe growth of R_0, which is caused by increasing anisotropy of dielectric permittivity ε_0. Indeed, in accord with Eq. (3), $\varepsilon_0^{\parallel}(x)$ increases from 9.9 up to 16.9 as a consequence of the $\varepsilon^*(x)$ parameter growth from $\varepsilon_v = 1$ up to 1.7.

In accord with the expression

$$a_{exc}\,[nm] = 0.053 \cdot \varepsilon_0(x)/\mu, \quad (6)$$

where a_{exc} is the exciton radius, the exciton diameter grows linearly with $\varepsilon_0(x)$ from 9.51 nm up to 11.43 nm, which is equivalent to increasing the exciton diameter d_{exc} = 11.3 up to 13.5 crystalline layers (while the layer thickness in InSe crystal is d_{layer} = 8.44 Å). Thereof, for m = 13, accordingly to Eq. (9) for x = 0.5 (P = 2.3 MPa) and T = 80 K, one can obtain ε_i = 1.04.

Analogous calculations for x = 1 give the following results: ε_0 = 14.25 and the parameter ε^* = 2.5. In this case, R_0 should decrease further down to the value 6.45 мeB, and the exciton diameter should increase up to d_{exc} = 12.9 nm (d_{exc} = 15.3 crystalline layers). For m = 15 (x = 1, P = 4.7 MPa and T = 80 K), one obtains ε_i = 1.062.

For x = 2, one can obtain ε_0 = 16.67 and ε^* = 4.0, R_0 = 5.73 meV, d_{exc} = 15.1 nm (d_{exc} = 17.9 crystalline layers). For the value m = 18 (x = 2, P = 9.4 MPa and T = 80 K), one can obtain ε_i = 1.081.

The performed calculations show that the value ε_1 also grows with x and begin to reach experimental values inherent to dielectric permittivity of liquid nitrogen (Malkov, 1985), which is equal to 1.253 and 1.230 for the temperatures 14 and 20.5 K, respectively.

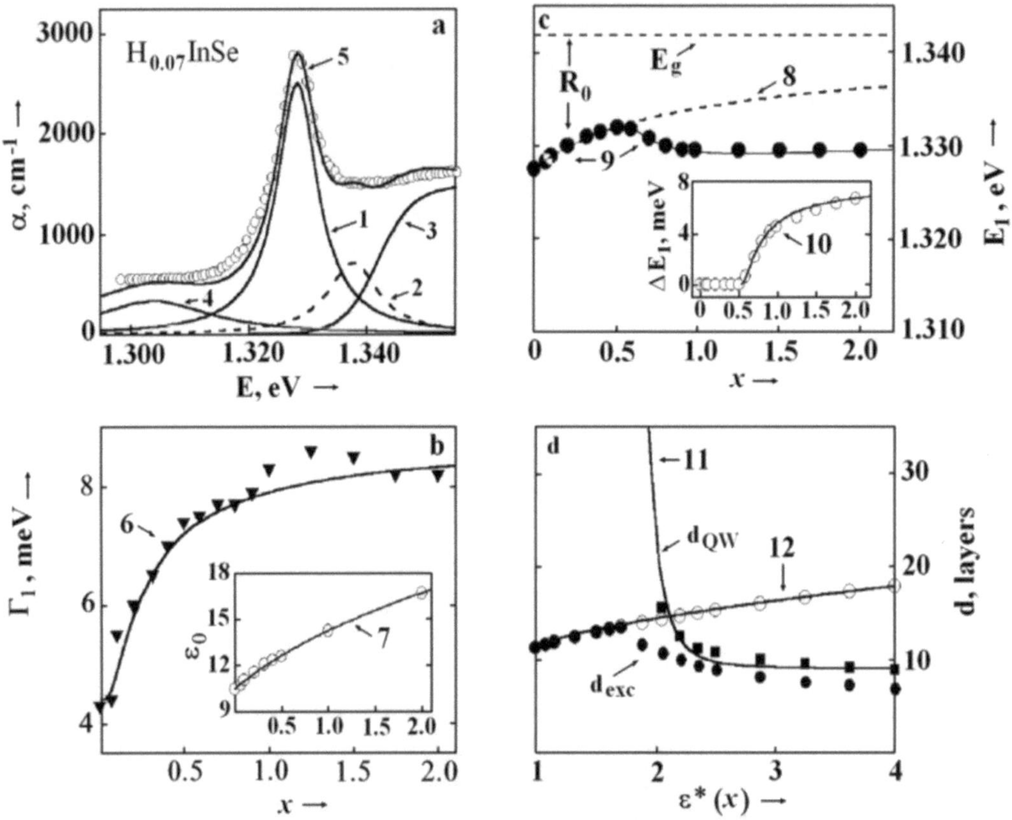

Figure 4. a – open circles and curve 5 – experimental and calculated spectra of exciton absorption for the crystal $H_{0.07}InSe$ at T = 80 K; b – triangles – experimental dependences for the halfwidth Γ_1 on $\boldsymbol{x}$. In the insert – open circles and experimental dependences of ε_0 on $\boldsymbol{x}$; c – circles – experimentally observed shift of $\boldsymbol{n}$ = 1 maximum of the exciton absorption band as a dependence E_1 on $\boldsymbol{x}$. The straight line E_g shows a stable position of the bandgap energy E_g with varied $\boldsymbol{x}$. R_0 – illustrates lowering the exciton binding energy with varied $\boldsymbol{x}$ in the case when localization of excitons is not taken into account. In the insert – circles designate experimental dependences ΔE_1 on $\boldsymbol{x}$; d – squares – calculated values for the dependence of the quantum well thickness $\boldsymbol{d_{QW}}$ on the parameter $\varepsilon^*(\boldsymbol{x})$. Open and filled circles – experimental dependences of the exciton diameter $\boldsymbol{d_{exc}}$ on the parameter $\varepsilon^*(\boldsymbol{x})$ with and without account of exciton localization, respectively. The curves 1 to 4 – absorption bands for $\boldsymbol{n}$ = 1, 2 exciton states, band-to-band transition and shallow defect level, the curve 6 is plotted in accord with Eq. (1), curve 7 – Eq. (5), curve 8 – Eq. (2), curve 9 – with account of exciton localization inside the plane of layers, curve 10 – Eq. (8), curves 11 and 12 are built in accord with Eqs. (10) and (6).

It is noteworthy that the numeric values of parameters ε^* and ε_1 versus x are well described by the following analytical dependences:

$$\varepsilon^*(x) = \varepsilon_v + 1.5 \cdot x \quad \text{and} \quad \varepsilon_1(x) = \varepsilon_v + 0.06 \cdot x^{1/2}, \tag{7}$$

where ε_v=1. Thereof, it is seen that growth of the dielectric permittivity in the interlayer space and within the range $0<x=0.5$ corresponds to experimentally observed decrease in R_0, however, within the range $0.5<x<1$ the shift of $E_1(x)$ dependence changes its sign to the opposite one, and for $1<x<2$ E_1 position becomes stabilized.

This $E_1(x)$ behavior allows to assume that the parameter $\varepsilon^*(x)$ characterizing the degree of anisotropy of dielectric permittivity in the part of medium comprised by one exciton increases in such a manner that for $\varepsilon^* = 1.7$ ($x = 0.5$) motion of excitons becomes localized within the plane of a crystalline layer. Therefore, further growth of $\varepsilon_i(x)$ and $\varepsilon^*(x)$ results in lowering the exciton radius and width of the quantum well where its localization takes place. This conclusion is confirmed by the experimental data for the dependence $\Delta E_1(x)$ shown with open circles in the insert to Fig. 4c and described with the expression

$$\Delta E_1(x)\ [\text{meV}] = 13/\exp[x/2\cdot(x-0.5)] \tag{8}$$

represented by the curve 10.

In Fig. 4d, squares show the dependence for the width of quantum well where the exciton is localized on the parameter $\varepsilon^*(x)$, this dependence being obtained in accord with the classic exprerssion

$$E = \frac{\pi^2 \hbar^2}{2 d_{QW}{}^2 M_{exc}}\ , \tag{9}$$

and experimental data $\Delta E_1(x)$. For convenience, as a dimensional unit of the quantum well width d_{QW} we took the thickness of one crystalline layer, and $M_{exc} = m_e + m_h$ is the translational mass of excitons. Open circles show the dependence of the exciton diameter d_{exc} on the parameter $\varepsilon^*(x)$ without account of its localization inside the layer plane, while the squares do it with account of localization.

As seen in Fig. 4d, within the range of parameter values $1<\varepsilon^*(x)<1.7$, the quantum well does not arise ($d_{QW} = \infty$), here one can observe the case of ordinary growth of the exciton diameter d_{exc} (open circles), which is caused by growth of $\varepsilon_i(x)$. Within the range of values $1.7<\varepsilon^*(x)<2.5$, strong localization of excitons inside the plane of crystalline layers, and the quantum well is narrowed from 107 down to 12.8 crystalline layers. Simultaneously, the exciton diameter d_{exc} decreases from 13.5 down to 8.8 crystalline layers.

In accord with Eq. (7), further x growth results in increasing the parameter $\varepsilon^*(x)$ from 2.5 up to 4.0, which reduces the quantum well width from d_{QW} = 12.8 down to 10.55 layers as well as the exciton diameter – from d_{exc} = 8.8 down to 6.8 layers. The values R_0 and d_{exc} are finally stabilized, and exciton motion is fully localized inside the layer plane. The experimental data for the dependence $d_{QW}(\varepsilon^*)$, where (see Eq. (7)) $\varepsilon^* = f(x)$, are well described with the following analytical expression

$$d_{QW}(\varepsilon^*)\ [\text{layers}] = 9 + 0.4/(\varepsilon^*(x) - 1.7)^3 \tag{10}$$

that is plotted with the curve 11 in Fig. 4d.

Let us note that numeric values for $\boldsymbol{d}_{exc}$ in the case $\varepsilon^*(\boldsymbol{x}) > 1.7$ were obtained in approximation that growth of the anisotropy parameter should be accompanied with the condition $\varepsilon_1(\boldsymbol{x})^{\boldsymbol{m}} \leq 1.7$. As far as the value $\varepsilon_1(\boldsymbol{x})$ growing with $\boldsymbol{x}$ becomes higher than 1.7, the amount of layers $\boldsymbol{m}$ comprised by the exciton should decrease.

This statement is confirmed by the experimental dependence for the halfwidth Γ_1 of $\boldsymbol{n}$=1 exciton absorption band on $\boldsymbol{x}$. This dependence is shown in Fig. 4b. Indeed, growth of $\boldsymbol{d}_{exc}$ in the range $\boldsymbol{x}$<0.5 results in increasing probability of exciton scattering by point defects and growth of initial values for homogeneous widening the main (n=1) and excited excitonic states (n>1) at T = 0 K $\Gamma'_n(0) = g^2[\hbar\Omega(R_0/n^2 - \hbar\Omega)]^{1/2}$. In what follows, in the range $\boldsymbol{x}$>0.5 when the exciton dimension is stabilized inside the quantum well, the experimental dependence of the halfwidth $\Gamma_1(\boldsymbol{x})$ goes to saturation. The curve 6 in Fig. 4b, in accord with Eq. (1), shows the dependence $\Gamma_1(\boldsymbol{x})$ for Γ_1 (T = 80 K) = 4.4 meV.

4.2. GaSe hydrogen intercalates

The investigation of the influence of hydrogen intercalation on optical properties of layered crystals was extended to GaSe crystals at T = 4.5 K, as at this temperature exciton scattering is only caused by lattice defects. In this case, hydrogen molecules can be considered as some defects relatively to the proper system of elementary excitations in the crystal. Their concentration grows with $\boldsymbol{x}$, where $\boldsymbol{x}$ is the amount of hydrogen atoms per one formular unit of the crystal.

The spectra of excitonic absorption for GaSe crystals intercalated with hydrogen in concentrations $\boldsymbol{x}$ = 0…5 are depicted in Fig. 5 in geometry $\vec{E} \perp$ C-axis of the crystal, at the temperature T = 4.5 K (with account of reflection from the frontal face of the crystal). To take into account the light reflectivity, the reflection index $n_o = [n_0^{\perp 2} n_0^{\parallel}]^{1/3} = 3.1$ was determined directly from the values $\varepsilon_0^{\perp,\parallel}$ (Zhirko, 2000), where superscripts $\perp$ and $\parallel$ corresponds to directions relatively to crystal optical **C**-axis oriented along the normal to the plane of crystalline layers.

As can be seen in Fig. 5a, near the absorption edge of non-intercalated single crystal GaSe ($\boldsymbol{x}$ = 0) at T = 4.5 K, one can observe the absorption bands of the main n=1 and first excited n=2 excitonic states, which energy positions of peaks are E_1 = 2.1096 eV and E_2 = 2.1247 eV, and the obtauned via the wellknown relation $E_n = R_0/n^2$ (Elliott, 1957) the exciton binding energy R_0 = 20.1±0.15 meV corresponds to the classic value R_0 = 20.0 meV (Zhirko, 2000) for a non-intercalated crystal GaSe.

Increasing the hydrogen concentration up to the values $\boldsymbol{x}$ = 1.5 does not result in visible changes in absorption spectra. When the hydrogen concentration is further increased, essential transformation of the contour describing the n=1 exciton absorption band takes place. Besides, as seen in Figs. 5e and 5f, for $\boldsymbol{x}$ > 1.5 one can observe widening and energy shift of the band peak for n=1 exciton state; when $\boldsymbol{x}$ = 2.5 (Fig. 5c) the main excitonic state is widened, and an additional band with the peak at the energy E'_1 = 2.1123 eV arises near the shortwave edge of n=1 absorption band. The splitting value between E_1 and E'_1 states is equal to 2.7 meV. This formation of the additional band for the main excitonic state, when x grows, is related with creation of quasi-two-dimensional localized excitonic states splitted from the main excitonic state.

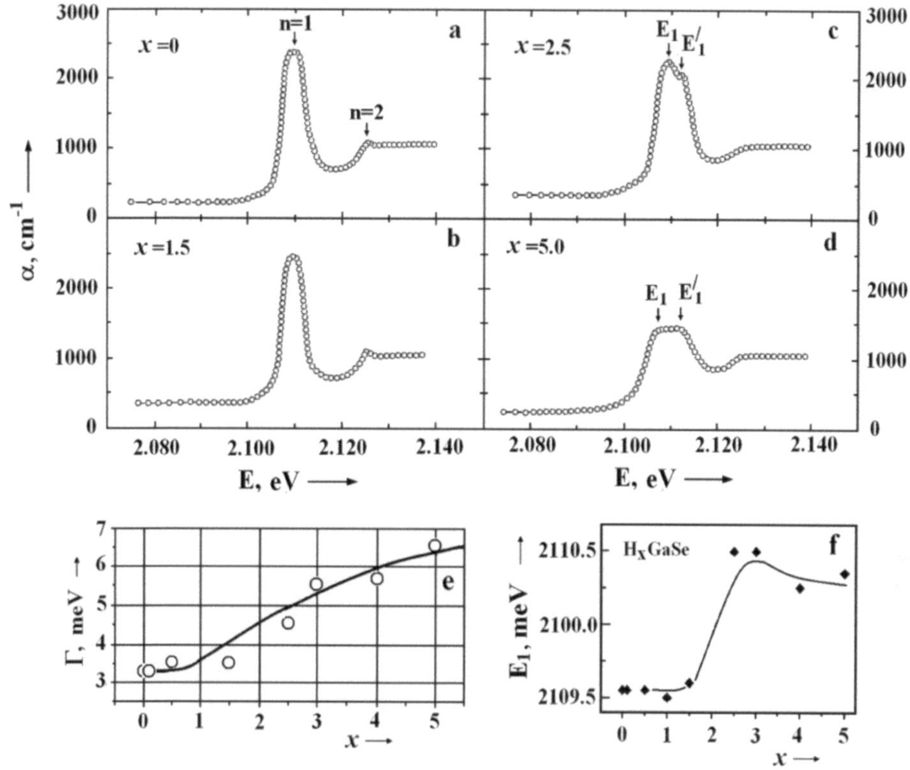

Figure 5. a-d –excitonic absorption spectra of GaSe crystalas intercalated with hydrogen in various concentrations; e, f – dependence of the halfwidth and shift of the peak for n=1 exciton absorption band in GaSe crystal on the hydrogen concentration

When x=5.0, the splitting of the main excitonic state is $E'_1 - E_1 = (2.1117 - 2.1078) = 3.9$ meV. Our attention was attracted by the fact that the positions of peaks for the doublet main excitonic state are practically at the same distance from the peak of main excitonic state of non-intercalated crystal (x=0). Using the known relation for the energy positions of the main and excited states in the case of quasi-two-dimensional exciton $E_n = R_0/(2n-1)^2$ as well as experimental data for values $E'_1 = 2.1117$ eV and $E_2 = 2.1256$ eV and $x = 5.0$, one can estimate the binding energy of quasi-two-dimensional exciton states in the crystal $H_{5.0}GaSe$ as R_0 = 15.6 meV.

Moreover, as seen from Fig. 5f, the energy position of the mass center of absorption band for n=1 excitonic state possesses a non-linear dependence on the concentration of embedded hydrogen: when the concentration increases up to x = 2.5, one can observe the shortwave shift, but in the range of higher concentrations the mass center begin to shift to low-energy side, and at x>4.0 reaches saturation. Simultaneously, as seen in Fig. 5e widening the excitonic absorption band takes place.

It should be noted that the same dependences $E_{n=1}(x)$ and $\Gamma_1(x)$ are observed in InSe crystals intercalated with hydrogen. However, in the case of GaSe crystals they are less pronounced and observed for higher hydrogen concentrations. For instance, in H_xGaSe crystals widening the main excitonic state with x growth can be approximated by the following dependence

$$\Gamma_1(x)\ [\text{meV}] = \Gamma_1 \cdot \{1+1.7/\exp[1/(0.1 \cdot \Gamma_1 \cdot x)]\}. \quad (11)$$

for Γ_1 (T = 4.5 K) = 3.3 meV.

This difference is caused by several reasons. The main one is as follows: the effective Bohr radius of excitons in non-intercalated GaSe crystals is 3.7 nm, and this exciton comprises 9.3 crystalline layers, while the exciton in non-intercalated InSe comprises 11.3 crystalline layers. It results in the fact that, for availability of these effects, it requires a higher hydrogen concentration in the interlayer space. The second and important reason lies in the following difference: in InSe $\mu^{\perp}\varepsilon_0^{\perp}/\mu^{\parallel}\ \varepsilon_0^{\parallel} = 1$, while in GaSe $\mu^{\perp}\varepsilon_0^{\perp}/\mu^{\parallel}\ \varepsilon_0^{\parallel} > 1$. It means that excitons in GaSe are not spherical but a littile oblate in the plane of crystalline layers. It lowers the amount of layers comprised by excitons and results in lower probability of their scattering.

At the same time it was shown (Zhirko, 2011) that temperature dependences of n=1 exciton band half-width in H_xGaSe intercalates as well as in pure and doped GaSe and InSe single crystals attributed with exciton scattering on $A_1^{/}$-homopolar optical phonon and observed temperature increase of integral intensity of n=1 exciton absorption band attributed to polariton damping by optical phonons in exciton-like states.

As a whole, one can state that intercalation of InSe and GaSe crystals results in essential transformations of their excitonic spectra. As it was shown using optical investigations, these transformations are caused by changes in the dielectric permittivity of the van der Waals gap, when it is filled with hydrogen. They manifest themselves in changing the exciton binding energy and, for sufficient degree of hydrogen intercalation, lead to further localization of exciton motion inside the plane of crystalline layers.

To check this hypothesis, we performed direct investigations of the hydrogen state in InSe and GaSe crystals by using the method of NMR spectroscopy. The respective results have been represented in the following chapter.

5. NMR investigations of GaSe and InSe monocrystal hydrogen intercalates and their powders

In (Zhirko et al., 2007b) the previous investigation of MNR spectra of $H_{1.0}GaSe$ powder hydrogen intercalates were done. It was shown that NMR spectra of $H_{1.0}GaSe$ powder under room temperature contain three bands (Fig. 6a, curve 1) associated with molecular hydrogen. In accord with the model offered in chapter 3 and with account of the band energy shift caused by the influence of matrix crystalline field on H_2 molecules, these bands were identified with H_2 (Fig. 6, insert) present in: i) crystal layers, where strong ion-covalent bonds are active (***L***-band); ii) interlayer space with weak van der Waals bonds (***I***-band); and iii) the regions between crystalline grains of the powder (***G***-band).

In this case, as shown in chapter 4, at low concentrations ($x \leq 2.0$) molecular hydrogen is predominantly located in interlayer space, while with growing x, when this space is filled, it comes into intra-layer space. Indeed, as seen from NMR spectra of powders the integrated intensities of ***G, I, L*** bands for $x = 1.0$ are in proportion 1 : 9 : 2.5 between each other. It has been also shown in (Zhirko et al., 2007b) that growth of the hydrogen concentration results in increasing a_0 and C_0 lattice parameters of the crystal.

In this part of our work we present a further refining of the mechanism and dynamics of hydrogen introduction into the layered crystals, which is based on additional temperature (T = 295…380*K*) and polarized radiospectroscopy investigations of GaSe single crystals intercalated with hydrogen in concentrations $x = 0…4.0$. And at the end of this part some extrapolate conclusion on H_xInSe intercalates were done and some complement experimental NMR spectra of $H_{1.0}InSe$ were presented.

5.1. Experimental part

Presented there radiospectroscopic investigation of H_xGaSe intercalates were performed using the NMR spectrometer Bruker AvanceTM^{400}. As seen from Fig. 6a, the spectra of $H_{1.0}GaSe$ single crystals (curve 2) contain only clearly pronounced ***L***- and ***I***-bands inherent to molecular hydrogen located in layer and interlayer spaces, and in external magnetic field these bands are split by doublets $L^{\uparrow}$, $L^{\downarrow}$ and $I^{\uparrow}$, $I^{\downarrow}$. The fact that the split value $\Delta L > \Delta I$ also confirms conclusion (Zhirko et al., 2007b) that H_2 molecules in H_xGaSe single crystals are located in two different crystal fields.

As it was shown in chapter 2 (Fig 2c), deintercalation of molecular hydrogen from layered crystals takes place at the temperature 110 °C and permanent pumping down. To study temperature influence on H_2 behavior in layered crystals, we performed temperature investigations of NMR spectra for $H_{1.0}GaSe$. The respective data have been depicted in Fig. 6, where doublet $I^{\uparrow}$, $I^{\downarrow}$ is associated with H_2 located in van-der-Waals gap and (renamed here for next discussion) $L_1^{\uparrow}$, $L_1^{\downarrow}$ doublet with H_2 located in layer call.

To further refine the considered model for introduction of hydrogen into a layered crystal, we carried out the investigations of the influence of H_2 concentrations on NMR spectra. Depicted in Fig. 6a are the NMR spectra of H_xGaSe for the concentrations $x = 1.0$ and $x = 4.0$ (T = 295 *K*). Obtained experimental data maximum of $I^{\uparrow}$, $I^{\downarrow}$ and $L_1^{\uparrow}$, $L_1^{\downarrow}$ bands shifting with temperature (T) and hydrogen concentrations (x) was collected in Table 1.

5.2. Discussion

It is known that molecular hydrogen at T≥ 295K consist of para-hydrogen (25%) and ortho-hydrogen (75%) molecules. At T=0 all molecules are in para-hydrogen state. Para-hydrogen possesses the total molecular spin $S_{p\text{-}H2} = S(+1/2) + S(-1/2) = \uparrow + \downarrow = 0$, while ortho-hydrogen - $S_{o\text{-}H2} = S(+1/2) + S(+1/2) = \uparrow + \uparrow = 1$. In external magnetic field ***B*** the energy level of a free molecules are split by three states with spins projection $S_{H2} = (+1, -1, 0)$.

x	T (K)	$I^{\uparrow}$ (ppm)	$I^{\downarrow}$ (ppm)	ΔI (ppm)	$L^{\uparrow}$ (ppm)	$L^{\downarrow}$ (ppm)	ΔL (ppm)
4.0	295	1.0	8.4	7.4	3.3	14.05	10.75
1.0	295	1.19	8.19	7.0	3.19	12.42	9.23
	310	1.22	8.10	6.88	3.0	12.12	9.02
	330	0.95	7.57	6.62	2.73	12.0	9.27
	360	0.48	6.96	6.48	2.0	11.05	9.05
	380	0.50	6.74	6.24	2.0	10.84	8.84

Table 1. Data for temperature and concentration dependences of the energy shift and splitting inherent to ***L***- and ***I***-bands observed for H_xGaSe in external magnetic field of the spectrometer Bruker Avance***TM*** 400

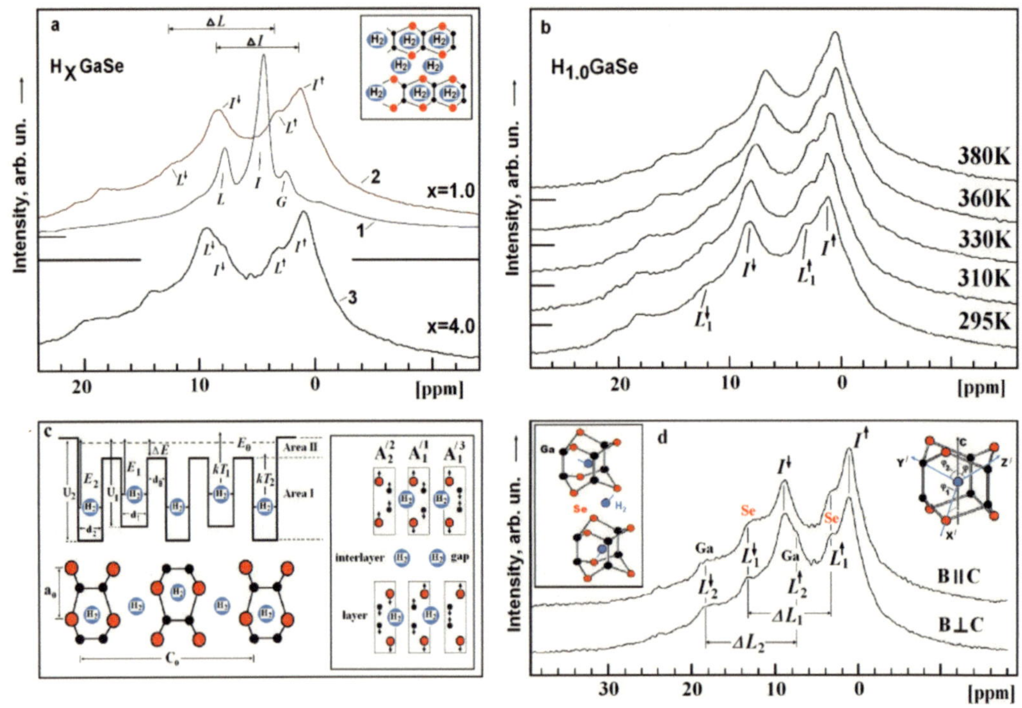

Figure 6. a) - NMR spectra of $H_{1.0}GaSe$ single crystals (curve 2) and its powder (curve 1) intercalates at room temperature. Curve 3 – for $H_{4.0}GaSe$. Insert. 2D sketch of molecular hydrogen location in $H_{1.0}GaSe$ powders; b) - temperature dependences of the shift and splitting of molecular hydrogen ***I***- and L_1-bands for intercalates in $H_{1.0}GaSe$ crystals; c) - 2D sketch and scheme of H_2 1D-localization in QW's of interlayer and layer spaces of GaSe crystal at T=0. Insert. Full symmetric vibrations in ε-GaSe crystal; d) - NMR spectra of $H_{1.0}GaSe$ single crystal at T=295K for longitudinal and transverse orientation of applied external magnetic field ***B*** according to crystal axis ***C***. Inserts. 2D sketches presenting slight spin orientation of ortho-hydrogen molecules in a layer and intra-layer crystal space.

Any hydrogen molecule in crystal experiences action of a crystalline field that causes limitation of its free precession relatively to external magnetic field ***B***, which eventually results in lowering its energy state. When measuring powder NMR spectra of H_xGaSe (analogue to the so-called «magic angle»), optical axes of crystalline grains (of 1-μm diameter, in our case) are chaotically average oriented relatively to external magnetic field ***B***, and in this case the total spin of all ortho-hydrogen dipole molecules in powder was equal to zero that corresponds with a para-hydrogen molecules. The bands ***G***, ***L*** and ***I*** observed in NMR spectra of H_xGaSe powder (Fig. 6a, curve 1) are related with resonant absorption of hydrogen molecules with total spin equal zero which are located in three different crystalline surroundings, namely: between grains, between layers and inside layers.

In the case of H_xGaSe single crystals, pronounced in the NMR spectra are two bands of molecular hydrogen located in interlayer and intra-layer spaces. In external magnetic field ***B***, these bands are split by doublets corresponding to ortho-hydrogen which spins oriented parallel and anti-parallel to ***B***. In our experiments, the field ***B*** is oriented in parallel to the optical axis ***C*** that is directed along the normal to layer planes in GaSe.

As seen from Fig. 6b with increasing the temperature from 295 up to 380K, one can observe the following features in NMR spectra of $H_{1.0}GaSe$:

- Some decrease in splitting the doublets of L_1- and ***I***-bands,
- The very doublets of L_1- and ***I***-bands are shifted to the side of lower energy,
- Lowering the integral intensity of $L_1^{\uparrow}$-band relatively to $I^{\uparrow}$- band.

It is also worth to note that the shift of these bands and their splitting value are in linear proportion to the inverse temperature [≈ ***const***·(1/T)] and the constant of the temperature shift for $L_1^{\uparrow}$ and $I^{\uparrow}$ bands of ortho-hydrogen oriented along the field ***B*** is somewhat lower than that one for opposite orientation.

This behavior of NMR spectra is indicative of the increasing mobility of hydrogen molecules in intra-layer and interlayer spaces with increasing the temperature, which results in a reduced influence of matrix crystalline field on hydrogen. As shown by calculations within the framework of the ideal gas model (Zhirko et al., 2007a), in GaSe crystals (for $x = 1.0$ and T = 300K) molecular hydrogen creates the pressure in the interlayer space equal to P = 17.7 MPa, which results in growth of the lattice parameter C_0. This fact is also confirmed by direct X-ray investigations (Zhirko et al., 2007b).

When $x = 1.0$ and T = 380K, the H_2 pressure value in the interlayer space is 22.4 MPa. In accord with our estimation (see chapter 2), at T = 384K (110 ^{0}C) and weak pumping down hydrogen leaves crystal: for the low concentration $x = 0.1$, its yield is approximately 60%, and for $x = 4.0$, when molecular hydrogen occupies both interlayer and intra-layer spaces, its yield reaches 85-90%.

It seems obvious to expect that growth of the hydrogen concentration resulting in the pressure increase in interlayer space and, respectively, in growing the lattice parameters a_0 and C_0, should result in the increasing mobility of hydrogen in intra-layer and interlayer spaces, i.e., to drop in splitting of L_1- and ***I***-bands as well as to their shift into the range of lower ***ppm*** values. However, as it can be seen in Fig. 6a, with growing the hydrogen concentration x the splitting of doublets in L_1- and ***I***- bands is increased, and $L_1^{\uparrow}$- , $L_1^{\downarrow}$- and $I^{\downarrow}$-bands (except $I^{\uparrow}$-band) are shifted to the range of higher ***ppm*** values.

It is indicative of the fact that, with increasing the hydrogen concentration, the influence of matrix crystalline field on the hydrogen molecule grows, and fixed orientation of the molecule relatively external magnetic field ***B*** is kept. It should be also noted in accord with (Zhirko et al., 2007a) that for $x \geq 2.0$ hydrogen, having filled all the translationally ordered states in interlayer space, goes into intra-layer space by virtue of quantum-dimensional effects. It also leads to a growing pressure inside crystalline layers. In this case, as seen in Fig. 6a the splitting value for L_1-band is larger than that for ***I***-band.

5.3. Model for hydrogen intercalation in layered GaSe and InSe crystals

Note that even at high concentrations (x=4.0) molecular hydrogen in the interlayer space at room temperature can be considered as condensed gas that, with the concentration growth, comes more and more into crystalline layers, which due to increased pressure enhances lattice parameters.

Performed in (Zhirko et al., 2007a) was an estimation of the level energy value for hydrogen molecule localization inside a one-dimensional well inherent to the interlayer space. In accord with the known expression

$$E_0 = \frac{\pi^2 \hbar^2}{2M_{H2} d_z^2} \tag{12}$$

as well as values for hydrogen molecule mass M_{H2} =3673 a.u. and width of the interlayer space d_z = 0.308 nm for InSe crystal (Olguin et al., 2003), obtained was the energy of the localization level E_0 = 1.1 meV. Note that the considered one-dimensional case describes appearance of the level in the interlayer space in a right manner, since motion of the molecule in the interlayer plane has a quasi-continuous spectrum.

However, the estimation of E_0 value performed in (Zhirko et al., 2007a) explains localization of H_2 in interlayer space only qualitatively. It is quite sufficient to consider processes of two-dimensional exciton movement in the plane of crystalline layers when hydrogen fills the interlayer space. But to explain behavior of NMR spectra for molecular hydrogen with increasing the temperature and concentration as well as its localization in inter- and intra-layer spaces, it seems no longer sufficient. Note that the localization energy for hydrogen molecules E_0=1.1meV obtained in (Zhirko et al., 2007a) is much less than the kinetic energies which necessary for H_2 molecule to deintercalating from crystal (kT for T = 384K is equal to 33.0 meV).

With this aim, taking into account the data obtained in this work, let us refine the offered in (Zhirko et al., 2007a; Zhirko et al., 2007b) model describing intercalation and deintercalation of hydrogen in layered crystals.

For this purpose (see Fig. 6c), let us consider GaSe single crystal consisting of three cells from crystalline layers and two interlayer spaces that contain H_2. In this case, there takes place one-dimensional localization of molecules in the direction normal to layers in the interlayer space with the width d_1. While in the planes of layers, molecular motion will be quasi-free, i.e., free for low hydrogen concentrations and strongly localized when the rest of the phase space will be practically filled with other hydrogen molecules.

For the molecule H_2 present in the cell of layer space, there takes place a three-dimensional potential well with the energy of a ground localization level (111) in the case of cubic geometry (linear sizes of the well in three directions x, y, z are the same, i.e. $d_2^X = d_2^Y = d_2^Z \equiv d_2$) that is equal

$$E_0 = \frac{3\pi^2\hbar^2}{2M_{H2}d_2^2}. \tag{13}$$

Thus, we have a set of two type wells, namely: one- and three-dimensional ones, where the hydrogen molecule can be present, being limited with potential barriers of finite width and height. To shorten the description of H_2-molecule behavior, let us simplify the model by changing the 3D well with 1D one. As seen from Eqs. (12) and (13), at equal energies of the ground state the only well width should be changed.

With this simplification, let us construct a set of five 1D wells (Fig. 6c) separated with potential barriers. Since during intercalation the molecular hydrogen penetration into the intra-layer space is more difficult than into the interlayer one, three deeper wells correspond to localization of H_2 inside crystalline layers and two more shallow - to localization of H_2 in interlayer space. This area of quantum wells with localized there H_2 molecules is marked in the figure as Area I. Depicted on a Fig. 6c case corresponds to the concentration $x = 4.0$, when T = 0K; $\boldsymbol{B} = 0$. The rest over-barrier phase space of the layered crystal is designated as Area II and was considered latter.

Before proceeding to calculation of well parameters in Area I, note some additional requirements:

1. When considering well geometrical parameters d_1 and d_2, it is necessary to take into account the fact that the diameter of a hydrogen molecule $\mathbf{D}_{H2} = 0.148$ nm is comparable with the width of interlayer space d_1. Moreover, covalent radii of selenium and gallium atoms $\mathbf{R}_{Se} = 0.116$ nm and $\mathbf{R}_{Ga} = 0.126$ nm in the layered crystal GaSe possess sizes comparable with $\mathbf{D}_{H2}$, too. The above particle sizes should be taken into account when determining the efficient geometrical sizes of wells

$$\delta_1 = d_1 - \mathbf{D}_{H2} \text{ and } \delta_2 = d_2 - \mathbf{D}_{H2}, \tag{14}$$

as well as d_B – the width of the potential barrier between them

$$d_B = [(C_0/2) - 2\mathbf{D}_{H2} - \delta_1 - \delta_2]/2, \tag{15}$$

where $C_0 = 1.595$nm (Kuhn et al., 1975) is the lattice parameter of the GaSe crystal.

2. Since the molecule spin is equal to integer (0, 1), molecular hydrogen obeys Bose statistics. Then, levels in each of the well sets are degenerated and possess the ground

level E_1 for interlayer space and E_2 for the intra-layer one. Assume also that in every well the hydrogen molecule possesses only localization level and $E_2 > E_1$. The depths of wells have finite values, and $U_2 > U_1$. In this case, the localization levels of H_2 molecules can be found as solutions of the quadratic equation

$$E_{(1,2)} \cong U_{(1,2)} - \frac{M_{H2}\delta_{(1,2)}^2}{2\pi^2\hbar^2} U_{(1,2)}^2 \tag{14}$$

3. Since at T = 384K molecular hydrogen comes out of the interlayer space, let the level energy will be as $E_1 = kT_1 = 33.0$ meV.
4. At the same time, at T = 384K and $x = 4.0$ with a weak pumping down approximately 85% of hydrogen comes out of the crystal. Therefore, transfer of hydrogen from the intra-layer space to the interlayer one should take place at T ≤ 384K, i.e. $kT_2 \le kT_1$.
5. We assume that at temperatures above 330*K* molecular H_2 is in a combined state (Area II), i.e., with increasing the temperature up to 380*K*, one can observe the only interlayer hydrogen present in a large common well. Thus, the difference between levels is $\Delta E = E_2 - E_1 = 3.0$ meV, where $E_2 = 36.0$ meV.

Indeed, as seen from the NMR spectra (Fig. 6b), at temperatures T > 330K $L_1^{\uparrow}$- band becomes to vanish, and at T = 380K one can observe the only $I^{\uparrow}$-band that practically stops its shifting to the range of lower ***ppm*** values. While $L_1^{\downarrow}$ - and $I^{\downarrow}$- bands caused by resonance absorption related with molecular ortho-hydrogen oriented anti-parallel to the external magnetic field ***B*** and being in more low-energy states prolong to shift with increasing temperature.

Thereof, in accord with the set H_2 localization levels E_1 and E_2 and using the expression (14), let us determine the parameters of quantum wells in absence of external magnetic field ***B***:

- The interlayer well depth is $U_1 = 64$ meV, efficient width $\delta_1 = 0.0557$ nm, $d_1 = 0.2037$nm,
- The intra-layer well depth is $U_2 = 69$ meV, $\delta_2 = 0.0533$ nm, $d_2 = 0.2013$ nm,
- And in accord with (13a), the barrier width between them is $d_B = 0.1473$ nm.

Note that parameters δ_1 and δ_2 obtained without any account of the experimental fact that, with increasing the temperature and hydrogen concentration in the layered crystal, a_0 и C_0 grow, too, which enhances the well width d_1, d_2 and changes the barrier thickness d_B in places where hydrogen is present. By other words $d_1, d_2, d_B = f(T, x)$.

5.3.1. The case B=0, T>0, x>0

As it was noted above, growth of the temperature forces the molecular hydrogen to come out of quantum wells to Area II and fill the phase space of the layered crystal, and for sufficient concentrations when $x \rightarrow 4.0$ creates in it some level more E_0 with the energy ≤ 1.0 meV. For $x \rightarrow 0$, the level energy $E_0 \rightarrow 0$: H_2 behavior is close to the quasi-free one limited by the crystal bulk and atoms of its composition. By this cause, at low concentrations the hydrogen amount leaving the crystal does not exceed 60…70%, while at high concentrations it reaches 80…90%.

With increasing the temperature, the molecule kinetic energy in the well can be enhanced due to scattering at potential well walls with absorption of vibration energy quantum from crystalline lattice. When T = 0, the potential barrier walls in the first approximation are rigid and fixed (we do not take into account zero vibrations of atoms). But increase in temperature means that phonons of the crystalline lattice become active, the density of population n_i for phonon branches Ω_i grows with temperature in accord with the law

$$n_i = \frac{1}{e^{\hbar\Omega_i/kT} - 1}. \quad (15)$$

Periodical shifts of atoms in lattice that are caused by presence of phonons, results in changing the potential barrier width d_B, which cannot but influence the energy of hydrogen molecule localization: for inelastic H_2 reflection from the barrier wall (with absorption of a phonon) the molecule can pass to Area II.

The process of H_2 molecule scattering by lattice vibrations will be the most efficient when: i) the molecule kinetic energy in the well is close to the phonon energy, and ii) the momentum of the molecule reflected from the barrier wall coincides (is summed up) with the barrier wall momentum (phonon momentum), which is possible when lattice atoms in the course of vibrations change the width of the potential well and barrier.

In the case of GaSe crystal, the unit cell of which consists of 8 atoms located in two crystal layers, there exist 24 normal vibrations of the lattice (Belenkii & Stopachinskii, 1983).

Three acoustic vibrations of them for $\boldsymbol{k} = 0$, where $\boldsymbol{k}$ is the wave vector of the crystal reciprocal lattice, possess the energy equal to zero and do not directly contribute to the molecule kinetic energy. All the doubly-degenerated *E*-vibrations also do not essentially contribute to the molecule kinetic energy, as they do not take part in significant changing d_1, d_2 and d_B width. Only three transverse (totally symmetrical) *A*-vibrations can provide direct contribution to changing the molecule kinetic energy because they can change quantum well width. Motion of atoms during this vibration is shown on Insert of Fig. 6c.

The first totally symmetrical $A_2^{\,l2}$-vibration (interlayer) is not valid here, because: i) it possesses the low energy (40 cm^{-1}) to scatter molecule from quantum well; ii) interlayer space is changed, but the thickness of the crystalline layer (barrier width d_B and d_2) remains constant.

The second $A_1^{\,l1}$-vibration (half-layer with energy 133 cm^{-1}) can change width d_1, d_2 of quantum wells but remain unchanged barrier width d_B between them. It was an active at T = 300K to take part in molecule scattering from both type quantum wells.

The third $A_1^{\,l3}$-vibration (intra-layer with energy 307 cm^{-1}) are realized with changing both the volume of the crystalline layer cell and the thickness of the interlayer space, but $A_1^{\,l3}$-vibration is still insufficiently active even at temperatures close to 380K.

Therefore, we assume that the totally symmetrical half-layer $A_1^{\,l1}$ -vibration takes main part in the processes of hydrogen molecule scattering in the wells of two kinds, besides even at sufficiently low temperatures (150 to 200K) it promotes H_2 molecule tunneling from one well to another. Also, this vibration takes very active part in the processes of exciton decay (Zhirko, 2000).

5.3.2. The case B≠0, T>0, x>0

As seen from Figs. 6d and 7, the application of external magnetic field ***B*** results in vanishing degeneration and splitting of the ortho-hydrogen molecule level in every well: there arise doublets of L_1 and ***I***-bands for ortho-hydrogen molecules in NMR spectra of H_xGaSe crystals, these molecules being in two different crystalline fields. Energy levels of para-hydrogen molecules with zero spin (not observed in NMR spectra of single crystal) remain the same.

At the same time as molecular hydrogen in a gaseous-like state possess diamagnetic properties the amount of ortho-hydrogen molecules oriented in parallel and anti-parallel direction to applied external magnetic field ***B*** must be equal one to another. Really on the Fig. 6d one can see that the summarized integral intensities of $I^{\uparrow}$ and $L^{\uparrow}$ bands for ortho-hydrogen oriented along ***B*** are in well coincidence with summarized integral intensities of $I^{\downarrow}$ and $L^{\downarrow}$ bands for ortho-hydrogen oriented opposite ***B***.

Finally the NMR investigations of $H_{1.0}GaSe$ single crystals at T=295K with different orientation of external magnetic field $B \perp C$ and $B \| C$ (where ***C*** is a crystal axis oriented normally to crystal payer plane) were conducted. Experimental data (see Fig. 6d) cannot find energetic shift of H_2 absorption bands maximums with changing ***B*** orientation. But at the same time slight redistribution of integral intensities for ***I*** and L_1 bands occurred. Thus in $B \perp C$ geometry $I^{\downarrow}$-band increased and are compatible (in intensities) with $I^{\uparrow}$-band but $L_1^{\downarrow}$-band decreased. For $B \| C$ geometry their mutual redistribution are reciprocal.

Also in geometry $B \perp C$ we can clearly observe additional one doublet of $L_2^{\uparrow}$ and $L_2^{\downarrow}$ bands. Note that these additional $L_2^{\uparrow}$and $L_2^{\downarrow}$ bands also observed in Figs. 6a and 6b but they are not so pronounced because of some intermediate $B \angle C$ geometry. It is important that L_2 doublet has greater shift than L_1 doublet and $\Delta L_2 > \Delta L_1$.

Two L_1 and L_2 absorption doublets for hydrogen molecule in a crystal layer evidenced about presence of two different position of H_2 inside a crystal cell. Really in layer space (see inserts of a Fig. 6d) for H_2 molecule located in center of a cell there are two environments consisting from atoms Ga and Se. The nearest sub-lattice consisted from Ga atoms give the greatest contribution of a crystal field into chemical shift and is identified by us to a doublet L_2. The Se atoms are removed further from the centre of a cell: therefore contribution of Se atoms in chemical shift is less. We identify the contribution of Se atoms sub-lattice to a doublet L_1. For everyone sub-lattice there are three equivalent dislocations of a ortho-hydrogen molecule along an ***C*** axis and three in opposite.

With nearest-neighbour distances ***d*** obtained in (Olguin et al., 2003) for InSe: $d_{In\text{-}In}$ = 2.79Å; $d_{In\text{-}Se}$ = 2.65Å; $\angle\varphi$ = 119°30′, layer thickness d_l = 5.36Å, the distance between layers d_i = 3.08Å and their interlayer distance $d_{Se\text{-}Se}$ = 3.80Å., one can find an angle ∠ between optical axis ***C*** and nearest Se ($\angle\varphi_1$) and Ga ($\angle\varphi_2$) atoms by values $\angle\varphi_1 = \pm 40^0 45'$ and $\angle\varphi_2 = \pm 59^0$ respectively.

To define orientation of hydrogen molecule in interlayer space it is essential that each layers are shift one to another and angle of Ga-Ga-Se bound in layer is about 120^0. In this circumstance angles in Ga and Se directions are degenerated, equal to $\angle\ \varphi = \pm 52^0$ and have an intermediate value between $\angle\ \varphi_1$ and $\angle\ \varphi_2$ angles.

As it is seen on Fig. 6d intra-layer doublets L_1 and L_2 also differently react to a direction of an external magnetic field $\boldsymbol{B}$ concerning axis $\boldsymbol{C}$. So in geometry $\boldsymbol{B} \perp \boldsymbol{C}$ doublet L_2 are more pronounced, and for $\boldsymbol{B} \| \boldsymbol{C}$ geometry L_1 bands are more appreciable. It allows us to make a conclusion that spin orientation of H_2 inside a layer in sub-lattices Ga and Se try to compensate each other in external magnetic field $\boldsymbol{B}$. It is possible only in that case, when H_2 spins in Ga and Se sub-lattices have opposite directions concerning axis $\boldsymbol{C}$. Thus in absence of an external magnetic field $\boldsymbol{B}$ a corner of a spin inclination concerning an axis $\boldsymbol{C}$ in direction to Se atom is negative $\angle\varphi_1 = -\ 40^0 45'$ and in direction to Ga atom is positive $\angle\varphi_2 = +59^0$ respectively. In this case as it is seen on an inserts of a Fig. 6d corner between H_2 spins in Ga and Se sub-lattices with good accuracy are equal to 90^0 and their summarized projection on axis $\boldsymbol{C}$ are negative.

In interlayer space at $\boldsymbol{B}$=0 an angle between spin of ortho-hydrogen molecule and axis $\boldsymbol{C}$ was positive and $\angle\ \varphi = +52^0$. In the whole three various directions of H_2 spins orientation in GaSe made a right-hand rectangular system of coordinates, which allow compensate H_2 spins in absence of external magnetic field $\boldsymbol{B}$. Presented on Fig. 6d schema of H_2 molecule spin projection clearly shown the reason of $\boldsymbol{I}$, L_1 and L_2 – band doublets redistribution with changing $\boldsymbol{B}$ orientation according to axis $\boldsymbol{C}$.

Finally, using the data obtained as a result of these investigations, in Fig. 7 we have shown the scheme of splitting the level of a hydrogen molecule in H_xGaSe single crystals and powders in external magnetic field $\boldsymbol{B}$ with changing temperature and hydrogen concentration.

In insert the scheme of hydrogen molecule localization in layer and interlayer quantum wells with account of two Ga and Se sub-lattices of a layer cell are proposed. The given scheme consists of 1D interlayer QW, projection of a 3D crystal cell QW and potential barrier between them, which becomes transparent enough at high $\boldsymbol{k}$T.

The scheme allows to:

i. show splitting of H_2 ground state in 3D QW at $\boldsymbol{B}$=0 and $\boldsymbol{B}$>0 due to different crystal field of a nearest Ga and Se atomic sub-lattices constituting crystal cell,
ii. differentiate phase spaces of hydrogen molecule located in 1D and 3D QW and,
iii. demonstrate process of hydrogen exit from 3D to 1D QW with growing $\boldsymbol{k}$T.

In this case when H_2 molecule exit from 3D to 1D QW it loses energy and localized in interlayer 1D QW. Accordingly the process of hydrogen entry in crystal cells occurs at high enough temperatures and concentration that allows to overcome a potential barrier created by atoms of a crystal cell. Nevertheless some part of hydrogen molecules at deintercalation remains localized in crystal cells.

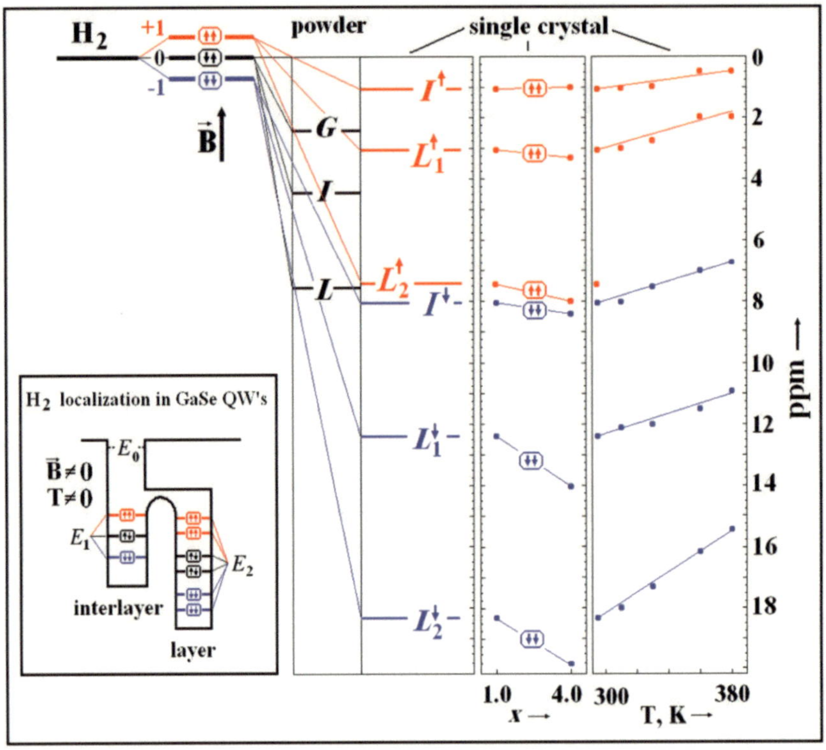

Figure 7. Scheme of splitting the level of a hydrogen molecule in HxGaSe single crystals and powder in external magnetic field (*B*) with changing the hydrogen concentration (*x*) and temperature (*T*).

Also at large hydrogen concentration and high temperatures, when hydrogen molecules are not located in QW's and posses high kinetic energy in volume of interlayer space appearance of an additional quazi-local level with energy E_0 is possible.

In Introduction, we adduced the lattice parameters for InSe and GaSe crystals. As in InSe crystals the lattice parameter a_0 and the width of the van der Waals gap are larger than those in GaSe crystals, it should be expected that the increase of the lattice parameter and interlayer space taking place in InSe as compared to GaSe crystals will result in increasing the halfwidth of H_2 bands in NMR spectra and to reduction of their chemical shift and splitting.

Indeed, as seen in Fig. 8 adduced for $H_{1.0}InSe$ single crystal there observed is practically 30% decrease in splitting and simultaneous growth of the width of NMR bands corresponding to hydrogen molecules. It is indicative of a decreased matrix influence on the hydrogen dissolved in it.

6. Electrophysical and EPR investigations of hydrogen intercalates in InSe and GaSe crystals

In what follows, we shall describe in brief some electrical properties of InSe and GaSe crystals, their hydrogen intercalates and related with it electron paramagnetic resonance (EPR) investigations of hydrogen in these crystals.

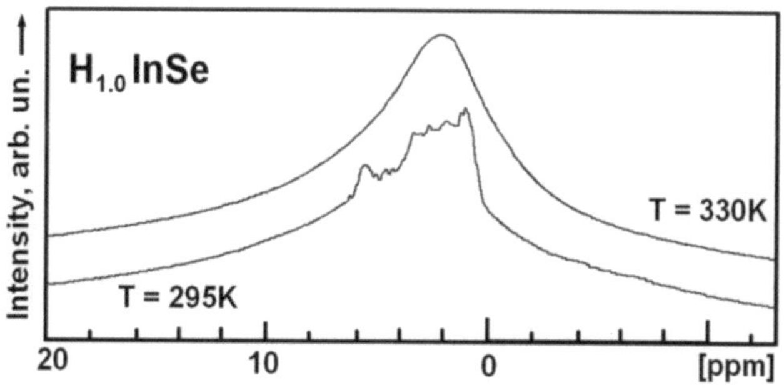

Figure 8. NMR spectra for $H_{1.0}InSe$ crystal at T= 295K and T=330K.

Contrary to GaSe crystals that are high-ohmic semiconductors of ***p***-type conductivity, InSe crystals are low-ohmic and possess ***n***-type conductivity. It enables to determine the carrier concentration in InSe crystals by using the Hall effect. As it was shown in (Zhirko et al., 2010) (see Fig. 9a), the concentration of free electrons in ***n***-InSe at T = 80 K increases from n=1.5·10^{15} cm^{-3} up to 1.4·10^{16} cm^{-3} even for comparatively low hydrogen concentrations (x=0.05).

In addition, contrary to pure InSe crystals, the concentration of carriers does not grow with temperature in $H_{0.05}InSe$ intercalates. Moreover, in these intercalates at T ≈ 290 K one can observe the so-called "exhaustion" of the level caused by transition of electrons from the donor level into the crystal conduction band. Thereof, using the expression

$$(T_d / 300)^{3/2} = (m_0 / m_e^*)^{3/2} \cdot 4 \cdot 10^{-18} N_d, \tag{16}$$

that relates the carrier concentration with the temperature of their exhaustion T_d (290 K for $H_{0.05}InSe$), and the experimentally registered electron concentration N_e = 1.4·10^{16} cm^{-3}, one can determine the effective electron mass value m_e^* = 0.152m_0 that is in satisfactory coincidence with the value $m_e^* = (0.143m_{0\parallel} \cdot 0.156m_{0\perp}^2)^{1/3} = 0.15m_0$ based on experimental data (Kuroda & Nishina, 1980).

As the electron concentration does not visibly grow in H_xInSe at T > 80 K, one can estimate the activation energy of the donor level in this crystal as 5 to 7 meV. Besides, in InSe crystals at 80 K the electron concentration is less by one order than that in intercalates and, as seen in Fig. 9a, grows with temperature. It allows drawing the conclusion that introduction of hydrogen causes a decrease in the activation energy for the shallow donor level (see Fig. 9c) that is located in InSe by 16 to 20 meV below the conduction band bottom (Martinez-Pastor et al., 1992), while the concentration of its carriers reaches 10^{15} cm^{-3}.

Further investigations of free carriers in hydrogen intercalates of ***n***-InSe and ***p***-GaSe crystals were performed with the method of EPR spectroscopy (Bruker X-Band Fourier Transform EPR Spectrometer ELEXSYS E 580-10/12). The experimental data have been summarized in Fig. 9b. It is seen that for T = 120 K the halfwidth of EPR-line and g-factor of the hydrogen unpaired electron in H_xGaSe intercalates ($g = 2.40$, $\Delta H = 1500$ Gs) are higher than those in the H_xInSe ones ($g = 1.87$, $\Delta H = 60$ Gs). It is indicative of the fact that hydrogen passivation of donor type point defects in *n*-InSe and those of acceptor type in *p*-GaSe differs to some extent. For comparison, represented in Fig. 9d are the simplest defects by Schottky, which results in creation of donor and acceptor centers in *n*-InSe and *p*-GaSe crystals, respectively.

To explain this difference between EPR spectra, let us assume that during intercalation some part of hydrogen atoms located within the range of point and spatial defects of the crystal (where present are broken chemical bonds caused by absence of atoms) do not recombine to H_2 molecules but are sorbed inside the range of these defects, which results in passivation of them with their proton H^+ or electron *e*.

It is noteworthy that, contrary to graphen where hydrogen atoms break the double carbon C=C bond, which transforms it to graphan, hydrogen does not break the double In(Ga)=In(Ga) bond inside crystalline layer, as it does not form chemical compounds with In and Ga atoms.

Therefore, we assume that one of the main reasons for this difference between EPR spectra of hydrogen unpaired electrons in $H_{1.0}InSe$ and $H_{1.0}GaSe$ crystals is the charge state of defects. On the Fig.9d one can see the simplest defects of a donor and acceptor type in GaSe and InSe crystals. In the case of a donor the crystalline matrix within the

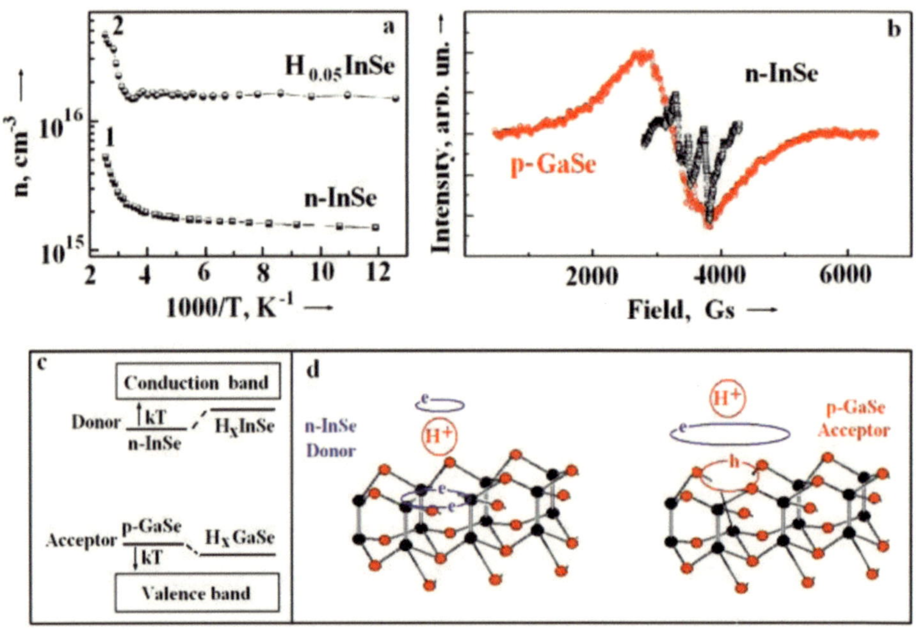

Figure 9. a) - Temperature dependence of free electron concentration for undoped InSe and hydrogen interalated InSe with x=0.05; b) - EPR spectra for $H_{1.0}GaSe$ (red circleles) and $H_{1.0}InSe$ (black squares) crystals obtained at 120K; c) - Change of donor and acceptor energy level displacement in forbidden gap of n-InSe and p-GaSe crystals after hydrogen intercalation; d) – A simple defect of donor and acceptor type in n-InSe and p-GaSe crystals and scheme of their hydrogen passivation.

defect range (due to absence of a Se atom) has two excess electrons that attracts by their Coulomb potential a positively charged proton from hydrogen, while the unpaired electron of the latter being captured in the proton field in accord with the scheme: donor (2e) - proton – electron. In the case of acceptor, the excess hole (absence of Ga atom) of the crystal due to its Coulomb potential attracts the hydrogen unpaired electron, on the opposite side of which the proton rotates in accord with the scheme: acceptor (h) – electron – proton.

Since the proton mass is 1836 times higher than that of electron, it is obvious that, in the case of donor, the proton (and, in general, the hydrogen atom) should be less mobile than in the acceptor case. The second reason for the difference in EPR spectra can be possibly as follows. Geometrical dimensions of the acceptor can be higher than those of donors, which is able to result in increasing the mobility of hydrogen atom inside the defect range.

The described above mechanism for hydrogen passivation of point defects (free radicals) as it is shown on a Fig. 9c should also lead to decreasing the activation energy for carriers (electrons, holes) in the crystal and increasing the concentration of carriers at low temperatures, which follows from the experimental data for $H_{0.05}InSe$ crystals as compared to non-intercalated InSe crystals.

In addition, our calculations show that the concentration of atomic hydrogen adsorbed by defects of crystalline layers is 4…5 orders less than that of molecular hydrogen, and it is comparable with the concentration of intrinsic defects in the crystalline structure.

7. Conclusion

The performed complex electron-microscopic SEM, EDS, HKL, electrophysical (Hall effect), microwave (EPR, NMR) and optical-and-spectral (exciton absorption) investigations of processes for hydrogen intercalation in layered crystals InSe and GaSe enabled us to ascertain the following features.

In the course of intercalation atomic hydrogen recombine to the molecular state, occupies interlayer space of a crystal, and due to quantum-dimensional effects with increase of hydrogen concentration comes into the interstitial space, occupies centers of crystalline layer cells where its recombination to molecular state takes place.

It results in localization of hydrogen molecules, which, in its turn, leads to growth of lattice parameters a_0 and C_0. At the same time, very small fraction of hydrogen (≈0.001%) in H_xInSe and H_xGaSe intercalates, being in atomic states, passivate point and spatial lattice defects, too. Simultaneously, due to decreasing the defect activation energy the concentration of free carriers increases by one order.

It is shown that, in absence of an external magnetic field ***B***, molecular hydrogen as diamagnetic gas (owing to symmetry of the crystal and nearest nuclear environment) forms in GaSe and InSe crystals three spin sub-lattices where spins are located mutually-perpendicularly to each other and form right-hand rectangular system of coordinates. At $\boldsymbol{B} = 0$, the total projection of two spins inside the layer is as a whole located against the optical axis $\boldsymbol{C}$, and between layers along the axis $\boldsymbol{C}$.

Offered is the scheme of 1D and 3D quantum wells (QW's) that is capable to explain localization of hydrogen molecules in layered crystals in the course of intercalation as well as mechanism of their deintercalation from the crystals with participation of totally symmetrical lattice $A_1{}^{/1}$-phonons taking part in scattering of molecules and in formation of oscillating parameters of potential barriers and QW's. In addition, described is the mechanism of H_2 scattering by $A_1{}^{/1}$-phonons in QW's.

Increasing the lattice parameter and width of the van der Waals gap results in decreasing the matrix influence on behavior of molecular hydrogen dissolved in it. It can be observed as a decreasing chemical shift and splitting the level of ortho-hydrogen molecule in magnetic field as well as widening the absorption bands.

Our optical investigations enabled us to ascertain that appearance of molecular hydrogen in the interlayer space of InSe crystals results in non-monotonous dependence of the exciton binding energy on the hydrogen concentration as a consequence of growth of the crystal dielectric permittivity ε_0. It has been shown that, for low hydrogen concentrations, the growth of the anisotropy parameter for the dielectric permittivity $\varepsilon^*(x)$ both along layers and along the normal to them manifests itself as a decrease in the exciton binding energy R_0. In the case of further growth of the parameter $\varepsilon^*(x)$ up to the critical value $\varepsilon^* \leq 2$, there arises 2D localization of the exciton motion inside the plane of crystalline layers, which is accompanied by growth of R_0 and decrease and, in what follows, stabilization of linear dimensions both of the exciton and quantum well. As a consequence of the difference between exciton diameters in GaSe and InSe crystals ($\boldsymbol{d}_{exc}{}^{GaSe} < \boldsymbol{d}_{exc}{}^{InSe}$), non-linear behavior of the exciton binding energy in GaSe is slower and weaker pronounced.

The performed investigations have shown that the degree of hydrogen electrochemical intercalation of layered crystals GaSe and InSe can reach the concentration $x = 5.0$, where $\boldsymbol{x}$ is the amount of hydrogen atoms embedded to the crystal per one its formular unit. Deintercalation of molecular hydrogen from H_xInSe and H_xGaSe crystals is realized for 3 to 9 hours at the temperature $T = 110\,^0$C and continuous pumping down. The degree of deintercalation grows practically linearly from 60% at $\boldsymbol{x}\to 0$ up to 80…85 % at $\boldsymbol{x}\to 4$. Our investigations of intercalation-deintercalation processes in powders of these crystals with sizes of faces 1…20 μm have shown that the hydrogen concentration can reach the values $\boldsymbol{x} = 6.0$, while the degree of deintercalation can reach 90 %. The repeated cycles of soft intercalation-deintercalation regimes do not lead to essential worsening the physical parameters of InSe and GaSe crystals. At the same time hard regimes of intercalation caused destruction of crystals up to powder-like states.

8. Afterword

The investigations performed by us enabled to draw the conclusion that these crystals and their powders are able to take and return hydrogen in its molecular form very well, keeping at the same time their initial physical-and-chemical properties. The calculations performed in (Zhirko et al., 2007a) have shown that the pressure of molecular hydrogen inside the van der Waals gap at room temperature for $x = 2.0$ is close to 35.4 MPa (354 atm.), which is quite comparable with the hydrogen pressure in the vessels of high pressure (400…500 atm. at -30 … +40 ^{0}C). Moreover, an important advantage of all the hydrogen solid-state storages is their safety when storing, as compared to the above vessels.

The hydrogen content in layered intercalates of InSe and GaSe crystals is also comparable in mass fractions with that in vessels of high pressure. For example, when the pressure in these vessels reaches 500 atm., the mass fraction of hydrogen is close to 11%, while in $H_{5.0}GaSe$ – 3.3% and in $H_{5.0}InSe$ – 2.6%. These values are comparable with those in metal hydrides: BeH_2 – 18.28%, LiH – 12.7%, MgH_2 – 7.6%, NaH – 4.2%, as well as in hydrides of intermetallic compounds: $LaNi_5H_6$ – 1.5%, $TiFeH_2$ – 1.8%, Mg_2NiH_2 – 3.8%, $CeCo_3H_{4.5}$ – 1.4%.

An important factor promoting hydrogen intercalation in layered crystals GaSe and InSe is that molecular hydrogen at temperatures below its melting temperature (<14 K), being in the solid state, possesses dense hexagonal lattice of the spatial group D^4_{6h} (C6/mmc) with the following parameters: a_{H2}=3.75Å and C_{H2}=6.12Å (Ormont, 1950). Note that the parameter a_{H2}=3.75 Å of dense hexagonal stacking for crystalline molecular hydrogen practically coincides with the parameter a_0 =3.755 Å (Kuhn et al., 1975) inherent to hexagonal stacking for the GaSe crystal lattice and is only a little less than that parameter a_0 =4.001 Å (Goni et al., 1992; Olguin et al., 2003) for InSe crystal.

It provides good conditions for location of hydrogen molecules inside centers of hexagonal cells within crystalline layers and interlayer spaces in GaSe and InSe crystals. However, the parameters $C_{GaSe} = C_0/2 = 7.98$ Å for GaSe crystal and $C_{InSe} = C_0/3 = 8.44$ Å for InSe crystal exceed the parameter C_{H2} for crystalline hydrogen by 1.3 and 1.4 times, respectively. Besides, with account of the dielectric permittivities of crystals InSe ($\varepsilon_0 = 10.5$) and GaSe ($\varepsilon_0 = 9.8$), one can assume that availability of the matrix crystalline layer between two molecular monolayers of hexagonal densely stacked hydrogen in the gap as well as its presence inside centers of crystalline layer hexagonal cells should result in screening the interaction of hydrogen molecules. As a consequence, it should cause growth of the parameter C_{H2} and its approaching to the parameters C_{GaSe} and C_{InSe} inherent to these crystals.

The performed investigations also provide the possibility to search more suitable layered structures with the aim to apply them for keeping hydrogen in various aggregate states.

Author details

Yuriy Zhirko
Institute of Physics of the National Academy of Sciences (NAS) of Ukraine, Ukraine

Volodymyr Trachevsky
Institute for Physics of Metal NAS of Ukraine, Ukraine

Zakhar Kovalyuk
Chernivtsy Department of Institute of Material Science Problems NAS of Ukraine, Ukraine

Acknowledgement

The authors express their sincere gratitude to Dr. Tkach V. N., Dr. Klad'ko V.P., Ph.D. Shapovalova I.P., Ph.D. Pyrlja M.M., Vorsovskii A.L., Zaslonkin A.V., Boledzyuk V.B. & Mel'nik A.K. who took part in performing the experiments discussed in this work.

9. References

Anderson, S.H. & Chung, D.D.L. (1984). Exfoliation of intercalated graphite. *Carbon*, Vol.22, No3, pp.253-263

Belenkii, G.L. & Stopachinskii V.B. (1983). Electronic and vibration spectra of A^3B^6 layered semiconductors. *Uspekhi Fizicheskih Nauk (in Russian)*, Vol.140, No2, pp. 233-270

Elliott, R.J. (1957). Intensity of optical absorption by excitons. *Phys. Rev.*, Vol.108, pp.1384-1389

Ghosh, P.H. (1983). Vibrational spectra of layer crystals. *Applied Spectroscopy Rewiev*, Vol.19, No2, pp.259-323

Goni, A.; Cantarero, A.; Schwarz, U.; Syassen, K. & Chevy, A. (1992). Low-temperature exciton absorption in InSe under pressure. *Phys. Rew. :B*, Vol.45, No8, pp.4221-4226

Grigorchak, I. I.; Zaslonkin, A. V.; Kovalyuk, Z. D. et al. (2002). Patent of Ukraine, Bulletin No5, No46137

Kozmik, I.D.; Kovalyuk, Z.D.; Grigorchak I.I. & Bakchmatyuk, B.P. (1987). Preparation and properties of hydrogen intercalated gallium and indium monoselenides. *Isvestija AN SSSR Inorganic materials. (in Russian)*, Vol .23, No5, pp.754-757

Kuhn, A.; Chevy, A. & Chevalier, R. (1975). Crystal structure and interatomic distance in GaSe. *Physica Status Solidi(b)*, Vol.31, No2, pp. 469-473

Kuroda, N. & Nishina, Y. (1980). Rezonance Raman Scattering Study of Exciton and Polaron Anisotropies in InSe. *Sol. St. Commun.* Vol.34, No6, pp.481-484

Lebedev, A.A.; Rud', V.Yu. & Rud' Yu.V. (1998). Photosensitivity of geterostructures porous silicon-layered $A^{III}B^{VI}$ semiconductors. *Fizika i Tekhnika Poluprovodnikov (in Russian)*, Vol.32, No3, pp.353-355

Malkov, M.P. (Ed.) (1985). Hand-book on physical-chemical fundamentals of cryogenics. (in Russian), Energoatomizdat, Moscow, USSR

Martines-Pastor, J.; Segura, A. & Valdes, J.L. (1987). Electrical and photovoltaic properties of indium-tin-oxide/p-InSe/Au solar cells. *J. Appl. Phys.*, Vol.62, No4, pp.1477-1483

Olguin, D.; Cantarero, A.; Ulrich, C. & Suassen, K. (2003). Effect of pressure on structural properties and energy band gaps of γ-InSe. *Physica Status Solidi (b)*, Vol.235, No2, pp.456-463

Ormont, B.F. Structures of inorganic substances, Moscow, Gosudarstvennoje Izdatel'stvo Techniko Teoreticheskoj Literatury, 1950

Polian, A.; Kunc, K. & Kuhn, A. (1976). Low-frequency lattice vibrations of ∂-GaSe compared to ε- and γ-polytypes. *Solid State Communications*, Vol.19, No8, pp.1079-1082

Shigetomi, S. & Ikari, T. (2000). Electrical and photovoltaic properties of Cu-doped *p*-GaSe/*n*-InSe heterojunction. *J. Appl. Phys.*, Vol.88, No3, pp.1520-1524

Toyozawa, Y. (1958). Theory of line-shapes of the exciton absorption bands. *Progr. Theor. Phys*. Vol.20, No1, pp.53-81

Zhirko, Yu.I. (1999). Investigation of the light absorption mechanisms near exciton resonance in layered crystals. N=1 state exciton absorption in InSe. Physica Status Solidi (b), Vol.213, No1, pp.93-106

Zhirko, Yu.I. (2000). Investigation of the light absorption mechanisms near exciton resonance in layered crystals. Part 2. N=1 state exciton absorption in GaSe. Physica Status Solidi (b), Vol.219, No1, pp.47-61

Zhirko, Yu. I.; Kovalyuk, Z. D.; Pyrlja, M. M. & Boledzyuk, V. B. (2007a). Application of layered InSe and GaSe crystals and powders for solid state hydrogen storage. In: *Hydrogen Materials Science and Chemistry of Carbon Nano-materials*, N.T. Vezirogly et al. (Ed.), 325-340, © 2007 Springer

Zhirko, Yu.I.; Kovalyuk, Z.D.; Klad'ko, V.P. et al. (2007b) Investigation of hydrogen intercalation in layered crystals InSe and GaSe, *Proceedings of HTM-2007 15 International Conference "Hydrogen Economy and Hydrogen Treatment of Materials"*, Vol.2, pp. 606-610., Donetsk, Ukraine, May 21-25, 2007

Zhirko, Yu.; Kovalyuk, Z.; Zaslonkin, A. & Boledzyuk, V. (2010). Photo and electric properties of hydrogen intercalated InSe and GaSe layered crystals, *Proceeding of Solar'10 International Conference "Nano/Molecular Photochemistry and Nanomaterials for Green Energy Development"* pp.48-49, Cairo, Egypt , February 15 -17, 2010

Zhirko, Yu.I. (2011). Optical properties of GaSe and InSe layered crystal hydrogen intercalates. Comparative study. *Proceeding of International Symposium on Reactivity of Solids (ISRS17)*. p.14, Bordeaux, France, 27 June – 1 July, 2011

Zhirko, Yu. I.; Skubenko, N.A.; Dubinko,V. I. et al. (2012). Influence of Impurity Doping and γ-Irradiation on the Optical Properties of Layered GaSe Crystals. *Journal of Materials Science and Engineering B*, Vol.2, No2, pp.91-102.

Chapter 10

Hydrogen Storage for Energy Application

Rahul Krishna, Elby Titus, Maryam Salimian, Olena Okhay, Sivakumar Rajendran, Ananth Rajkumar, J. M. G. Sousa, A. L. C. Ferreira, João Campos Gil and Jose Gracio

Additional information is available at the end of the chapter

http://dx.doi.org/10.5772/51238

1. Introduction

The rising population and increasing demand for energy supply urged us to explore more sustainable energy resources. The reduction of fossil fuel dependency in vehicles is key to reducing greenhouse emissions [1-2]. Hydrogen is expected to play an important role in a future energy economy based on environmentally clean sources and carriers. As a fuel of choice it is light weight, contains high energy density and its combustion emits no harmful chemical by-products. Moreover, hydrogen is considered as a green energy, because it can be generated from renewable sources and is non-polluting [3-5].

Nevertheless, there is a still remaining significant challenge that hinders the widespread application of hydrogen as the fuel of choice in mobile transportation, namely, the lack of a safe and easy method of storage. Vehicles and other systems powered by hydrogen have the advantage of emitting only water as a waste product. The efficient and safe storage of hydrogen is crucial for promoting the "hydrogen economy" as shown in Figure1 [6-9].

The United States' Department of Energy (DOE) has established requirements that have to be met by 2015; regarding the reversible storage of hydrogen according to which the required gravimetric density should be 9 wt % and the volumetric capacity should be 81 g of H_2/L [10-11]. To fulfill such requirements there are main problems for hydrogen storage such as:

- reducing weight and volume of thermal components is required;
- the cost of hydrogen storage systems is too high;
- durability of hydrogen storage systems is inadequate;
- hydrogen refuelling times are too long;
- high-pressure containment for compressed gas and other high-pressure approaches limits the choice of construction materials and fabrication techniques, within weight, volume, performance, and cost constraints.

For all approaches of hydrogen storage, vessel containment that is resistant to hydrogen permeation and corrosion is required. Research into new materials of construction such as metal ceramic composites, improved resins, and engineered fibbers is needed to meet cost targets without compromising performance. Materials to meet performance and cost requirements for hydrogen delivery and off-board storage are also needed [10].

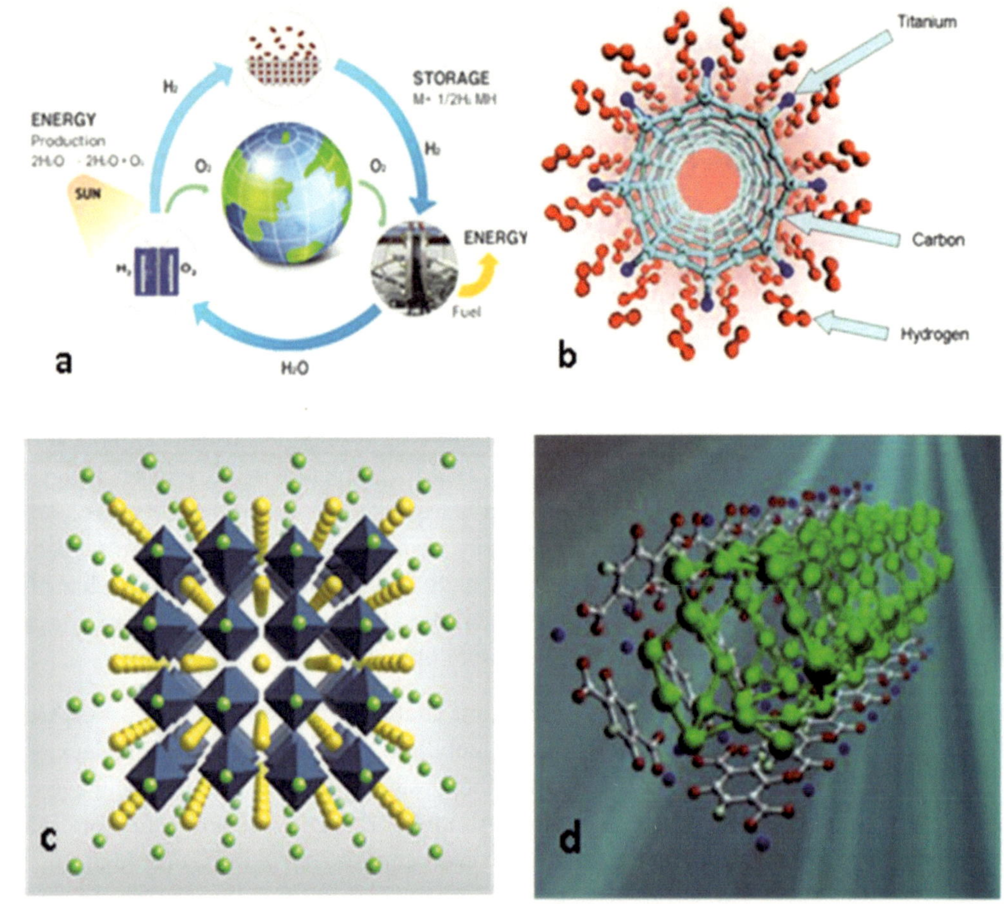

Figure 1. (a) Hydrogen production and storage by renewable resource [6], (b) hydrogen storage in metal doped carbon nanotubes [7], (c) storage in mesoporous zeolite: by controlling the ratio of different alkali metal ions (yellow and green balls), it is possible to tailor the pressure and temperature at which hydrogen is released from the material [8], (d) hydrogen storage in metal–organic framework (MOF)-74 resembles a series of tightly packed straws comprised mostly of carbon atoms (white balls) with columns of zinc ions (blue balls) running down the walls. Heavy hydrogen molecules (green balls) adsorbed in MOF-74 pack into the tubes more densely than they would in solid form [9].

Moreover, for all methods of hydrogen storage thermal management is a key issue. In general, the main technical challenge is heat removal upon re-filling of hydrogen for compressed gas and onboard reversible materials within fuelling time requirements. Onboard reversible materials typically require heat to release hydrogen. Heat must be provided to the storage media at reasonable temperatures to meet the flow rates needed by the vehicle power plant, preferably using the waste heat of the power plant. Depending upon the chemistry, chemical hydrogen approaches often are exothermic upon release of hydrogen to the power plant, or optimally thermal neutral. By virtue of the chemistry used, chemical hydrogen approaches require significant energy to regenerate the spent material and by-products prior to re-use; this is done off the vehicle.

2. Methods and problems of hydrogen storage

At the moment, several kinds of technologies of hydrogen storage are available. Some of them will be briefly described here.

1. The simplest is compressed H_2 gas. It is possible at ambient temperature, and in- and out-flow are simple. However, the density of storage is low compared to other methods.
2. Liquid H_2 storage is also possible: from 25% to 45% of the stored energy is required to liquefy the H_2. At this method the density of hydrogen storage is very high, but hydrogen boils at about -253ºC and it is necessary to maintain this low temperature (else the hydrogen will boil away), and bulky insulation is needed.
3. In metal hydride storage the powdered metals absorb hydrogen under high pressures. During this process heat is produced upon insertion and with pressure release and applied heat, the process is reversed. The main problem of this method is the weight of the absorbing material – a tank's mass would be about 600 kg compared to the 80 kg of a comparable compressed H_2 gas tank.
4. More popular at this time is carbon absorption: the newest field of hydrogen storage. At applied pressure, hydrogen will bond with porous carbon materials such as nanotubes.

So, it can be summarized that even mobile hydrogen storage is currently not competitive with hydrocarbon fuels; it must become so in order for this potential environmentally life-saving technology to be realized on a great scale.

3. High pressure hydrogen storage

The most common method of hydrogen storage is compression of the gas phase at high pressure (> 200 bars or 2850 psi). Compressed hydrogen in hydrogen tanks at 350 bar (5,000 psi) and 700 bar (10,000 psi) is used in hydrogen vehicles. There are two approaches to increase the gravimetric and volumetric storage capacities of compressed gas tanks. The first approach involves cryo-compressed tanks as shown in Figure 2 [12]. This is based on the fact that, at fixed pressure and volume, gas tank volumetric capacity increases as the tank temperature decreases. Thus, by cooling a tank from room temperature to liquid nitrogen temperature (77 K), its volumetric capacity increases. However, total system volumetric capacity is less than one because of the increased volume required for the cooling system. The limitation of this system is the energy needed to compression of the gas. About 20 % of the energy content of hydrogen is lost due to the storage method. The energy lost for hydrogen storage can be reduced by the development of new class of lightweight composite cylinders. Moreover, the main problem consisting with conventional materials for high pressure hydrogen tank is embrittlement of cylinder material, during the numerous charging/discharging cycles [13-14].

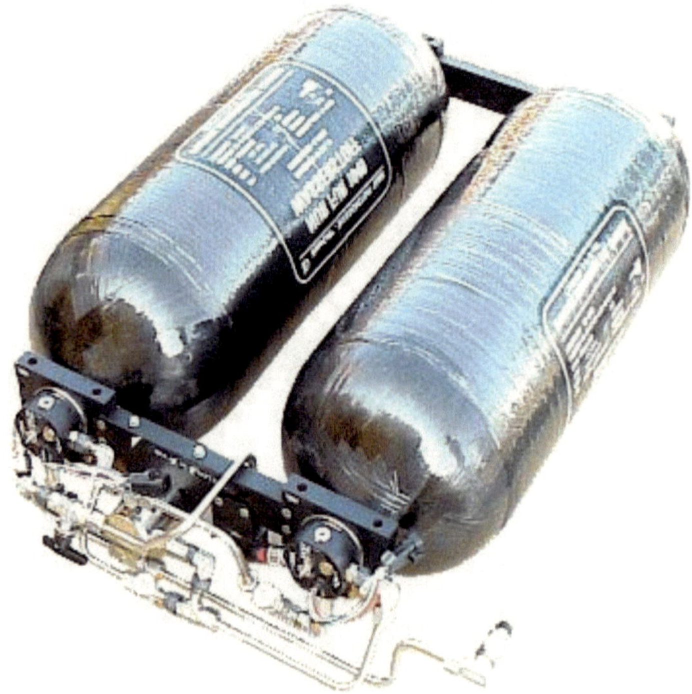

Figure 2. Hydrogen storage in tanks presently used in hydrogen-powered vehicles [12].

4. Liquefaction

The energy density of hydrogen can be improved by storing hydrogen in a liquid state. This technology developed during the early space age, as liquid hydrogen was brought along on the space vessels but nowadays it is used on the on-board fuel cells. It is also possible to combine liquid hydrogen with a metal hydride, like Fe-Ti, and this way minimize hydrogen losses due to boil-off.

In this storage method, first gas phase is compressed at high pressure than liquefy at cryogenic temperature in liquid hydrogen tank (LH_2). The condition of low temperature is maintained by using liquid helium cylinder as shown in Figure 3 [15]. Hydrogen does not liquefy until -253 °C (20 degrees above absolute zero) such much energy must be employed to achieve this temperature. However, issues are remaining with LH_2 tanks due to the hydrogen boil-off, the energy required for hydrogen liquefaction, volume, weight, and tank cost is also very high. About 40 % of the energy content of hydrogen can be lost due to the storage methods. Safety is also another issue with the handling of liquid hydrogen as does the car's tank integrity, when storing, pressurizing and cooling the element to such extreme temperatures [10,16-19].

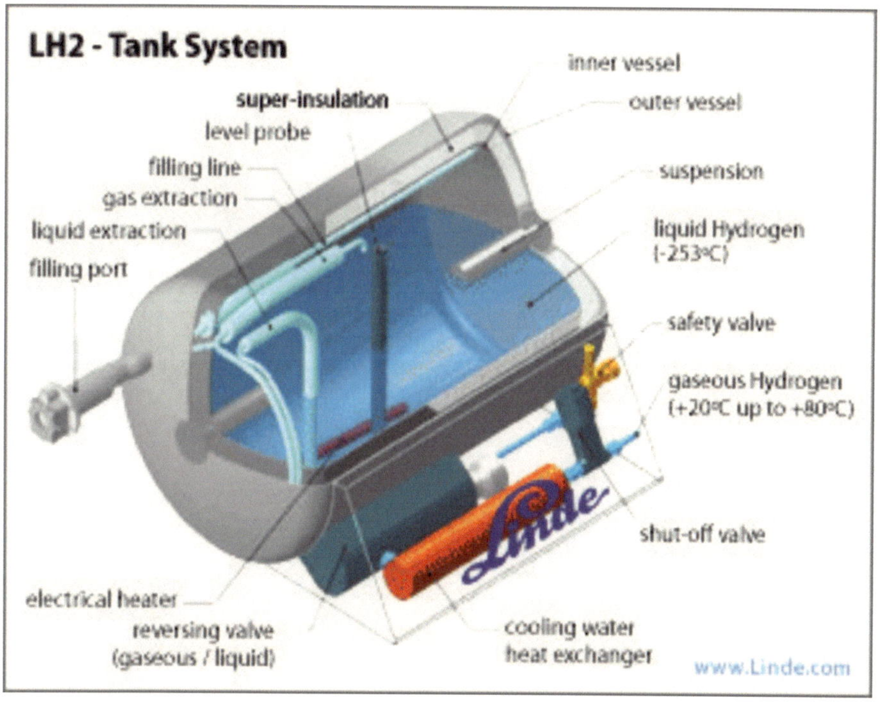

Figure 3. Liquid hydrogen storage tank system, horizontal mounted with double gasket and dual seal [15].

5. Solid state hydrogen storage

As mentioned above, certainly some practical problems, which cannot be circumvented, like safety concerns (for high pressure containment), and boil-off issues (for liquid storage), both are challenging for hydrogen storage. There is a third potential solution for hydrogen storage such as *(i)* metal hydrides and *(ii)* hydrogen adsorption in metal-organic frameworks (MOFs) and carbon based systems [10,17,18].

In these systems, hydrogen molecules are stored in the mesoporous materials by physisorption (characteristic of weak van der Waals forces). In the case of physisorption, the hydrogen capacity of a material is proportional to its specific surface area [20-22]. The storage by adsorption is attractive because it has the potential to lower the overall system pressure for an equivalent amount of hydrogen, yielding safer operating conditions. The advantages of these methods are that the volumetric and cryogenic constraints are abandoned. In recent decades, many types of hydrogen storage materials have been developed and investigated, which include hydrogen storage alloys, metal nitrides and imides, ammonia borane, etc.

Currently, porous materials such as zeolites, MOFs, carbon nanotubes (CNTs), and graphene also gained much more interest due to the high gravimetric density of such materials [23-24]. The corresponding hydrogen storage capabilities of these materials are displays in Figure 4 [25].

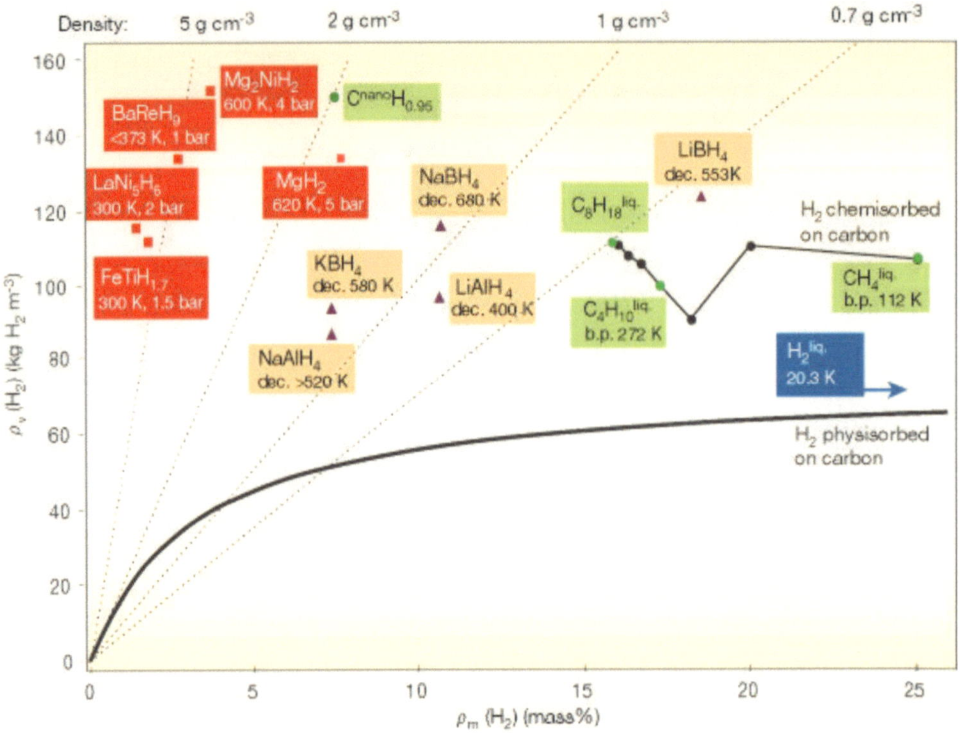

Figure 4. A complete survey plot of hydrogen storage in metal hydrides and carbone-based materials [25].

5.1. Hydrogen storage in metal hydrides

Initially, metal alloys, such as $LaNi_5$, TiFe and MgNi [10] were proposed as storage tanks since by chemical hydrogenation they form metal hydrides as previously shown in Figure 2 (compressed tank) [12]. Latter, hydrogen can be released by dehydrogenation of metal hydrides with light elements (binary hydrides and complex hydrides) because of their large gravimetric H_2 densities at high temperature [10,17]. Regarding vehicle applications, metal hydrides (MHs) can be distinguished into high or low temperature materials [26]. This depends on the temperature at which hydrogen absorption or desorption is taking place. Normally, in MHs hydrogen uptake and release kinetics is considered as above or below of 150 °C, respectively [27]. La-based and Ti-based alloys are examples of some low temperature materials with their main drawback as they provide very low gravimetric capacity (<2 wt %). The corresponding hydrogen storage capabilities of metal hydrides are displays in Figure 5 [13].

The analysis of above plot $LiAlH_4$ (LAH) (Fig.5) shows that the gravimetric weight ratio of hydrogen is 10.6 wt%; thereby LAH seems a potential hydrogen storage medium for future fuel cell powered vehicles. But, in practice the hydrogen storage capacity is reduced to 7.96 wt% due to the formation of LiH + Al species as the final product. Due to this, a substantial research effort has been devoted to accelerating the decomposition kinetics by catalytic doping in the MHs [10]. The high hydrogen content, as well as the discovery of reversible hydrogen storage is reported in Ti-doped $NaAlH_4$. In order to take advantage of the total hydrogen capacity, the intermediate compound LiH must be dehydrogenated as well. Due to its high thermodynamic stability this requires temperatures higher than 400 °C which is not considered feasible for transportation purposes [13]. Another problem related to hydrogen storage is the recycling back to $LiAlH_4$ due to its relatively low stability, requires an extremely high hydrogen pressure in excess of 10000 bar [27].

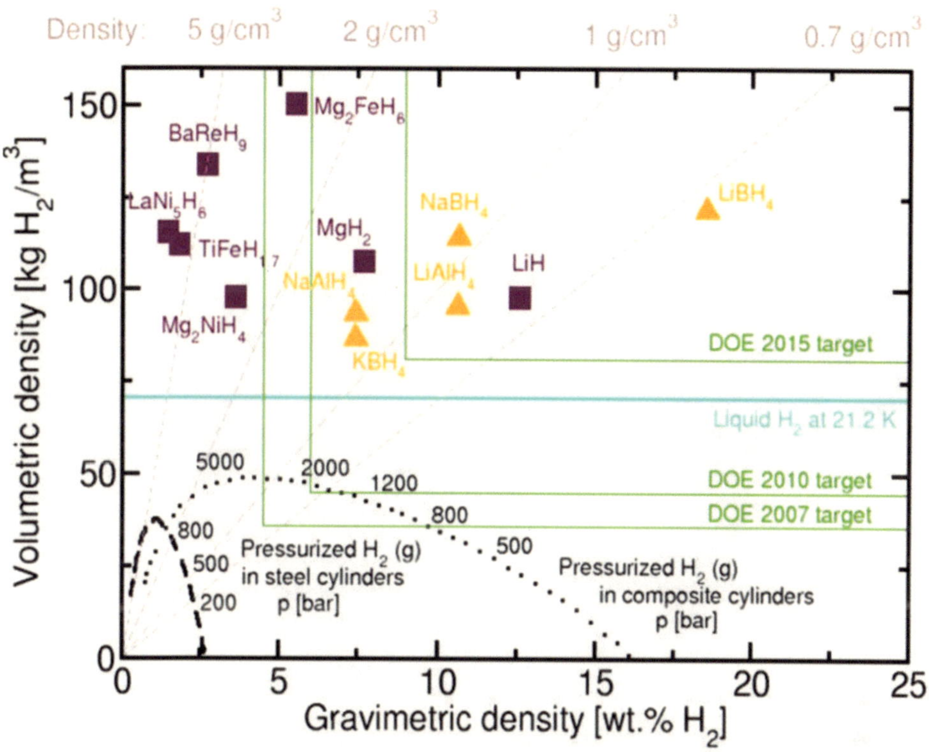

Figure 5. Volumetric and gravimetric hydrogen storage densities of different hydrogen storage methods. Metal hydrides are represented with squares and complex hydrides with triangles. $BaReH_9$ has the highest know hydrogen to metal ratio (4.5), Mg_2FeH_6 has the highest known volumetric H_2 density, $LiBH_4$ has the highest gravimetric density. Reported values for hydrides are excluding tank weight. DOE targets are including tank weight, thus the hydrogen storage characteristics of the shown hydrides may appear too optimistic [13].Controversially, high temperature materials like Mg-based alloys can reach a theoretical maximum capacity of hydrogen storage of 7.6 wt%, suffering though from poor hydrogenation / dehydrogenation kinetics and thermodynamics [10,18,23].

Thus, the high work temperature and the slow reaction rate (high activation energy) limit the practical application of chemical hydride systems. Those properties can be improved by the nanocomposite materials (Fig.6) [28].

The nanocomposite materials for hydrogen storage encompass a catalyst and composite chemical hydrides at the nanometer scale. The catalyst increases reaction rate. The thermodynamic stability of the nano-composite materials can be controlled by the composite chemical hydrides having protide (hydride) ($H\delta^{-}$) and proton ($H\delta^{+}$). In addition, the hydrogen absorption kinetics is accelerated by the nanosize materials and they may change the thermodynamic stability of the materials [28].

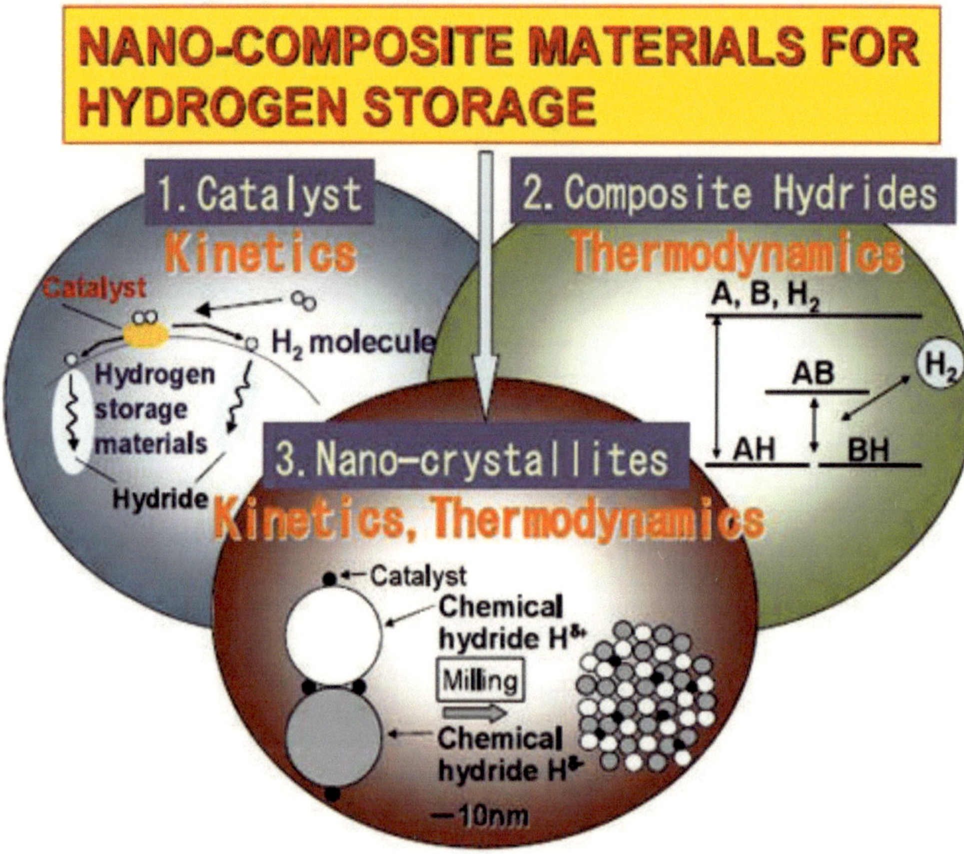

Figure 6. Desigh concept of nano-composite materials for hydrogen storage [28].

Another strategy to increase the interaction energy refers to the modifications of the chemical properties: surface treatment, ion exchange, doping, etc. It was observed that the capacity of hydrogen adsorption in metal oxide increased remarkably when Pt was introduced into mesoporous nickel oxide and magnesium oxide [29].

The analyse of hydrogen storage property of Ni-nanoparticles with cerium shell structure showed that the quantity of hydrogen released gradually increased with increasing temperature up to a maximum of 400 °C, then gradually decreased at temperatures above 400 °C. The catalytic activity of nanoparticles in the gas phase hydrogenation of benzene was related to their hydrogen storage properties and it reached the maximum value when the maximum amount of bulk H was released. By comparing the catalytic activity of nano-Ni particles with and without Ce shell structure, the higher activity of nano-Ni with Ce shell was attributed to the property of hydrogen storage and the synergistic effect of Ce and Ni in the shell structure, which was also illustrated by the result of gas phase benzene hydrogenation over supported nano-NiCe particles [30].

5.2. Hydrogen storage in nanostructured / porous material

It is well known that there are three categories of porous materials: microporous with pores of less than 2 nm in diameter, mesoporous having pores between 2 and 50 nm, and macroporous with pores greater than 50 nm. The term "nanoporous materials" has been used for those porous materials with pore diameters of less than 100 nm. Many kinds of crystalline and amorphous nanoporous materials such as framework silicates and metal oxides, pillared clays, nanoporous silicon, carbon nanotubes and related porous carbons have been described lately in the literature. It will be focused here on microporous zeolites, nanoporous metal organic frameworks (MOFs) and carbon-based materials.

5.2.1. Zeolites

Zeolite is a type of microporous solid used commercially in catalysis and gas separation [31]. Zeolites are prominent candidates for a hydrogen storage medium, due to their structural and high thermal stability, large internal surface area, low cost and adjustable composition [32].

Zeolites contain well defined open-pore structure, with often tunable pore size, and show notable guest-host chemistry, with important applications in catalysis, gas adsorption, purification and separation [33]. Additionally, this material is cheap and has been widely used in industrial processes for many decades. The extensive experimental survey depicts the hydrogen storage capacity of zeolites to be <2 wt% at cryogenic temperatures and <0.3 wt% at room temperatures and above [34]. Figure 7 shows that the structure of these minerals is most commonly based on a framework of alternating AlO_4 and SiO_4 species, with charge balancing (hydroxyl or cationic) entities, forming networks of cavities, channels and openings of varying dimensions [35].

The specific structural configuration of the zeolite provides great influence for their properties with respect to adsorption, selectivity and mobility of the guest molecules. An important property of zeolites is their high ion-exchange capacity, which allows for the direct manipulation of the available void space inside the material, as well as the chemical properties of the binding sites, greatly influencing their storage capacity [36]. Theoretical modelling also provides a close insight of the hydrogen storage capacity of zeolite materials [37-38].

A hydrogen capacity equal to 1.28 wt% was obtained at 77 K and 0.92 bar for H-SSZ-13 zeolite [39]. Also at cryogenic temperatures and 15 bar gravimetric storage capacities of 2.19 wt% was reported for Ca exchanged X zeolite [32]. However, hydrogen adsorption on zeolites at room temperature and 60 bar was reported less than 0.5 wt% [34,40]. It was shown that the amount of hydrogen adsorbed on zeolites depended on the framework structure, composition, and acid–base nature of the zeolites [32].

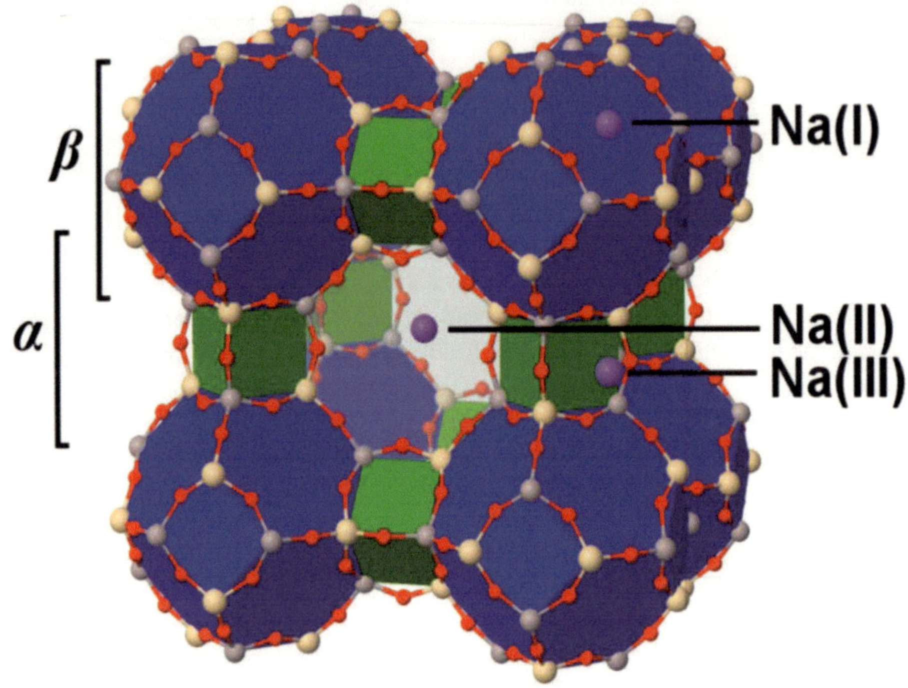

Figure 7. An unit cell of sodium zeolite with cage and cavity [32].

5.2.2. Metal organic frameworks

Lately, novel nanoporous materials like metal organic frameworks (MOFs) is have been targeted to the hydrogen storage problem [14,24,41]. MOFs are porous materials constructed by coordinate bonds between multidentate ligands and metal atoms or small metal-containing clusters. MOFs can be generally synthesized via self-assembly from different organic linkers and metal nodules. Due to the variable building blocks, MOFs have very large surface areas, high porosities, uniform and adjustable pore sizes and well-defined hydrogen occupation sites. These features make MOFs (some type of MOF is called IRMOF) promising candidates for hydrogen storage. As usually, MOFs are highly crystalline inorganic-organic hybrid structures that contain metal clusters or ions (secondary building units) as nodes and organic ligands as linkers (Fig.8) [42].

When guest molecules (solvent) occupying the pores are removed during solvent exchange and heating under vacuum, porous structure of MOFs can be achieved without destabilizing the frame and hydrogen molecules will be adsorbed onto the surface of the pores by physisorption. Compared to traditional zeolites and porous carbon materials, MOFs have very high number of pores and surface area which allow higher hydrogen uptake in a given volume [43]. The research interests on hydrogen storage in MOFs have been growing since 2003 when the first MOF-based hydrogen storage material was introduced [44]. Since there are infinite geometric and chemical variations of MOFs based on different combinations of secondary building units (SBUs) and linkers, many researches explore what combination will provide the maximum hydrogen uptake by varying materials of metal ions and linkers as shown in Figure 9 [45].

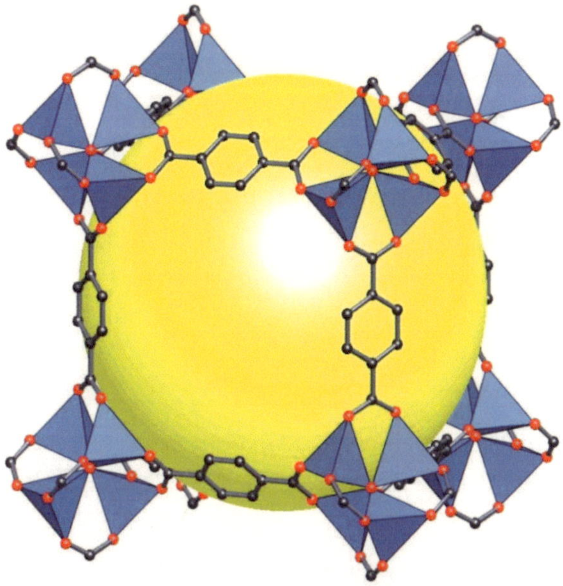

Figure 8. Eight units surround this pore (yellow ball represents space available in pore) in the metal-organic framework called MOF-5. Each unit contains four ZnO_4 tetrahedra (blue) and is connected to its neighboring unit by a dicarboxylic acid group [42].

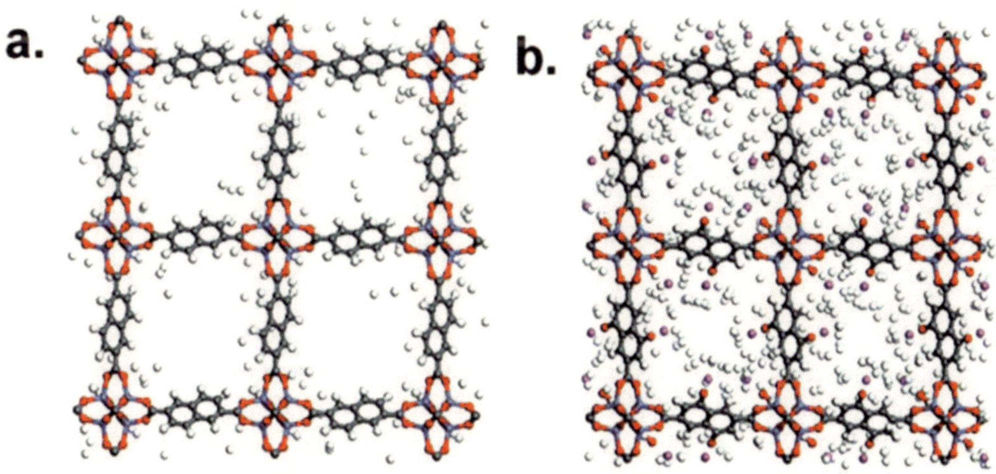

Figure 9. Snapshots of the IRMOF-8 cage with adsorbed H_2 at 77 K and 1 bar: (a) unmodified IRMOF-8 and (b) Li-alkoxide modified IRMOF-8 [45].

The effects of surface area, pore volume and heat of adsorption on hydrogen uptake in MOFs were discussed and in result the extensive work was directed toward the synthesis of MOFs with high surface areas and pore volumes [46].

As example, it was reported about storage capacity of 3.1 wt% at 77 K under 1.6 MPa in the formated 1D lozenge-shape tunnels, with pores of 8.5 Å and average surface area of 1100 m²/ g in the nanoporous metal-benzenedicarboxylate M (OH) ($O_2C–C_6H_4–CO_2$) ($M=Al^{3+},Cr^{3+}$), MIL-53 systems (Fig. 10) [44].

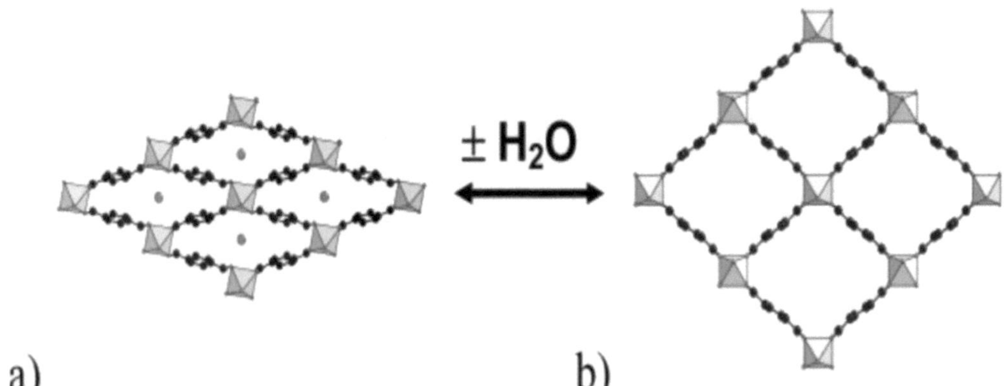

Figure 10. Representation of the structure of MIL-53 showing the expansion effect due to the removal of a water molecule, (a) hydrated (left), (b) dehydrated (right); octahedra: $MO_4(OH)_2$, $M=Al^{3+},Cr^{3+}$. The dehydrated form of MIL-53 was tested for the hydrogen adsorption experiment [44].

A series of isoreticular (meaning having the same underlying topology) metal organic frameworks, $Zn_4O(L)$, were constructed by changing the different linking zinc oxide clusters with linear carboxylates L, so as to get a high porosity. Recently, MOF-177 ($Zn_4O(BTB)_2$) was formed by linking the same clusters with a trigonal carboxylate. MOF-177 was claimed to have a high Langmuir surface area of 5640 $m^2\ g^{-1}$, and the highest hydrogen storage of 7.5 wt% H_2 at 77 K and 70 bar. The more meaningful surface area, BET surface area (calculated by well known Brunauer-Emmett-Teller method), for MOF-177 is around 3000 $m^2\ g^{-1}$. More recently, it was reported a nanoporous chromium terephthalate-based material (MIL-101) with the highest Langmuir surface area (4500–5900 $m^2\ g^{-1}$) among all MOFs [44]. It was reported that the hydrogen storage capacities in this material at 8 MPa were 6.1 wt% at 77 K, and 0.43 wt% at 298 K. Although these MOFs have remarkable hydrogen capacities at 77 K, no significant hydrogen storage capacities were obtained with the MOFs at room temperature [11].

5.2.3. Carbon-based materials

Among the vast range of materials, carbon-based systems have received particular research interest due to their light weight, high surface area and chemical stabilities. Early experimental data for hydrogen storage in carbon nanomaterial's was initially promising, indicating high hydrogen storage capacities exceeding DOE targets [17,18]. Therefore, hydrogenation of carbon-based materials e.g., activated carbon, graphite, carbon nanotubes and carbon foams, have gained large technological and scientific interest for hydrogen storage and were included in the group of hydrogen storage materials as shown in Figure 4 [25].

It was already mentioned stored hydrogen in the liquid form requires large energy consumption for liquefaction at 20 K and it also suffers from the "boil-off" problem. At the same time, carbon materials, as well as MOFs and other nanostructured and porous materials, have high surface areas and have exhibited promising hydrogen storage capacities at 77 K. However, among the currently available candidate storage materials, none is capable of meeting the DOE criteria for personal transportation vehicles at moderate temperatures and pressures. The hydrogen adsorption capacities at the ambient temperature on all known sorbents are below 0.6–0.8 wt% at 298 K and 100 atm. This is true for all sorbent materials including the MOFs and templated carbons [11].

Hydrogen storage by spillover has been proposed as a mechanism (i.e., via surface diffusion) to enhance the storage density of carbon-based nanostructures as well as MOF structures. This approach relies on the use of a supported metallic catalyst to dissociate molecular hydrogen, relying on surface diffusion through a bridge to store atomic hydrogen in a receptor.

The term hydrogen spillover was coined decades ago [47] to describe the transport of an active species (e.g., *H*) generated on one substance (activator, *Act*) to another (receptor, *Rec*) that would not normally adsorb it. Common in heterogeneous catalysis, the activator is metal and the receptor can be a metal or a metal oxide, $H_2 - Act \rightarrow 2H$, $H + Rec \rightarrow H@Rec$ [47].

The number of adsorbed H atoms can exceed that of the activator by orders of magnitude and approach the number of receptor atoms. This feature makes the spillover attractive for H storage: if a receptor is made from light elements, notably C, then the gravimetric fraction of the "spilled" H may be large and approach the DOE's goals in its use as an onboard energy source [48].

Obviously, in order for the spillover process to take place, next processes need to occur: *(i)* H atoms migrate from the catalyst to the substrate via physical adsorption (physisorption), when the adsorbate molecules are attracted by weak van der Waals forces towards the adsorbent molecules, or chemical adsorption (chemisorption), when the adsorbate molecules are bound to the surface of adsorbent by chemical bonds; *(ii)* H atoms diffuse from the adsorption sites at the vicinity of the catalyst to the sites far away from the catalyst. Thus the process of physisorption results in the H_2 molecule remaining intact. And, in opposite, chemisorption leads to H_2 bond dissociation, with the resulting H atoms forming chemical bonds with the storage substrate [49]. Physisorption can occur as a preliminary state to chemisorption.

It was already reported that a hydrogen spillover induced increase of hydrogen storage capacity for activated carbon and single-walled carbon nanotubes by factors of 2.9 and 1.6, respectively [50]. It has been suggested that the stored hydrogen atoms in the carbon-based materials are "loosely adsorbed" on surfaces of substrates via spillover upon H_2 dissociation on nickel catalyst. It was explained that at a given pressure of H_2 gas, the H_2 molecules undergo dissociative chemisorption upon interacting with a supported transition metal catalyst, e.g. nanoparticles of platinum. The generated H atoms then migrate from the catalyst particles to the storage material through a "bridge" built of carbonized sugar molecules and further diffuse throughout the entire bulk solid. Obviously, in order for the spillover process to occur to any significant extent in solid materials, it is essential that the H atoms should be able to move from the vicinity of catalyst particles to substrate sites far from where the catalysts reside (Fig.11) [50].

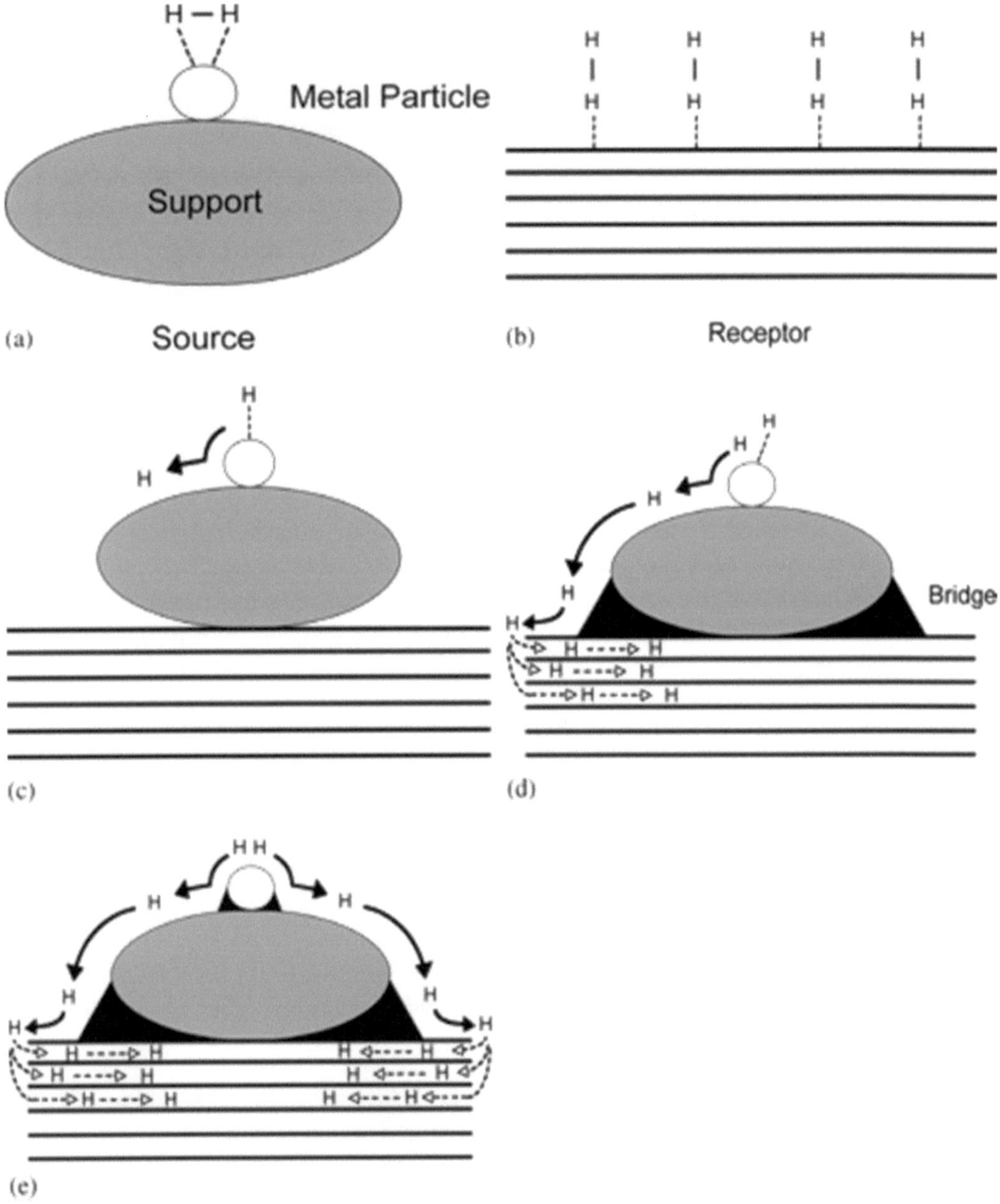

Figure 11. Hydrogen spillover in a supported catalyst system: (a) adsorption of hydrogen on a supported metal particle; (b) the low-capacity receptor; (c) primary spillover of atomic hydrogen to the support; (d) secondary spillover to the receptor enhanced by a physical bridge; (e) primary and secondary spillover enhancement by improved contacts and bridges [50].

While active research efforts are being made to understand the hydrogen spillover processes, to date, there has been no general consensus on the spillover mechanisms that can satisfactorily explain the observed large storage capacity (up to 4 wt% of H_2) or facile hydrogen desorption kinetics from the carbon-based storage compounds at near-ambient temperatures. Hydrogen spillover on solid-state materials is not a newly discovered phenomenon. The concept of hydrogen spillover has its genesis in fundamental studies with heterogeneous metal catalysts, particularly with such systems as are used for chemical hydrogenation reactions.

On another way, in catalytic processes, the metal has the role of "activating" hydrogen by reversibly dissociating H_2 into metal-H atom (hydride) species on its surface. It has been observed that, for instance, by heating Pt dispersed on carbon at 623 K, Pt/Al_2O_3 at 473–573 K, Pd/C at 473 K, and Pt/WO_3, under hydrogen pressure the amount of H_2 absorbed is in excess of the known H_2-sorption capacity of the metal alone. While the concept of hydrogen spillover is normally associated with solid-state materials such as activated carbon (or transition metal oxides), there have been a small number of reports of the solid-state hydrogenation of organic compounds that appear to implicate hydrogen spillover as the mechanism for hydrogenation [49].

Thus, the hydrogen spillover process includes three consecutive steps. In the first step, H_2 molecules undergo dissociative chemisorption upon interacting with metal catalysts. Subsequently, H atoms on the catalyst surfaces migrate to the substrate in the vicinity of the catalyst particles. Finally, H atoms near the catalysts diffuse freely to the surface or bulk sites far from where the catalyst particles reside [49].

So, as it was mention before, hydrogen can be stored in the light nanoporous carbon materials by physisorption and the interaction between H_2 and host material is dominated only by weak van-der Waals forces, and only a small amount of H_2 can be stored at room temperature. Thus, high surface area and appropriate pore size are key parameters for achieving high hydrogen storage. The nanoporous carbon structures fulfill this criterion, placing them since the beginning among of the best candidates for hydrogen storage media.

5.2.3.1. Carbon nanotubes

Because of unique hollow tubular structure, large surface area, and desirable chemical and thermal stability carbon nanotubes (CNTs) are considered as a promising candidate for gas adsorption (Fig.12) [51].

The experimental results on hydrogen storage in carbon nanomaterials scatter over several orders of magnitudes. It was reported in 1997 that single-walled CNTs (SWCNTs) could store ~10 wt% hydrogen at room temperature, and predicted a possibility to fulfill the benchmark set for on-board hydrogen storage systems by DOE [26].

Soon after this work, other optimistic results of hydrogen storage in CNTs were reported [52,53]. A few years later, very low hydrogen storage capacity of CNTs started to emerge, in particular, those experimentally obtained at room temperature: lower than 0.1 wt% at room temperature and 3.5 MPa [54]. It was also pointed that "the application of CNTs in hydrogen storage is clouded by controversy" [55], the reproducibility is poor, and the mechanism of how hydrogen is stored in CNTs remains unclear. In summary, it was mention that amount of hydrogen can be stored in CNTs, the reliable hydrogen storage capacity of CNTs is less than 1.7 wt% under a pressure of around 12 MPa and at room temperature, which indicates that CNTs cannot fulfill the benchmark set for onboard hydrogen storage systems by DOE [56]. However, higher values hydrogenation of carbon atoms in the SWCNTs (approximately 5.1±1.2 wt%) was also demonstrated as well as reversibility of the hydrogenation process [57].

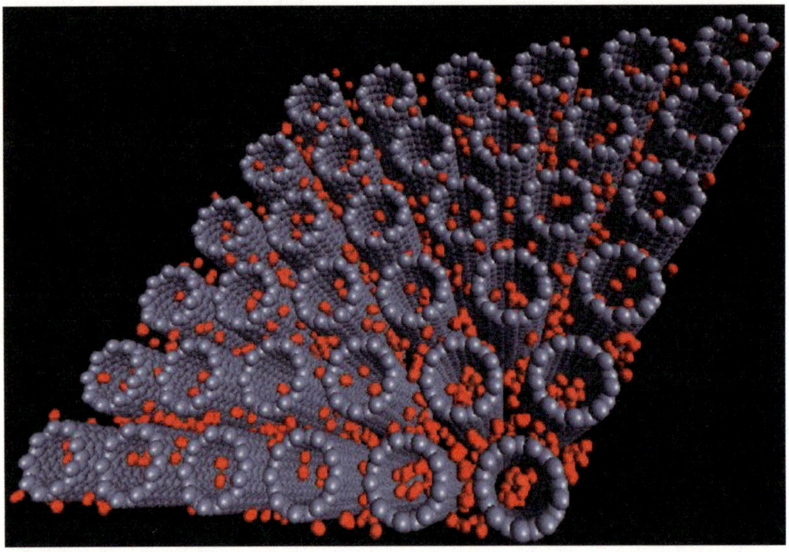

Figure 12. Hydrgen gas (red) adsorbed in an array of carbon nanotubes (grey). The hydrogen inside the nanotubes and in the interstitial channels is at a much higher density than that of the bulk gas [51].

All these results suggest that pure CNTs are not the best material for investigating hydrogen uptake. However, CNTs can be an effective additive to some other hydrogen storage materials to improve their kinetics. For example, the hydrogen storage capacity of CNTs can be increasing by the doping effect: potassium hydroxide (KOH) to increase the capacity of hydrogen storage on multi-walled carbon nanotubes (MWNTs) from 0.71 to 4.47 wt%, respectively, under ambient pressure [58]. Also, experimental results revealed that the structure of CNTs became destructive after being activated by KOH at 823 K in H_2 atmosphere [58].

Metal doping (Pt catalyst) influence on electronic structure and hydrogen storage of SWCNT was also studied [59]. It was represented that the spillover mechanism is responsible for hydrogenation of Pt-SWNT composites using molecular hydrogen (Fig.13). Moreover these materials store hydrogen by chemisorption, that is, the formation of stable C–H bonds. The hydrogen uptake indicates 1.2 wt% hydrogen storage for LB-film composites (monolayer assemblies of SWNT films prepared by the Langmuir-Blodgett method) and 1 wt% for chemical vapour deposition (CVD)-grown CVD composites. This maximum hydrogen storage was possible only after obtaining a uniform dispersion of monolayer thick unbundled SWNTs, doped with optimal size and density of Pt catalyst particles (~2 nm, which corresponds to nominal thickness of 6 Å) [59]. This storage capacity may be increased by improving the kinetics of atomic-hydrogen diffusion from the Pt catalyst to the SWNT support and also by employing nonbundled CNTs with an increased Pt-to-C dispersion [59].

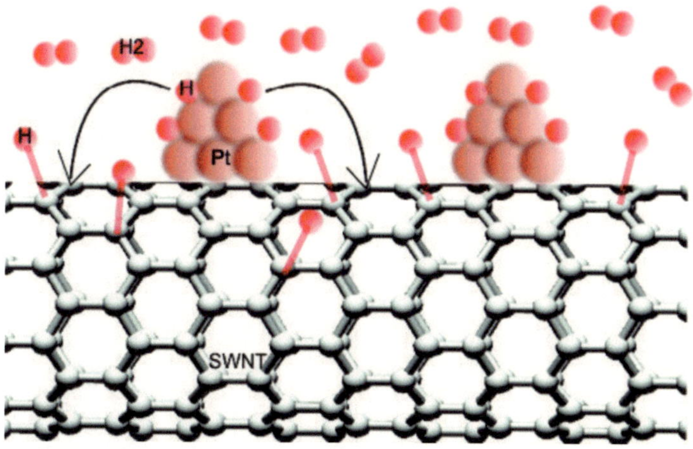

Figure 13. Schematic spillover mechanism: hydrogen uptake of single-walled carbon nanotubes is influenced by the addition of Pt nanoparticles [59].

5.2.3.2. Graphene and graphane

After the discovery of graphene scientists move in that direction. Graphene has one of the highest surface area-per-unit masses in nature, far superior to even carbon nanotubes and fullerenes [33-37]. It can store individual hydrogen atoms in a metallic lattice, through chemical bonding to a metallic host. Besides the physisorption, the chemisorption of hydrogen in graphene is even more interesting for catalysis and electronic purpose. For the first time, it was possible to isolate two dimensional mono layers of atoms. A suspended single layer of graphene is one of the stiffest known materials characterized by a remarkably high Young's modulus of ~ 1 TPa. The high thermal conductivity (3000 Wm^{-1} K^{-1}), high electron mobility (15000 cm^2 V^{-1} s^{-1}), and high specific surface area of graphene nanosheet are also few amazing characteristics of this material. As an electronic material, graphene represents a new playground for electrons in 2, 1, and 0 dimensions where the rules are changed due to its linear band structure (Fig.14) [60].

Recently, graphene has triggered enormous interest in the area of composite materials and solid state electronics. When incorporated into a polymer, its peculiar properties manifest as remarkable improvements in the host material. The mechanical and thermal properties of these materials rank among the best in comparison with other carbon-based composites. For energy applications graphene is a very suitable candidate to produce ultra high charge capacitor or super capacitor. Here, storage capacity must be realized by rapid charging and discharging of composite materials. Therefore, the charge and discharge kinetics must be very high. Recently, it was reported that by using of different activation procedures other than heat treatment, such as chemical reduction, to remove unreacted functional groups the O/C ratio could be reduced and improve the surface area and adsorption capacity of graphene oxide frameworks (GOFs) increases significantly [61]. Moreover, doping effect also can increase the capacity of hydrogen storage in graphene: the maximum hydrogen storage capacity in lithium ion doped pillared graphene material (Fig.15) around 7.6 wt% in the condition of 77 K and 100 bar pressure [10].

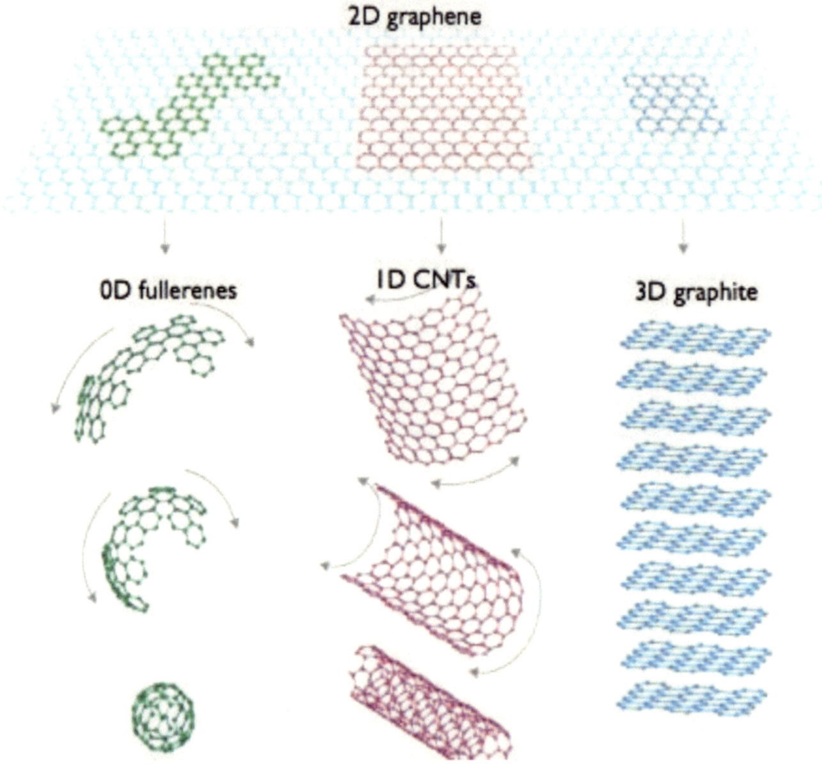

Figure 14. 2D Graphene can be transformed into CNTs, graphite, or fullerenes [60].

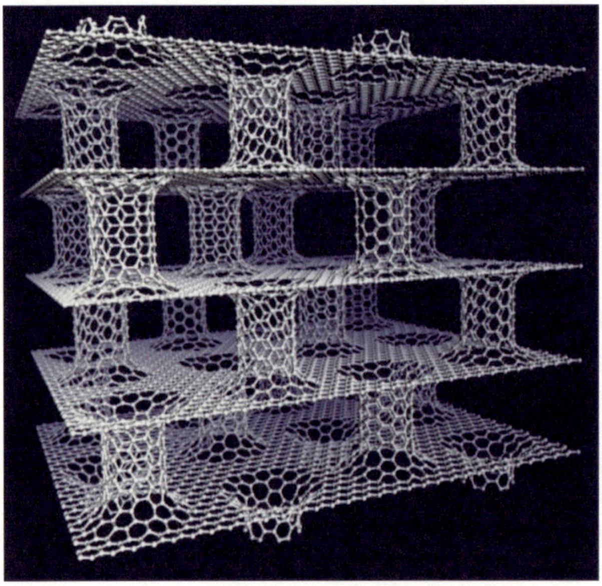

Figure 15. Pillared graphene structure, a theoretical approach [10].

Hydrogen storage is also dependent on number of layers of graphene. If graphene is single layer it can store more hydrogen as compare to bilayer or few layers. The analyse of multilayers of graphene on to the surface of Ru (0001) substrate by scanning tunnelling microscopy (STM) technique represented moiré patterns of superstructure as shown in Figure 16 [62]. Moiré pattern is generated due to the rotational misorientation between the layers. Graphene is one atom thick hexagonally arranged carbon material and whenever layers are stacking together (due to the van-der Walls forces) moiré pattern is observed. In the moiré pattern it was analyzed 12 stacking faults of graphene sheets with 2.46 Å typical lattices spacing of graphite [62].

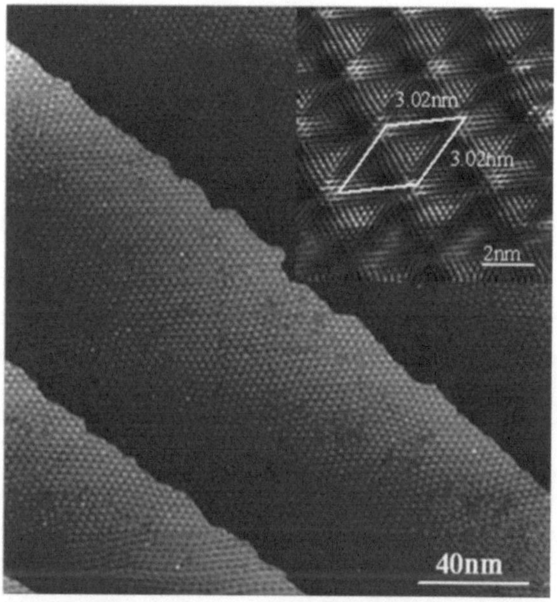

Figure 16. STM image (200 nm × 200 nm) of a monolayer graphene on Ru (0001). The moiré pattern with a periodicity of 30 Å is clearly visible. The inset shows the atomic structure of graphene/Ru (0001). The unit cell is marked as a rhombus [62].

The efficient hydrogenation of graphene can also be possible by catalytic conversion where catalyst provides a new reduction pathway of graphene through dissociation of graphene molecule [63]. The intercalated Ni nanoparticles inside of the alumina matrix with the flow of H_2 at 820°C demonstrated the catalytic hydrogenation of graphene films (Fig.17).

The degree of hydrogenation of graphene layers was also found 16.67 wt% and depended on the thickness of graphene [64]. Moreover, it was also reported about 100 % coverage of graphene by hydrogen atoms or each carbon atom has binding with hydrogen atoms. These results are very exciting that provides the future platform for hydrogenated graphene system [64].

The stability of a new extended two-dimensional hydrocarbon on the basis of first-principles total energy calculations was predicted [65]. The compound was called graphane and it was a fully saturated hydrocarbon derived from a single graphene sheet with formula CH. All of the carbon atoms are in sp³ hybridization forming a hexagonal network and the hydrogen atoms are bonded to carbon on both sides of the plane in an alternating manner (Fig.18). Graphane is predicted to be stable with a binding energy comparable to other hydrocarbons such as benzene, cyclohexane, and polyethylene.

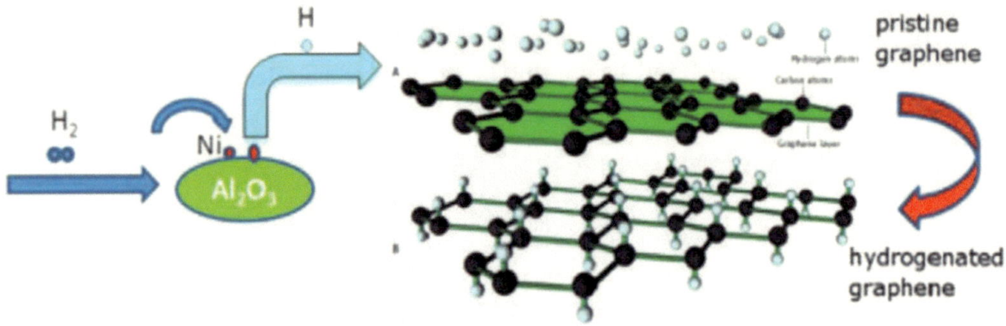

Figure 17. Schematic illustration for catalytic hydrogenation of graphene [63].

Thanks to its low mass and large surface area, graphane has also been touted as an ideal material for storing hydrogen fuel on vehicles. However, making graphane had proven to be difficult. The problem is that the hydrogen molecules must first be broken into atoms and this process usually requires high temperatures that could alter or damage the crystallographic structure of the graphene.

Figure 18. Structure of graphane in the chair conformation.
The carbon atoms are shown in gray and the hydrogen atoms in white.
The figure shows the hexagonal network with carbon in the sp³ hybridization [65].

However, a team led by Geim and Novoselov has worked out a way to make graphane by passing hydrogen gas through an electrical discharge (Fig.19). This creates hydrogen atoms, which then drift towards a sample of graphene and bond with its carbon atoms. The team studied both the electrical and structural properties of graphane and concluded that each carbon atom is bonded with one hydrogen atom. It appears that alternating carbon atoms in the normally-flat sheet are pulled up and down - creating a thicker structure that is reminiscent of how carbon is arranged in a diamond crystal. And, like diamond, the team found that graphane is an insulator - a property that could be very useful for creating carbon-based electronic devices [66].

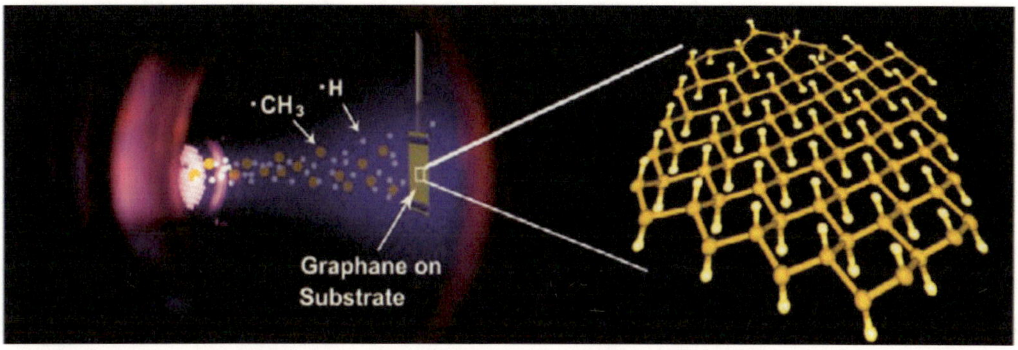

Figure 19. The growth of large-area graphane-like film by RF plasma beam deposition in high vacuum conditions. Reactive neutral beams of methyl radicals and atomic hydrogen effused from the discharged zone and impinged on the Cu/Ti-coated SiO_2/Si samples placed remotely. A substrate heating temperature of 650 °C was applied [67].

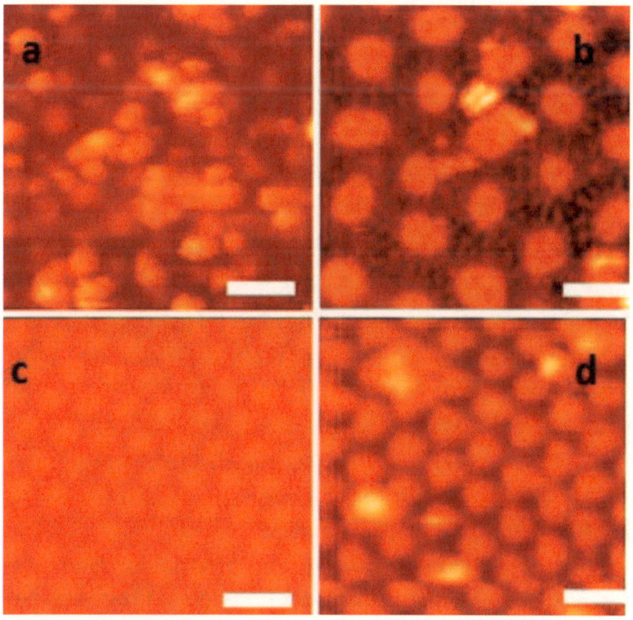

Figure 20. (a) STM images of graphane. The bright protrusions in the image are identified as atomic hydrogen clusters; (b) after annealing at 300 °C for 20 min; (c) after annealing at 400 °C for 20 min; (d) graphene recovered from graphane after annealing to 600 °C for 20 min. Scale bar 3 nm [67].

It was reported that the obtained graphane was crystalline with the hexagonal lattice, but its period was shorter than that of graphene. The reaction with hydrogen was found reversible, so that the original metallic state, the lattice spacing, and even the quantum Hall effect can be restored by annealing [67]. The degree of hydrogenation of graphene layers was analyzed by changing in moiré pattern and interconversion of graphane to graphene is presented in Figure 20.

6. Conclusion

The hydrogen revolution following the industrial age has just started. Hydrogen production, storage and conversion have reached a technological level although plenty of improvements and new discoveries are still possible. The hydrogen storage is often considered as the bottleneck of the renewable energy economy based on the synthetic fuel hydrogen. Different hydrogen storage methods and materials have been described already and need to be study more.

Author details

Rahul Krishna, Elby Titus*, Maryam Salimian, Olena Okhay,
Sivakumar Rajendran and Jose Gracio
Nanotechnology Research Division (NRD), Centre for Mechanical Technology and Automation (TEMA), Dept. of Mechanical Engineering, University of Aveiro, Portugal

J. M. G. Sousa, A. L. C. Ferreira and Ananth Rajkumar
Department of Physics& I3N, University of Aveiro, Portugal

João Campos Gil
Department of Physics, University of Coimbra, Portugal

7. References

[1] D.Leaf, H.J.H. Verlomec and W.F. Hunt, *J.Environ.Int.* 2003, *29*, 303
[2] J. M. Thomas and R. Raja, *Annu. Rev. Mater. Res.* 2005, 35, 315.
[3] Q. K.Wang, C. C. Zhu, W. H. Liu and T. Wu, *Int. J. Hydrogen Energy*, 2002, 27, 497.
[4] M. Bououdina, D. Grant, and G. Walker, *Int. J. Hydrogen Energy*, 2006, 31, 177.
[5] M. Kruk and M. Jaroniec, *Chem. Mater*, 2001, 13, 3169.
[6] http://hcc.hanwha.co.kr/english/pro/ren_hsto_idx.jsp
[7] http://nanopedia.case.edu/NWPage.php?page=hydrogen.storage
[8] http://www.bnl.gov/energy/AdvStorageSys.asp
[9] http://www.physorg.com/news126355316.html
[10] G.K. Dimitrakakis, E. Tylianakis and G. E. Froudakis, *Nano Lett.* 2008, *8*, 3166; Hydrogen, Fuel Cells & Infrastructure Technologies Program: Multiyear Research, Development and Demonstration Plan: Planned program activities for 2003–2010.

* Corresponding Author

Chapter 3.3, Revision 1. United States Department of Energy; http://www1.eere.energy.gov/hydrogenandfuelcells/mypp/pdfs/storage.pdf; https://www1.eere.energy.gov/hydrogenandfuelcells/pdfs/freedomcar_targets_explanations.pdf

[11] L. Wang and R. T. Yan, *Energy Environ.Sci.* 2008, *1*, 268

[12] http://news.softpedia.com/newsImage/New-Quantum-Alloy-Could-Improve-Hydrogen-Storage-for-Fuel-Cells-2.jpg/

[13] B. Bogdanovic and M. Schwickardi, *J.Alloys Comp.* 1999, *253*, 1; W. Grochala, P.P. Edwards, *Chem.Rev.* 2004, *104*, 1283

[14] Y. Li and R. T. Yang, *J. Am. Chem.Soc.* 2006, *128*, 726

[15] http://www1.eere.energy.gov/hydrogen and fuelcells/storage/hydrogen_storage.html

[16] M. Hirscher, *Handbook of hydrogen storage new materials for future energy storage*, Weinheim Wiley-VCH-Verl. 2010

[17] E. Perlt, J. Friedrich, M. Domaros and B. Kirchner, *Chem.Phys.Chem.* 2011, *12*, 3474

[18] Y. Kato, K. Otsuka and C. Y. Liu, *Chem.Eng.Res.Design*, 2005, *83*, 900

[19] G. Mpourmpakis, E.Tylianakis and E.F. George, *Nano Lett.* 2007, *7*, 1893

[20] M. Rzepka, P. Lamp, M.A. De La Casa-Lillo, *J.Phys.Chem.B* 1998, *102*, 10894

[21] A. Csaplinski, E. Zielinski, *Przem.Chem.* 1958, *37*, 640

[22] A. Herbst and P. Harting, *Adsorption* 2002, *8*, 111

[23] L. M. Viculis, J. J. Mack and R. B. Kaner, *Science* 2003, *299*, 1361

[24] J. Graetz, *Chem.Soc.Rev.* 2009, *38*, 73

[25] L. Schlapbach and A. Zütte, *Nature* 2001, *414*, 353

[26] A. C. Dillon, K. M. Jones, T. A. Bekkedahl, C. H. Kiang, D. S.Bethune and M. J. Heben, *Nature* 1997, *386*, 377

[27] http://en.wikipedia.org/wiki/Lithium_aluminium_hydride.

[28] Y. Kojima, *Materials Science Forum* 2010, *654-656*, 2935

[29] X. Sun, J.-Y. Hwang, and S. Shi, *J.Phys.Chem.C* 2010, *114*, 7178

[30] K.Z. Chen, Z.K. Zhang, Z.L. Cui, D.H. Zuo, D.Z. Yang, *Nanostructured Materials* 1997, *8*, 205

[31] R.T. Yang, *Adsorbents: Fundamentals and Applications*, Wiley, New York, 2003; M.E. Davis, *Nature* 2002, *417*, 813

[32] L. Regli, A. Zecchina, J. G. Vitillo, D. Cocina, G. Spoto, C. Lamberti, K.P. Lillerud, U. Olsbye and S. Bordiga, *Phys.Chem.Chem.Phys* 2005, *7*, 3197; H.W. Langmi, D. Book, A. Walton, S.R. Johnson, M.M. Al-Mamouri, J.D. Speight, P.P. Edwards, I.R. Harris and P.A. Anderson, *J. Alloys Comp.* 2005, *404–406*, 637

[33] S. Liu and X. Yang, J.*Chem.Phys* 2006, *124*, 244705

[34] J. Weitkamp, M. Fritz and S. Ernst, *Int.J.Hydrogen Ener.* 1995, *20*, 967

[35] J.M.G. Sousa, A.L.C. Ferreira, E. Titus, R. Krishna, D.P. Fagg and J. Gracio, accepted, *J.Nanoscience and Nanotech.* 2011

[36] D.W. Breck, *Zeolite molecular sieves: structure, chemistry, and use*, R.E. Krieger, 1984

[37] B. Smith and T.L.M. Maesen, *Chem.Rev.* 2008, *108*, 4125

[38] M. Hirscher, M. Becher, M. Haluska, F. von Zeppelin, X. Chen, U. Dettlaff-Weglikowska and S. Roth, *J. Alloys Comp.* 2003, *433*, 356

[39] A. Zecchina, S. Bordiga, J.G. Vitillo, G. Ricchiardi, C. Lamberti, G. Spoto, M. Bjrgen and K.P. Lillerud, *J.Am.Chem.Soc.* 2005, *127*, 6361
[40] S. B. Kayiran and F. L. Darkrim, *Surf.Interface Anal.* 2002, *34*, 100
[41] A.P. Cote, A.I. Benin, N.W. Ockwig, M. O'Keeffe, A.J. Matzger and O.M. Yaghi, *Science* 2005, *310*, 1166
[42] M. O'Keeffe, M. Eddaoudi, H. Li, T. Reineke, and O.M. Yagh, *J.Sol.State.Chem.* 2000, *152*, 3
[43] M. Dincă, J. R. Long, Angew, *Chem.Int.Ed.* 2008, *47*, 6766
[44] G. Férey, M. Latroche, C. Serre, and F. Millange, *Chem.Comm.* 2003, 2976.
[45] E. Klontzas, A. Mavrandonakis, E. Tylianakis, and G. E. Froudaki, *Nano Lett.* 2008, *8*, 1572
[46] R.Q. Snurr, *J.Phys.Chem.C*, 2007, *111*, 18794
[47] W.C. Conner and J.J.L. Falconer, *Chem.Rev.* 1995, *95*, 759; M. Boudart and G.D.-.Mariadassou, *Kinetics of Heterogeneous Catalytic Reactions*, Princeton University Press, Princeton, NJ, 1984
[48] Y. Lin, F. Ding, and B.I. Yakobson, *Phys.Rev.B* 2008, *78*, 041402 (R)
[49] H. Cheng, L. Chen, A. C. Cooper, X. Sha, and G.P. Pez, *Energy Environ. Sci*, 2008, *1*, 338
[50] A.J. Lachawiec, Jr., G. Qi, and R.T. Yang, *Langmuir* 2005, *21*, 11418
[51] http://puccini.che.pitt.edu/frameset3b.html
[52] P. Chen, X. Wu, J. Lin, K.L. Tan, *Science* 1999, *285*, 91-3
[53] C. Liu, Y.Y. Fan, M. Liu, H.T. Cong, H.M. Cheng, M.S. Dresselhaus, *Science* 1999, *286*, 1127-9
[54] G.G. Tibbetts, G.P. Meisner, C.H. Olk, *Carbon* 2001, *39*, 2291–301
[55] R.H. Baughman, A.A. Zakhidov, and W.A. de Heer, *Science* 2002, *297*, 787–92
[56] C. Liu, Y. Chen, C.-Z. Wu, S.-T. Xu, and H.-M. Chen, *Carbon* 2010, *48*, 452
[57] A. Nikitin, H. Ogasawara, D. Mann, R. Denecke, Z. Zhang, H. Dai, K. Cho, and A. Nilsson, *Phys.Rev.Lett.* 2005, *95*, 225507
[58] C. Chen, and C. Huang, *Int.J.Hydrogen Energy* 2007, *32*, 237
[59] R. Browmick, S. Rajasekaran, D. Friebel, C. Beasley, L. Jiao, H. Ogasawara, H. Dai, B. Clemens, and A. Nilsson, *J.Am.Chem.Soc.* 2011, *133*, 5580
[60] A.K. Geim and K.S. Novoselov, *Nature* 2007, *6*, 183
[61] J.W. Burress, S. Gadipelli, J. Ford, J.M. Simmons, W. Zhou, and T. Yildirim, *Angew.Chem.Int.Ed.* 2010, *49*, 8902
[62] Q. Liao, H.J. Zhang, K. Wu, H.Y. Li, S.N. Bao, and P. He, *Appl.Surf.Science* 2010, *257*, 82
[63] L. Zheng, Z. Li, S. Bourdo, F. Watanabe, C.C. Ryerson, and A.S. Biris, *Chem.Commun.* 2011, *47*, 1213
[64] Z. Luo, T. Yu, K. Kim, Z. Ni, Y. You, S. Lim, Z. Shen, S. Wang, and J. Lin, *ACS Nano* 2009, *3*, 1781
[65] J.O. Sofo, A.S. Chaudhari, and G.D. Barber, *Phys.Rev.B* 2007, *75*, 153401
[66] D.C. Elias, R.R. Nair, T. M. G. Mohiuddin, S.V. Morozov, P. Blake, M.P. Halsall, A.C. Ferrari, D.W. Boukhvalov, M.I. Katsnelson, A.K. Geim, and K.S. Novoselov, *Science* 2009, 30, 610
[67] Y. Wang, X. Xu, J. Lu, M. Lin, Q. Bao, B. Özyilmaz, and K.P. Loh, ACS Nano 2010, 4, 6146